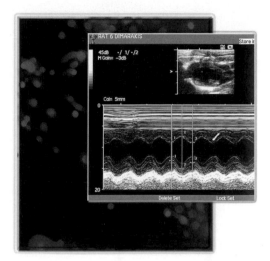

HANDBOOK OF CARDIAC STEM CELL THERAPY

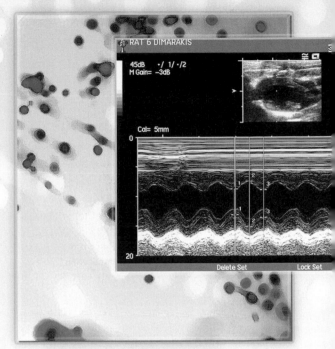

HANDBOOK OF CARDIAC STEM CELL THERAPY

edited by

Ioannis Dimarakis
Imperial College London, UK

Philippe Menasché
Hôpital Européen Georges Pompidou, France

Nagy A. Habib
Imperial College London, UK

Myrtle Y. Gordon
Imperial College London, UK

Imperial College Press

Published by

Imperial College Press
57 Shelton Street
Covent Garden
London WC2H 9HE

Distributed by

World Scientific Publishing Co. Pte. Ltd.
5 Toh Tuck Link, Singapore 596224
USA office: 27 Warren Street, Suite 401-402, Hackensack, NJ 07601
UK office: 57 Shelton Street, Covent Garden, London WC2H 9HE

British Library Cataloguing-in-Publication Data
A catalogue record for this book is available from the British Library.

ISBN-13 978-1-84816-256-3
ISBN-10 1-84816-256-1

Typeset by Stallion Press
Email: enquiries@stallionpress.com

Printed in Singapore by Mainland Press Pte Ltd

Contents

Contributors

Svend Aakhus, MD, PhD
Department of Cardiology
Rikshospitalet University Hospital
0027 Oslo, Norway

Adil Al Kindi, MD, MSc
Resident of Cardiac Surgery and Research Fellow
Divisions of Cardiothoracic Surgery and Surgical Research
Faculty of Medicine, McGill University
Montreal, Quebec, Canada

Muhammad Ashraf, PhD
Department of Pathology and
Laboratory of Medicine
University of Cincinnati
231-Albert Sabin Way
Cincinnati, OH 45267-0529, USA

Anthony Atala, MD
W.H. Boyce Professor and Director
Wake Forest Institute for Regenerative Medicine
Chair, Department of Urology
Wake Forest University School of Medicine
Medical Center Boulevard
Winston-Salem, NC 27157, USA

Kishore Bhakoo, PhD
Head, Stem Cell Imaging Group
MRC Clinical Sciences Centre
Imperial College London
Hammersmith Campus
Du Cane Road
London, W12 0NN, UK

Yen Chang, MD
Division of Cardiovascular Surgery
Veterans General Hospital-Taichung and
College of Medicine, National Yang-Ming University
Taipei, Taiwan, R.O.C.

Chun-Hung Chen, PhD
Department of Chemical Engineering
National Tsing Hua University
Hsinchu, Taiwan, R.O.C.

Shuhua Chen, PhD
Center of Physiology and Pathophysiology
Institute of Neurophysiology
Robert-Koch-Str. 39
50931 Cologne/Germany

Sung-Ching Chen, PhD
Department of Chemical Engineering
National Tsing Hua University
Hsinchu, Taiwan, R.O.C.

Huihua Kenny Chiang, PhD
Institute of Biomedical Engineering
National Yang-Ming University
Taipei, Taiwan, R.O.C.

Ray Chu-Jeng Chiu, MD, PhD, FRCSC, FACC
Professor of Surgery and Director Emeritus
Divisions of Cardiothoracic Surgery and Surgical Research
Faculty of Medicine, McGill University
Montreal, Quebec, Canada

Paolo De Coppi, MD, PhD
Adjunct Assistant Professor, Institute for Regenerative Medicine
Wake Forest University School of Medicine
Clinical Senior Lecturer and Consultant
Surgery Unit, University College London (UCL)
Institute of Child Health and Great Ormond Street Hospital
30 Guilford Street, London, WC1N 1EH, UK

Ioannis Dimarakis, MRCS, PhD
Division of Surgery, Oncology,
   Reproductive Biology and Anaesthetics (SORA)
Faculty of Medicine
Imperial College London
Hammersmith Hospital
Du Cane Rd
London, W12 0NN, UK

Michael Xavier Doss, PhD
Center of Physiology and Pathophysiology
Institute of Neurophysiology
Robert-Koch-Str. 39
50931 Cologne/Germany

Myrtle Y. Gordon, DSc, PhD, FRCPath
Department of Haematology
Investigative Science Division
Faculty of Medicine
Imperial College London
Hammersmith Hospital
Du Cane Rd, 4N5 Commonwealth Building
London, W12 0NN, UK

Danielle Gottlieb, MS, MD, MPH
Department of Cardiovascular Surgery
Children's Hospital Boston and Harvard Medical School
300 Longwood Avenue, Boston, MA 02115, USA

Nagy A. Habib, FRCS, MD
Department of Biosurgery and Surgical Technology
Division of Surgery, Oncology,
   Reproductive Biology and Anaesthetics (SORA)
Faculty of Medicine
Imperial College London
Hammersmith Hospital
Du Cane Rd
World of Surgery
BN1/17 B Block
London, W12 0NN, UK

Khawaja Husnain Haider, MPharm, PhD
Department of Pathology and
Laboratory of Medicine
University of Cincinnati
231-Albert Sabin Way
Cincinnati, OH 45267-0529, USA

Marcel Halbach, MD
Center of Physiology and Pathophysiology
Institute of Neurophysiology
Robert-Koch-Str. 39
50931 Cologne/Germany

Jürgen Hescheler, MD
Center of Physiology and Pathophysiology
Institute of Neurophysiology
Robert-Koch-Str. 39
50931 Cologne/Germany

Shiaw-Min Hwang, PhD
Bioresource Collection and Research Center
Food Industry Research and Development Institute
Hsinchu, Taiwan, R.O.C.

Naidu Kamisetti
Center of Physiology and Pathophysiology
Institute of Neurophysiology
Robert-Koch-Str. 39
50931 Cologne/Germany

Po-Hong Lai, PhD
Department of Chemical Engineering
National Tsing Hua University
Hsinchu, Taiwan, R.O.C.

Jonathan Leor, MD
Neufeld Cardiac Research Institute
Sheba Medical Center, Tel-Aviv University
Tel-Hashomer, Israel 52621

Nataša Levičar, PhD
Department of Oncology
Division of Surgery, Oncology,
    Reproductive Biology and Anaesthetics (SORA)
Faculty of Medicine
Imperial College London
Hammersmith Hospital
Du Cane Rd
World of Surgery
BN1/17 B Block
London, W12 0NN, UK

Mayasari Lim
Biological Systems Engineering Laboratory
Department of Chemical Engineering
Imperial College London
South Kensington Campus
London, SW7 2AZ, UK

Wei-Wen Lin, MD
Division of Cardiology
Veterans General Hospital-Taichung
Taiwan, R.O.C.

Ketil Lunde, MD, PhD
Department of Cardiology
Rikshospitalet University Hospital
0027 Oslo, Norway

Jun Luo, MD, PhD
Research Fellow
Divisions of Cardiothoracic Surgery and Surgical Research
Faculty of Medicine, McGill University
Montreal, Quebec, Canada

Athanasios Mantalaris, PhD
Biological Systems Engineering Laboratory
Department of Chemical Engineering
Imperial College London
South Kensington Campus
London, SW7 2AZ, UK

Elad Maor, MD
Biophysics Graduate Group
University of California, Berkeley
CA 94720-3200, USA

John E. Mayer, Jr., MD
Department of Cardiovascular Surgery
Children's Hospital Boston and Harvard Medical School
300 Longwood Avenue, Boston, MA 02115, USA

Philippe Menasché, MD, PhD
Assistance Publique-Hôpitaux de Paris
Hôpital Européen Georges Pompidou
Department of Cardiovascular Surgery;
Faculté de Médecine, Université Paris Descartes;
INSERM U 633, Paris, France

Arnon Nagler, MD
Director, Hematology Division
Chaim Sheba Medical Center
Tel Hashomer, Israel 52621

Georgios Nteliopoulos, Bsc (hons), MSc
Department of Haematology
Division of Investigative Science
Faculty of Medicine
Imperial College London
Hammersmith Hospital Campus
Du Cane Road
London, W12 0NN, UK

Nicki Panoskaltsis, MD, PhD
Department of Hematology
Imperial College London
Northwick Park & St. Mark's Campus
London, HA1 3UJ, UK

Agapios Sachinidis, PhD
Center of Physiology and Pathophysiology
Institute of Neurophysiology
Robert-Koch-Str. 39
50931 Cologne/Germany

Preface

Huge steps have been made in cardiovascular stem cell transplantation over the past decade. Basic scientists, interventional cardiologists, as well as cardiac surgeons have coordinated their efforts to further investigate the potential therapeutic value of stem cells in ischaemic heart disease. Although we may have moved from terms such as 'regeneration' towards the more widely accepted 'repair', it is evident that majority of scientific queries remain to be addressed.

This book is an attempt to encompass contributions spanning at various levels of stem cell research from bench top and bedside. Insight regarding the different stem cell types in experimental and clinical trials accounts for a significant part of this book. In addition, we have incorporated a group of basic science papers deciphering various mechanisms of stem cell function in cardiac studies. The role for stem cell therapy in heart valve disease, gene therapy, and bioengineering is also been discussed. At last, we have integrated a chapter on 'bioprocessing' in an attempt to make more transparent the potential of this tool in cardiac stem cell research.

Inarguably clinicians have acquired the procedural know-how of cell delivery and subsequent assessment of cardiac performance. As adult stem cell subpopulations and embryonic stem cells will be entering the clinical arena in the maybe not so distant future, this knowledge will prove invaluable in their clinical assessment. Hopefully, cardiac stem cell research and ultimately therapy will provide an additional strategy for the treatment of ischaemic heart disease.

The editors are indebted to all contributors who are leading authorities in their own research niches. Finally, all editors would like to take the opportunity to thank Mr Lance Sucharov, Senior Commissioning Editor at Imperial College Press for his invaluable help in bringing this project to life.

Johannes Winkler, PhD
Center of Physiology and Pathophysiology
Institute of Neurophysiology
Robert-Koch-Str. 39
50931 Cologne/Germany

Hua Ye, PhD
Institute of Biomedical Engineering
Department of Engineering Science
University of Oxford
Oxford, OX3 7DQ, UK

Tomo Šarić, PhD, MD
Center of Physiology and Pathophysiology
Institute of Neurophysiology
Robert-Koch-Str. 39
50931 Cologne/Germany

Saverio Sartore, PhD
Professor
Department of Biomedical Sciences
University of Padua
Viale G. Colombo, 3
I-35121 Padua, Italy

Dominique Shum-Tim, MD, MSc, FRCSC
Associate Professor of Surgery
Divisions of Cardiothoracic Surgery and Surgical Research
Faculty of Medicine, McGill University
Montreal, Quebec, Canada

Matthias Siepe, MD
Department of Cardiovascular Surgery
University Medical Center Freiburg
Hugstetterstr. 55
D-79106 Freiburg, Germany

Hsing-Wen Sung, PhD
Department of Chemical Engineering
National Tsing Hua University
Hsinchu, Taiwan, R.O.C.

Fraser W. H. Sutherland, MA, MD, FRCS(Eng), FRCS(C-Th)
Consultant Cardiac Surgeon
Golden Jubilee National Hospital
Beardmore Street
Clydebank G81 4HX, Scotland

Hao-Ji Wei, MD
Division of Cardiovascular Surgery
Veterans General Hospital-Taichung and
College of Medicine, National Yang-Ming University
Taipei, Taiwan, R.O.C.

1 The Meritocracy of Stem Cells for Therapy

Myrtle Y. Gordon and Nagy A. Habib

Recent developments in stem cell biology have indicated a remarkable differentiation plasticity of adult stem cells in many tissues in the body. These findings have led to hopes that stem cell therapy can solve the problem of degenerative disorders for which organ transplantation is inappropriate or there is a shortage of organ donors. Stem cell transplantation for regenerative medicine is still in its infancy, but in many ways may be comparable to the early years of clinical bone marrow transplantation for haematological disease. Moreover, it can be anticipated that the progress of stem cell therapy in this broader context will gain from the experience in bone marrow transplantation. For example, bone marrow transplantation benefited from the dissemination of information via the transplant registries, much is known of the interactions between grafted cells and host tissue, there is considerable experience in the conditioning necessary for acceptance of the transplanted stem cells and processing technologies that are compatible with Good Manufacturing Practice requirements and regulatory authority approval have been developed.

Stem cells with the potential for the treatment of a wide range of degenerative disorders may be obtained from a variety of sources but, for practical reasons, some of them are more likely to find earlier clinical application than others. The main types that have been studied in the context of stem cell therapy are embryonic, foetal, and adult stem cells. The bone marrow is an easily available source of mesenchymal and haemopoietic stem cells that have been proposed for application in the treatment of tissue damage and degeneration like Alzheimer's and Parkinson's diseases, heart failure, myocardial infarction, diabetes, and liver insufficiency. Both the haemopoietic and mesenchymal stem cell pools contain, or generate, subpopulations of stem cells that have been variously claimed to have therapeutic potential. Already, the results of preliminary clinical trials evaluating bone marrow-derived stem cells in the treatment of heart disease and liver failure have been reported and it is timely to consider the relative merits of some of the candidate stem cell populations.

Adult stem cells exist in many tissues and organs and these stem cells have differentiation potentials beyond those required to regenerate the tissue or organ in which they reside.[1] Clearly, some of these sources of adult stem cells, like brain stem cells, are less accessible than others, like bone marrow stem cells. The use of bone marrow-derived stem cells for tissue and organ repair has the additional advantage that there is considerable experience in the clinical application of bone marrow transplantation for the regeneration of the haemopoietic system, dating from the 1970s when relatively large clinical studies were initiated. By 2002, 20,207 haemopoietic stem cell transplants had been performed in Europe by 586 teams in 39 countries. This clinical experience has been accompanied by a wealth of laboratory studies on haemopoietic stem cells and transplantation biology.

In general, two distinct stem cell populations are thought to reside in haemopoietic tissue. These are the haemopoietic and the mesenchymal stem cells. Classically, the haemopoietic stem cells are the precursors of all of the blood cells,[2] and the mesenchymal stem cells are the source of the supporting stromal cells of the bone marrow, lineages[3] including the osteogenic, chondrogenic, and adipogenic lineages.[4] Multipotent adult progenitor cells (MAPC) are a sub-population of cells that arise in cultures of mesenchymal stem cells and seem to have a broader differentiation potential than the mesenchymal stem cells themselves.[5] However, many cell population doublings are required before MAPC arise in mesenchymal cell cultures and this has been associated with a potential for genetic instability. Other consequences are that it is not known if MAPC exist *in vivo* or what their *in vivo* phenotype may be, MAPC cannot be prospectively isolated from a tissue like the bone marrow and there is no quantative assay for MAPC so that it is not possible to predict with any accuracy how much tissue would be required to supply enough cells for a particular application. Similar considerations apply to other types of stem cell sub-populations with a mesenchymal cell origin that have been described.[6–8]

There are a number of key points that should be considered with regard to the clinical application of stem cells: ethical acceptability, ease of availability, numbers available, phenotypic identification for prospective isolation, physiological normality, and regulatory compliance. Ideal features of a stem cell population for therapy would be the identification of a homogeneous stem cell population from an ethically uncontroversial source that exists normally *in vivo* and which could be prospectively isolated from an easily available tissue like the blood or bone marrow. Ideally, prolonged tissue culture would not be necessary and the number of cells required would be attainable within a short period of time. Identification of a cell type in the bone marrow with these properties and with the ability to differentiate into multiple cell types would fulfil the immediate requirements for the early clinical application of stem cell therapy.[9]

The origin of embryonic stem cells is ethically unacceptable to many individuals since it requires the manipulation of pre-implantation embryos and compromises embryonic survival. Unacceptability is based on the belief that an egg fertilised *in vitro* has the same rights as a post-natal human being and has resulted in limitations of federal funding in the United States. Considerable efforts are being made to circumvent the ethical hurdles involved in the clinical application of embryonic stem cells.[10-11] In contrast, adult stem cells for clinical trials in regenerative medicine are obtained without severe intervention from sources like bone marrow and blood, and with informed consent.

The stem cells need to be obtained from an accessible source. Bone marrow and blood are obvious choices with the advantage that there is considerable experience in bone marrow harvesting and leukapheresis of G-CSF mobilised blood in the haematological community. Moreover, the identification of the stem cells in the source tissue is desirable since it allows prospective enumeration of the cells and also confirms that they are a normal physiological component. This requires knowledge of their phenotypic properties. Mesenchymal stem cells remain poorly characterised and stem cell populations like the MAPC[5] and USSCs[8] are only derived after long periods in culture. In contrast, the VSEL,[12] side population and Hoechst low rhodamine low cells[13] are identifiable by multiparameter flow cytometry and can be purified by FACS and the small lymphocyte-like stem cells we have described express CD34 and are separated from the bulk of the CD34-positive cells using their adherence properties. A stem cell population that is purified to homogeneity may be desirable because this permits a high concentration of the relevant cells to be transplanted locally into injured tissue. Regulatory compliance includes method of isolation and flow cytometry, for example, does not fulfil this objective. However, isolation of CD34+ cells from haemopoietic tissue has been used in stem cell transplantation and some devices for this manipulation are accepted by the regulatory authorities.

The number of stem cells available for clinical use can be a limiting factor. These numbers may be available immediately or require a period in culture to amplify them. The difficulties associated with amplifying haemopoietic stem cell numbers are notorious because stem cells tend to divide asymmetrically and this is incompatible with an increase in stem cell numbers.[14-17] Also, large numbers of cell divisions introduce the risk of genetic instability.[18] Some of these obstacles may be overcome by so-called therapeutic cloning although it attracts some of the controversies associated with embryonic stem cell research. Amplification in cell numbers can be achieved by genetic manipulation of the stem cells but this step seems unwarranted, except perhaps in extreme cases, in view of the inherent risks attached to transplanting permanently, or conditionally, self-renewing cells. Embryonic stem cells and, to a lesser extent, foetal stem cells have the potential to repair many types of tissues because they are totipotent.[19] Embryonic stem cells can be greatly increased in number in culture as cell lines *in vitro* and may be immuno-priviledged. These attributes mean that they can be used to treat multiple patients. However, their use has been confounded by the very real likelihood that, being immortal, they will form tumours after they have been transplanted into patients.[20] Undoubtedly, however, these barriers to widespread application will be overcome in the future.

In spite of the limited knowledge about the best source and type of stem cells to use for clinical applications, stem cell therapy for degenerative conditions is being applied in several settings. For example, stem cell-induced cardiac regeneration in patients with ischaemic heart failure has now been investigated by many groups with encouraging results.[21] The cells administered were obtained from bone marrow and injected by intramyocardial, intracoronary, and transendocardial routes. We have performed a phase I clinical trial of stem cell transplantation in patients with liver insufficiency. For this, autologous mobilised stem cells were procured and purified before they were injected into the portal vein or hepatic artery for local delivery into the damaged tissue.[9] This experience has demonstrated the safety and lack of toxicity of the procedure and has led to the initiation of a phase II clinical trial. These examples demonstrate that the revolution of using stem cells for regenerative medicine has begun. It is to be anticipated that the future will see ever expanding applications of this novel approach to conditions involving tissue damage and degeneration.

References

1. Lakshmipathy U, Verfaillie C. Stem cell plasticity. *Blood Rev* 2005;19:29–38.
2. Quesenberry PJ, Levitt L. Hematopoietic stem cells. *N Engl J Med* 1979;301(Pt 1–3): 755–760, 819–823, 868–872.
3. Friedenstein AJ, Chailakhjan RK, Lalykina KS. The development of fibroblast colonies in mono-layers or guinea-pig bone marrow and spleen cells. *Cell Tissue Kinet* 1970;3:393–403.
4. Pereira R, Halford K, O'Hara M, *et al.* Cultured adherent cells from marrow can serve as long-lasting precursor cells for bone, cartilage and lung in irradiated mice. *Proc Natl Acad Sci USA* 1995;92:4857–4861.
5. Jiang Y, Jahagirder BN, Reinhardt RL, *et al.* Pluripotency of mesenchymal stem cells derived from adult marrow. *Nature* 2002;418:41–49.
6. Colter DC, Sekiya I, Prockop DJ. Identification of a subpopulation of rapidly self-renewing and multipotential adult stem cells in colonies of human marrow stromal cells. *Proc Natl Acad Sci USA* 2001;98:7841–7845.
7. Smith JR, Pochampally R, Perry A, Hsu S-C, Prockop DJ. Isolation of a highly clonogenic and multipotential subfraction of adult stem cells from bone marrow stroma. *Stem Cells* 2004;22:823–831.
8. Kogler G, Sensken S, Airy JA, *et al.* A new human somatic stem cell from human placental cord blood with intrinsic pluripotent differentiation potential. *J Exp Med* 2004;200:123–135.
9. Gordon MY, Levicar N, Bachellier P, *et al.* Characterisation and clinical application of human CD34+ stem/progenitor cell populations mobilized into the blood by granulocyte colony-stimulating factor. *Stem Cells* 2006;24:1822–1830.
10. Dolgin JL. Embryonic discourse. *Issues Law Med* 2004;19:203–261.
11. Sclaeger TM, Lensch MW, Yaylor PL. Science aside: the trajectory of embryonic stem cell research in the USA. *Drug Discov Today* 2007;12:269–271.
12. Kucia M, Zuba-Surma E, Wysoczynski M, Dobrowolska H, Reca R, Ratajczak J, Ratajczak MZ. Physiological and pathological consequences of very small embryonic-like (VSEL) stem cells in adult bone marrow. *J Physiol Pharmacol* 2006;57(Suppl 5):5–18.
13. Challen GA, Little MH. A side order of stem cells: the SP phenotype. *Stem Cells* 2006;24:3–12.
14. Gordon MY, Blackett NM. Some factors determining the minimum number of cells required for successful clinical engraftment. *Bone Marrow Transplant* 1995;15:659–662.
15. Sherley JL. Asymmetric cell kinetics genes: the key to expansion of adult stem cells in culture. *Stem Cells* 2002;20:561–572.

16. Marley SB, Lewis JL, Gordon MY. Progenitor cells divide symmetrically to generate new colony-forming cells and clonal heterogeneity. *Br J Haematol* 2003;121:643–648.

17. Joseph NM, Morrison SJ. Towards an understanding of the physiological function of mammalian stem cells. *Dev Cell* 2005;9:173–183.

18. Miura M, Miura Y, Hesed M, *et al*. Accumulated chromosomal instability in murine bone marrow mesenchymal stem cells leads to malignant transformation. *Stem Cells* 2006;24:1095–1103.

19. Lerou PH, Daley GQ. Therapeutic potential of embryonic stem cells. *Blood Rev* 2005;19: 321–331.

20. Erdo F, Buhrle C, Blunk J, *et al*. Host-dependent tumorigenesis of embryonic stem cell transplantation in experimental stroke. *J Cereb Blood Flow Metab* 2003;23:780–785.

21. Dimarakis I, Habib NA, Gordon MY. Adult bone marrow-derived stem cells and the injured heart: just the beginning? *Eur J Cardiothorac Surg* 2005;28:665–676.

2 Overview of Adult Stem Cell Therapy in Cardiac Disease

Georgios Nteliopoulos, Nataša Levičar
and Ioannis Dimarakis

Introduction

Heart failure is the terminal stage for many diseases that affect the myocardium. Advanced stages are characterised by changes at the nuclear, cellular and finally organ level. It is a condition that is extremely prevalent, affecting an estimated five million people in the United States, and it underlies or contributes to the deaths of 286,700 people a year.[1] Coronary artery disease leading to ischaemia of the human myocardium is a major cause of heart failure, especially in males.[2] An estimated 565,000 new myocardial infarctions and 300,000 recurrent infarctions occur each year in the United States alone.[1] Timely coronary reperfusion using pharmacological agents and percutaneous coronary intervention reduces mortality resulting from myocardial infarction,[3] but adverse myocardial remodelling may still occur, in which changes in ventricular architecture result in impaired pump function and contribute to heart failure.[4]

At present the only radical treatment of end-stage congestive heart failure is cardiac transplantation. Shortage of donors poses a severe limitation of this approach without mentioning the complications of post-operative immunosuppression. Various techniques have been described and developed to reshape the dilated left ventricle either by endocardial patch plasty or by passive constraint and shape-change.[5,6] Left ventricular assist devices do exist, mainly to facilitate bridging patients to cardiac transplantation. Therefore, several alternative novel approaches to treat end-stage heart failure are being considered. One such approach is cellular transplantation, where impaired myocardium could be replaced by stem cells that would expand in the recipient heart.

Regenerative Potential of the Native Myocardial Tissue

It has been traditionally considered that human cardiomyocytes retreat from the cell cycle in early post-natal life. Thus, adult myocardial cells are regarded as

post-mitotic end-differentiated cells with no ability for division and proliferation following an ischaemic insult. Recent experimental evidence in addition to
clinical findings are questioning these assumptions. In the murine foetal heart the
retinoblastoma protein has been implicated in the observed cell withdrawal from
the cell cycle.[7] It has been shown that inactivation of the retinoblastoma protein
in terminally differentiated cells allows them to re-enter the cell cycle.[8] Endo and
Nadal-Ginard were able to show terminally differentiated myotubes completing
canonical mitosis and cytokinesis illustrating the proof of principal that even terminally differentiated cells retain the ability to re-enter the cell cycle.[9]

Synthesis of cardiomyocyte DNA in the mammalian heart has been demonstrated
without researchers being able to ascertain if this is part of a physiologic response.
Activation of cardiomyocyte DNA may require certain *in vivo* micro-environmental
alterations.[10] Kajstura *et al.* reported, by examining control human hearts by con-
focal microscopy, that 14 myocytes per million were seen to be in mitosis. A nearly
ten-fold increase in these figures was observed in end-stage ischaemic heart disease
and in idiopathic dilated cardiomyopathy.[11] A slow but persistent myocyte turnover
seems to exist providing a theoretical explanation of why certain pathologies linked
with myocyte death, such as diabetic cardiomyopathy, do not progress to myocardial
cellular "wipe-out".[12]

Conventional teaching has been that cardiomyocyte hypertrophy is the only
mechanism to compensate for loss of functional myocardium following an ischaemic
insult. Cellular proliferation has been attributed exclusively to endothelial and
fibroblast populations, with collateral circulation and scar formation being the
product of these proliferations, respectively. Recent studies, however, have shown
that a small number of cardiomyocytes re-enter the cell cycle following such an
insult.[13] This proliferation is not sufficient to compensate for the up to one billion
cardiomyocytes that are lost after myocardial infraction.[14]

The origin of primitive cells residing in the myocardium remains unclear.
Whether they belong to a resident cardiac stem cell population or are recruited
within the heart from the systemic circulation, ultimately being derived from the
bone marrow remains to be answered.

The finding of cardiac chimerism following peripheral blood and bone marrow
transplantation in sex-mismatched patients has shed new light on this topic. Bayes-
Genis *et al.* reported the presence of cardiac chimerism with the use of a sensitive
polymerase chain reaction assay for donor and recipient genotyping.[15] Furthermore,
homing of recipient progenitor cells in the myocardium has been shown in sex-
mismatched cardiac transplantations.[16,17] Based on their observations, Quaini *et al.*
hypothesise that the donor heart harbours a population of resident primitive cells as
well as attracting a second population of progenitor cells from the recipient. Further
support for this notion has been provided by the demonstration of non-haemopoietic
post-natal human bone marrow cells expressing early cardiac markers[18]; data from
our group has also confirmed this.[19]

Several cardiac progenitor cell (CPC) pools residing in post-natal hearts have been reported in different species including mice, rats, dogs, and humans. The degree of overlap between different CPC sub-populations remains unclear.

Stem cell factor receptor (c-kit) is widely used as a cell surface marker of pluripotent stem cells. Clusters of c-kit$^+$/Lin$^-$ cells expressing Ki67 as well as the transcription factors GATA4, Nkx2.5, and MEF2C have been identified lying in the intersticia between terminally differentiated myocytes.[20] These cells show pluripotency and ability for self-renewal *in vitro* and can differentiate into cardiomyocytes, smooth muscle and endothelial cells.[20] The authors do acknowledge that it is impossible to determine whether these cells are true stem cells or multipotent cells that have progressed further along the differentiation pathway. The c-kit$^+$/Lin$^-$ fraction of stem cells within the myocardium at any given time consists of a mixture of resident and bone marrow cells that are chemoattracted in relation to inflammatory stimuli.[21] Age, underlying pathology and the stage of disease are important parameters determining the above ratio.

Stem cell antigen-1 (Sca-1) is another murine cell surface marker used to identify primitive cells. In the adult murine heart, Sca-1$^+$ cells are identified as small interstitial cells adjacent to the basal lamina, typically co-expressing platelet–endothelial cell adhesion molecules (CD31) in close proximity with endothelial Sca-1$^-$ CD31$^+$ cells.[22] Based on the lack of haemopoietic and (pre)mature endothelial marker expression, the authors argue against a haemopoietic or endothelial phenotype. Interestingly, telomerase activity suggestive of self-renewal capacity was documented in cardiac Sca-1$^+$ cells in similar levels to neonatal myocardium; telomerase was not detected in cardiac Sca-1$^-$ cells. Although 0.03% of cardiac cells exert side population (SP) properties, approximately 93% of the SP fraction was Sca-1$^+$ cells. SP cells have the ability to efflux a Hoechst dye by a MDR-like protein and this feature has been applied to isolate resident CPCs as will be discussed later. While cardiac Sca-1$^+$ cells initially lack the expression of cardiac structural genes, these cells are capable of differentiating into cardiomyocytes when exposed to 5-azacytidine (DNA demethylation promoting agent) *in vitro*. Pfister *et al.* reported that among the SP cells the greatest potential for cardiomyogenic differentiation is restricted to CD31$^-$/Sca-1$^+$ cells.[23]

Cardiospheres which are myocardial stem cells capable of proliferating *in vitro* isolated from adult hearts of humans and mice.[24] In a supplement with EGF, basic FGF, cardiotrophin-1, thrombin and under low-serum conditions a cell population of the dissociating heart formed spheroids on poly-D-lysine-coated culture dish. These cells expressed c-kit, CD34 and kinase insert domain receptor (KDR). *In vitro*, murine cardiosphere-derived cells beat spontaneously, while human cells require co-culture with rat cardiac myocytes. Another category of cardiospheres are the SP derived neuronal crest stem cells, which reside mainly at the outflow tracts, express markers of undifferentiated neural precursor cells and differentiate into neurons, glia, smooth muscle cells, and cardiomyocytes.[25]

Cardioblasts express the LIM homeodomain transcription factor Islet-1 (Isl-1) and are not present in the SP fraction.[26] The LIM-homeodomain transcription factor islet-1 (isl-1) is linked with a cell population that makes a substantial contribution to the embryonic heart, comprising most cells in the right ventricle, both atria, the outflow tract as well as specific regions of the left ventricle. They can be expanded and show distinct cardiomyogenic potential *in vitro*. Although these cells may be directly traced in the myocardium from the early embryonic stage, their presence appears to be confined to the first days postpartum in humans. Further work from the same group has lead to the discovery of the multipotent Isl1$^+$ cardiovascular progenitor cell (MICP).[27]

The theoretical possibility that these endogenous cardiac stem cells could be expanded and differentiated to cardiomyocytes *in vitro* to generate sufficient numbers for the purpose of implantation does exist.[24] However, the required invasive nature of cell harvesting in combination with the small amount of cells that may be eventually isolated from myocardial biopsies make such a procedure at least impractical at present. Myocytes injected into an infarct scar would also need to have an adequate vascular supply to be able to survive in their new environment.

Adult Stem Cells

The main difficulty in using cellular transplantation is their lack of structure. The myocardium is formed from units, consisting of myofibres oriented in a particular direction, with a vascular network, an electrical conduction system, gap junctions connecting myocytes allowing electromechanical coupling, a nervous system, and an underlying fibrous skeleton.[28] Ideally, if we aim to use a stem cell population to implant into infarcted myocardium, these cells should replenish dying myocytes by giving rise to functional cardiomyocytes (myogenesis) and new blood vessels (angiogenesis). For this, cells have to acquire cardiac morphology and express cardiac specific markers and sarcomeric structures. More importantly, they must demonstrate automatically robust functional and electromechanical coupling with the host myocytes and work in a functional syncytium with the rest of the cardiac muscle.[28] Other desired characteristics should be obtained from the same patient (autologous) so as to avoid immunosuppressive therapy and to be easy to isolate in relatively large quantities.

Bone marrow-Derived stem cells

The plasticity of adult bone marrow-derived stem cells is the motivating factor for research in the field of myocardial regeneration. Minimal criteria of stem cell plasticity as outlined by Verfaillie include capability for self-renewal, ability of a single cell to differentiate into cells of the tissue of origin and at least one different cell type, and functional differentiation *in vivo* into cells of the tissue of origin and at least one cell type other than the tissue of origin.[29] Bone marrow and blood are attractive, readily accessible sources of stem cells. The multilineage differentiation and the multiorgan engraftment potential of bone marrow-derived stem cells have been previously demonstrated.[30] Other advantages of these cells are the facts that they are easy to prepare, simple to administer and do not require additional immuno-suppressive treatment as well as their use does not raise the ethical controversies associated with the use of embryonic stem cells. Multiple experimental as well as phase I/II clinical studies have been carried out to date assessing various types of adult bone marrow-derived stem cell delivery (discussed in Chapter 10).

At least two distinct populations of stem cells reside within the adult bone marrow: haemopoietic stem cells (HSCs) and mesenchymal stem cells (MSCs). The former are the precursors of all blood myeloid and lymphoid lineages and the latter give rise to stromal cells of the bone marrow including distinct mesodermal lineages such as osteoblasts, chondrocytes, and adipocytes.

Reconstitution of the haemopoietic system of a myeloablated host is the sole assay available at present for identification of HSCs. Surface markers in humans, such as CD34 or the more primitive CD133 (AC133) help in distinguishing sub-populations enriched in HSCs. Other haemopoietic markers are CD45 and c-kit (CD117), and stem cell factor (SCF) receptor. As murine progenitor cells do not express the same surface markers, the principle means to identify enriched popu-lations is lineage depletion (Lin⁻).[31] Another murine cell surface marker used to identify primitive cells is stem cell antigen-1 (Sca-1). In Orlic's landmark paper,[32] transplanted Lin⁻ c-kit⁺ were demonstrated to form new myocytes, endothelial cells and smooth muscle cells, leading thus to *de novo* myocardial regeneration; similar data has also been produced with human peripheral blood CD34⁺ cells.[33] Nonetheless, controversy remains as to whether HSCs are capable of transdif-ferentiation into cardiomyocytes[34,35] or even confer any functional improvement whatsoever.[36]

Other populations of cells that are enriched for HSCs despite the fact that they are isolated independently of classic antigenic markers are the SP cells. SP cells account for about 0.05% of the total bone marrow cell population and are identified by their ability to extrude Hoechst dye via the ABCG2 transporter on the membrane.[37] SP cells reside in various organs besides the bone marrow and there is conflicting data as to the origin of these cells with some evidence supporting the fact of actually being bone marrow-derived HSCs rather than tissue-specific.[38,39] Furthermore, SP cells isolated from different tissues share transcriptome signatures[40] and common "stemness functions".[41]

Endothelial progenitor cells (EPCs) or angioblasts differentiate into endothelial cells and are defined by their ability to participate in neovasularisation. Endothelial progenitor cells and HSCs are thought to share a common precursor, the haemangioblast.[42] They exist in the bone marrow as well as the systemic circu-lation. Studies have shown the connection between angioblasts residing within the bone marrow and neovascularisation.[43,44] EPCs are recruited to sites of ischaemic injury and promote angiogenesis; Fazel and colleagues suggest that this is achieved via regulation of the myocardial levels of various angiogenic cytokines such as the angiopoietins-1 and -2, and vascular endothelial growth factor.[21] Primitive bone marrow EPCs express CD133, CD34 and CD45 and can be indistinguishable from primitive HSCs. Mature EPCs express CD34, vascular endothelial growth factor receptor (VEGFR2), Tie-2, KDR, VE-cadherin, CD31, von Willebrand factor and are negative for CD45. Absolute numbers and functional activity of EPCs appear to be inversely correlated with risk factors for coronary artery disease[45,46]; an issue to be considered in the design of clinical studies if one bears in mind target population characteristics.

MSCs (also known as marrow stromal cells) are characterised by two distinguishing features, the ability to adhere to cell culture dishes as well as to differentiate into a variety of lineages under appropriate conditioning. They are rare in bone marrow, representing approximately one in 10,000 nucleated cells[47] and besides renewing marrow stroma and supporting haemopoiesis, are capable of differentiating into osteogenic, chondrogenic and adipogenic lineages. Surface proteins including CD105 (SH2), CD73 (SH3, SH4), CD29, CD44, CD71, CD90 (Thy-1), CD106 (vascular cell adhesion molecule-1), CD120a, and CD124, have been reported to be expressed in multipotent MSCs.[48] In contrast to HSCs these cells appear uniformly negative for typical haemopoietic surface antigens such as CD45 and CD34. Many groups have demonstrated the proof of principal by trans-differentiating marrow stromal cells into cardiomyocytes *in vitro*. This has mainly been accomplished by treating cells with chemicals promoting DNA demethylation, such as 5-aza-2′deoxycytidine,[49–51] which induces transcription of critical transcription factors by demethylating CpG island of promoter regions. All cells produced expressed cardiomyocyte specific genes and could be seen to spontaneously beat in the petri dish. Apart from innate plasticity, MSCs may exhibit immunomodulatory effects[52] that may allow for allogeneic *in vivo* transplantation with minimal risk of immune rejection.[53] Possible co-delivery of MSCs with bone marrow-derived mononuclear cells to maximise clinical effect has also been proposed.[54]

An interesting adult bone marrow-derived population of high plasticity reported by the Minnesota group are the multipotent adult progenitor cells (MAPCs).[55] Originally isolated from mesenchymal stem cell cultures (minute sub-population), this subset of progenitor cells has been shown to differentiate *in vitro* in cells of the three germ layers. Worth mentioning is the claim that single MAPCs injected into an early blastocyst, contribute to most, if not all, somatic cell types. While the phenotype of mMAPCs in fresh bone marrow remains unknown, the phenotype of cultured mMAPCs is CD34, CD44, CD45, c-Kit, and major histocompatibility complex (MHC) classes I and II negative; mMAPCs were shown express low levels of Flk-1, Sca-1, and Thy-1, and higher levels of CD13 and the stage-specific antigen I. A recent study did not yield any promising results regarding either global contractile function or evidence of differentiation into cardiomyocytes or engraftment with MAPCs in a rat model of myocardial transplantation.[56]

Yoon *et al.* have also reported a sub-population of stem cells within adult human BM, isolated at the single-cell level for myocardial regeneration.[57] This population expresses minimal levels (less than 3%) of CD90, CD115, and CD117. Authors claim that these cells are capable of inducing both therapeutic neovascularisation as well as endogenous and exogenous cardiomyogenesis.

Other adult stem cells

Besides the bone marrow several sources of adult stem cells are known. Zuk *et al.* have demonstrated that adipose tissue contains both HSCs and mesenchymal stem cell populations.[58] Planat-Benard *et al.* reported that cells from adipose tissue after culture under specific conditions give rise to beating cardiomyocytes, expressing cardiomyocyte-specific as well as sarcomeric markers.[59] However, up to date, it remains unknown if these cells are capable of engrafting into the adult heart and whether they will be correctly electromechanically coupled.

Skeletal myoblasts reside under the basal lamina of adult skeletal muscle and can be easily obtained by muscle biopsy. They may be expanded in culture, are ischaemia resistant and provide fatigue resistant, slow-twitch fibres. When compared to CD133[+] bone marrow-derived haemopoietic progenitor transplantation in an animal model similar results of functional improvement were recorded.[60] The main concern of introducing skeletal myoblasts within myocardial tissue is the lack of electrical communication with native cardiomyocytes.[61] There has been no demonstration of gap junction formation post-transplantation, which may potentially contribute to the generation of arrhythmias seen in patients receiving these cells. Genetic modification of myoblasts to express the gap junction protein connexin-43 has been proposed as an anti-arrhythmic measure[62]; unfortunately though, it seems that arrhythmias may be also attributed to other mechanisms beyond the lack of functional gap junction mediated intercellular communication.[63] Encouraging phase I clinical trial data exists incorporating trans-epicardial, trans-coronary venous, and trans-endocardial approaches.[64–69]

Other Cell Types

As briefly presented in Table 1 the armamentarium of potential donor cell types available to researchers is not confined to adult stem cells. Introduction of the majority of these cell groups in clinical trials is associated with serious drawbacks. Although the totipotent/pluripotent embryonic stem cells may seem as the ideal candidate, their clinical application is not foreseeable in the near future. Serious ethical considerations remain the main drawback; other issues complicating clinical adaptation include insufficient availability, associated disadvantages of allograft transplantation, unpredictable electrical behaviour as well as the risk of tumour formation. Undebatable proof for the ability of these cells to differentiate into cardiomyocytes cannot be other than the spontaneous contraction observed in embryoid bodies[70]; besides spontaneous "beating", cells from these areas exhibit typical immunopheno-typical, molecular, and electrophysiological properties of cardiomyocytes.[70,71] Following intramyocardial transplantation into rodent models of myocardial infarction, engraftment with subsequent cardiac-lineage differentiation has been reported by many groups; sustained improvement in cardiac function was also noted.[72,73] Recent data also suggests that embryonic stem cells are capable of homing to myocardial injury following peripheral delivery being chemoattracted to locally released cytokines.[74] Although a high degree of *in vitro* electromechanical coupling has been documented for embryonic stem cells and cardiomyocytes,[75] concerns regarding underlying arrhythmogenic potential[76] are still on the table.

Umbilical cord blood cells are easily obtainable from the placenta and may be cryopreserved for autologous or even allogeneic matched transplantation. Experience from the bone marrow transplantation front has shown that partial tissue mismatches are well tolerated; a smaller incidence of graft-versus-host disease is also associated with allogeneic cord blood transplantation compared to bone marrow transplantation. One hour after ligation of the left anterior descending artery, rats were injected with human cord cells in the peri-infarct border.[77] Long-term follow-up revealed substantial reduction in infarction size with improvement of left ventricular function parameters compared to untreated animals; interestingly animals were not given any immunosuppression in this study, confirming that cord blood cells may be immunoprivileged to a certain degree. When human cord cells were injected into the tail vein of control as well as NOD/SCID mice that were previously rendered ischaemic via LAD ligation, homing to the myocardium was only observed in the latter group.[78] Although there was no evidence of differentiation into cardiomyocytes, transplanted cells promoted neoangiogenesis, and exerted a beneficial influence on left ventricular remodelling.

Foetal cardiomyocyte transplantation has been successfully performed in various animal models. Most groups have shown formation of intercalated disks between grafted foetal cardiomyocytes and host myocardium. In an interesting study, Sakak-ibara *et al.* noted that foetal cardiomyocyte transplantation was able to prevent, but not to reverse, cardiac remodelling in myocardial infarction rats.[79] A Chinese group from Beijing has carried out human foetal cardiomyocyte transplantation (obtained from aborted human embryos) in a rat model of myocardial infarction with good functional results.[80] As with the previous group ethical issues, lack of availability, and the disadvantages of allograft transplantation apply to the foetal cell category. An exciting field of fetal tissue transplantation that may provide answers to many of these concerns is nuclear transplantation. Not only Lanza *et al.* demonstrated the feasibility of this technique, but also recorded myocardial regeneration significantly superior to that achieved with adult bone marrow cells.[81]

Due to an established phenotype and an unlimited availability cell lines have been thought to be a possible alternative for myocardial regeneration. Koh *et al.* transplanted cells from an established myoblast cell-line to murine ventricles.[82]

Table 1. Other potential donor cell groups.

Cell type	Advantages	Disadvantages
Embryonic stem cells	Totipotent/pluripotent Highly expandable Possibly immuno-privileged Cryopreservation for future use Immortal *in vitro*	Ethical issues Insufficient availability Risk of tumour formation

Umbilical cord stem cells	Pluripotent
	Cryopreservation for future use
	Easily available (autologous
	   transplantation in future)
		Ethical issues
		Allograft (applies only
		   to present)

Foetal cardiomyocytes	Established phenotype	Ethical issues Allograft Insufficient availability

Skeletal myoblasts	Established phenotype
	Easily available
	Ischaemia resistant
	Fatigue resistant,
	slow-twitch fibres
	Excellent expansion in culture
	Autologous transplantation
		No evidence of gap
		   junctions
		Potentially
		   arrhythmogenic
		Need for *ex vivo*
		   expansion

Cell line	Established phenotype Highly expandable Cryopreservation for future use Immortal *in vitro*	Allograft Risk of tumour formation

The absence of a cardiomyocyte cell-line along with the innate characteristics of cell lines (allograft transplantation, potential for tumour development) does not make this group favourable to scientists.

Potential donor cells other than bone marrow-derived are listed in Table 1.

Homing to Ischaemic Myocardium

Complex mechanisms, unknown to a great degree, are responsible for homing and holding haemopoietic progenitor cells within the bone marrow. Among these, and of clinical relevance, is the CXCR4/SDF-1 chemotactic axis. Stromal cell-derived factor-1 (SDF-1/CXCL12) is a marrow-stromal and bone-derived chemokine with two isoforms, α and β. It is thought to form a decreasing gradient from the extravascular toward the intravascular compartment within the bone marrow. SDF-1 plays an important role during foetal development in bringing about the relocation of foetal haemopoiesis to the bone marrow from the liver.[83,84] In the adult it promotes engraftment of transplanted HSCs within the recipient's bone marrow.[85] The chemotactic effects of SDF-1 are mediated by its receptor CXCR4 which is responsible for the observed chemoattraction of primitive bone marrow CD34$^+$CD38$^-$ populations *in vitro*.[86]

Mobilisation of stem cells is a dynamic process involving chemokines, cellular surface receptors, metalloproteinases, and adhesion molecules in addition to remodelling of the extracellular matrix. Ischaemia is known to cause migration of bone marrow-derived progenitors within ischaemic areas. SDF-1/CXCL12 and VEGF appear to be involved in this process.[87,88]

Ceradini *et al.* have demonstrated that SDF-1 gene expression is controlled by the transcription factor hypoxia-inducible factor-1 (HIF-1) translating to increased expression in the presence of tissue hypoxia.[89] Heissig *et al.* have shown in a murine model that bone marrow mobilisation takes place via a mechanism initially activating metalloproteinase-9 (MMP-9).[90]

By blocking the binding of SDF-1 to receptor CXCR4 in a rodent myocardial infarction model, homing of progenitor cells in injured myocardium was inhibited.[91] The establishment of a positive gradient of SDF-1 concentration between ischaemic tissue and bone marrow is suggested to promote mobilisation of progenitor cells and subsequent homing at sites of injury.[92] This would be consistent with the fact of cytokine induced bone marrow mobilisation with granulocyte colony-stimulating factor (G-CSF) being achieved via decreasing bone marrow levels of SDF-1.[93]

Locally delivered SDF-1 can augment neovascularisation in ischaemic hindlimb models.[94] VEGF gene expression is also upregulated in myocardial infarction[95,96] HIF-1 is an early transcription factor for VEGF during ischaemia-related tissue injury. This has been shown in human ventricular biopsies taken during surgical revascularisation following myocardial infarction.[97]

Further to these two factors, ischaemia induces the production of a variety of other factors and mediators that may participate in stem cell homing.[98] Even if they are not directly involved in generating a homing signal their significance lies in creating the appropriate microenvironment required for integration of the homed cells.

Mechanisms of Cardiac Function Improvement

Even if cell trafficking to ischaemia-related homing signals was fully understood, what takes place once the "homed" cells reached or are directly introduced to their new microenvironment is a great mystery and topic of dispute. The functional improvement observed in animal and human studies may constitute manifestations of either direct mechanical reinforcement of the left ventricular wall, incorporation and integration of implanted stem cells into native myocardial fibres, improvement in neovascularisation, or even be a result of paracrine factor production and delivery exerted by the implanted cells.

By introducing a volume of cell suspension into the myocardium, a local thickening occurs. Thus, taking advantage of the law of Laplace, less tension is exerted in the left ventricular wall avoiding further dilatation and haemodynamic deterioration.

The histopathological evidence of engraftment reflects the plastic character of the introduced cells. Possible theoretical factors accounting for this attributed plasticity include:

- the existence of myocardial specific stem cells within the bone marrow or other non-myocardial tissue,
- transdifferentiation of introduced cells into cardiac and/or endothelial cells,
- cellular fusion between the transplanted cells and host cardiomyocytes,
- injected cells (or certain sub-populations) may retain a pluripotent character enabling them to reconstruct *de novo* myocardial tissue,
- restoration of CSC niches, and
- combinations of the above.

Transdifferentiation refers to stem cells adopting the characteristics of other tissue-specific cell types. Orlic *et al.* indicated that HSCs after transplantation into infarcted mice differentiated into new myocytes, endothelial, and smooth muscle cells.[32] This claim for transdifferentiation has been challenged by many other groups and remains a controversial issue to date.[34-36]

Fusion between transplanted and host cells is another proposed mechanism. According to this, cell fusion causes the transfer of cell contents, including genetic material from transplanted to host cells. Cell fusion is a tightly regulated process restricted to a few types of normal somatic cells in humans.[99] Several reports showed that BMSCs engrafted into the adult heart myocardium by fusing with pre-existing cardiac myocytes.[100-102] It has also been suggested by investigators that what was supposed to be transdifferentiation was in fact cell fusion. Subsequent studies have excluded the contribution of cell fusion to cardiac regeneration and as it happens in such a low frequency, it is unlikely that cell fusion effects are significant enough to affect contractile function.[103]

Another mechanism by which stem cell therapy may have beneficial effects is via paracrine effect. The concept is that transplanted cells release cytokines, growth factors and signalling proteins that induce positive effects on resident cell populations stimulating thus proliferation and repair within host tissue. This positive effect on resident myocytes includes neoangiogenesis, proliferation of endogenous CPCs and prevention of apoptosis. It is highly likely the resident stem cells unable to cope with extensive myocardial injury are supported by distant stem cell recruitment, which may in turn convey a "boosting" effect to the resident stem cell niches.

Neovascularisation as a result of bone marrow-derived or circulating endothelial progenitor cells has been documented.[43,104,105] Many reports have demonstrated that different sub-populations of bone marrow-derived cells have the ability to secrete angiogenic factors, such as vascular endothelial growth factor, fibroblast growth factor, hepatocyte growth factor and angioprotein-1.[106,107] The formation of new blood vessels results in decreased apoptosis of myocytes in the border zone of the infarct limiting the remodelling process in this way.

Besides promoting neovascularisation via secretion of angiogenic factors as discussed previously, bone marrow-derived cells also secrete an array of other cytokines and growth factors. These include interleukin-1, interleukin-6, tumour necrosis factor-α and transforming growth factor-β that play a major role in tissue responses to injury, placental growth factor, stem cell derived factor-1, metalloproteinases, and granulocyte-macrophage colony-stimulating factor. In a mouse model of hindlimb ischaemia, cell-free MSC conditioned medium were shown to have beneficial effects on blood flow and limb function.[106] Collateral perfusion has been shown to be improved by angiogenic factors and cytokines produced by the transplanted cells.[108]

Finally, an interesting theory attempting to explain the observed improvement in cardiac outcome is the dying stem cell hypothesis.[109] As a great proportion of delivered stem cells die from apoptosis, it is proposed that down-regulation of both innate and adaptive immunity take place as apoptotic cells act as inhibitors of inflammation. This is mainly mediated via the production of anti-inflammatory cytokines by macrophages as well as immature dendritic cells, eventually leading to reduced scar formation and improved cardiac outcome.

Final Comments

As seen above majority of data accumulated to date originates from *in vivo* studies with plasticity of transplanted cells being attributed mainly to the mechanisms of milieu dependent differentiation or cell-to-cell fusion. This may provide researchers with optimism, but all underlying mechanisms continue to remain within the realm of medical hypothesis. The development of *in vitro* protocols to direct stem cell differentiation towards the cardiomyocyte lineage will not only assist in deciphering molecular signal pathways of differentiation, but also augment in creation of safer and more efficient stem cell transplantation models. Soluble chemokines, cell surface receptors, extracellular matrix substrata as well as intercellular gap junctions, are all potential co-factors.[110,111] To be able to elucidate this process at a cellular and molecular levels *in vitro* would translate to the ability to initiate and direct the differentiation process to a predefined point prior to clinical transplantation.

The degree of stem cell differentiation required prior to transplantation should also be established. An immature, more plastic cell may be more effective than an *ex vivo* pre-differentiated, committed cell and *vice versa*. Bittira *et al.* have actually shown that in the acute setting unmodified marrow stromal cells facilitate myocardial angiogenesis and myogenesis, whereas converting scar into myogenic tissue may be augmented by cell preprogramming before implantation.[112] Lu and co-workers have shown that following myocardial infarction in a rat model the local cues for homing of circulating cells at the site of injury are restricted to days 3 and 7 post-infarction.[113] They concluded that the potential for bone marrow-derived progenitor cell regeneration resides within the two first weeks. The need for local injury has been shown in animal studies. Cells delivered in control animals with no injury did not show any sign of engraftment.[114] On the other hand intracoronary infusion of cell suspension in healthy canines lead to micro-infarctions, as no local cues were available for cell homing and attraction, and there probably was no distinguishing factor between these cell boluses and systemic emboli.[115]

The main argument when debating over crude or pre-selected bone marrow-derived stem cell populations remains that of synergy versus specification. By definition crude preparations are less concentrated in potent stem cells and encompass a variety of different cell types. This in turn may expose subjects to unjustifiable procedure-associated risks with potentially unwanted side-effects.[116,117] On the other hand pre-selected populations may require to undergo multiple population doublings *in vitro* prior to delivery in order to achieve sufficient cell counts.[57] Prolonged passaging in culture is not without concerns though; replicative senescence,[118] changes in multipotentiality,[119] and spontaneous transformation[120] have all been associated with long-term *in vitro* culture. Finally, data concerning the arrythmogenic potential of transplanted MSCs has emerged from *in vitro* experimental work.[121]

Definition of delivered cell type(s) may also allow cell-type specific complications to be recognised and attributed. As both extremes of whole marrow preparations as well as clonally expanded sub-populations have been associated with potential drawbacks, it is clear that potential "ideal" sub-populations may well exist within this grey zone. For example, researchers have claimed that within the CD34$^+$ cell population, the CD34$^+$ KDR$^+$ fraction is responsible for the visible improvement in cardiac haemodynamics and hence represents the candidate active CD34$^+$ sub-population.[122] When designing an experimental or clinical trial serious consideration must be given to the starting number of stem cells to be delivered per subject. Iwasaki *et al.* reported a dose-dependent augmentation of vasculogenesis/cardiomyogenesis as well as a dose-dependent inhibition of ventricular fibrosis after CD34$^+$ cell transplantation after myocardial infarction in a rodent model.[123]

The significance of this approach lies in the fact that stem cell delivery is linked with cellular death due to the inherent hostility of the environment these cells are introduced into. In an effort to overcome this and improve cell survival during this stage many strategies have been developed. These include angiogenic pretreatment with injection of an adenovirus encoding VEGF,[124] prevascularisation with gelatin microspheres containing various growth factors.[125] As well as *ex vivo* retroviral transduction of cells to overexpress the prosurvival gene Akt1.[126]

Research is conducted at both the level of experimental animal and clinical human transplantations, with a variety of stem cell types being investigated. An aspect of data interpretation that may influence the future of cardiac stem cell therapy is identifying patient subgroups appearing to exert benefit from certain cell types. Focus should be targeted on assessing the potential of a particular stem cell treatment in clinical subgroups besides analysing outcome(s) for treatment arms within studies. On the other hand the long-term follow-up of treated patients will shed light on the possibility of stem cell therapy having a temporary effect on clinical improvement. In that case it may of importance to develop supportive strategies to improve cell survival (immunosuppression, concomitant attempt to revascularisation, genetic manipulation of cells to reduce apoptosis, etc.) or even develop protocols for repeated administration of stem cells.

References

1. Thom T, Haase N, Rosamond W, Howard VJ, Rumsfeld J, Manolio T, *et al*. Heart disease and stroke statistics – 2006 update: a report from the American Heart Association Statistics Committee and Stroke Statistics Subcommittee. *Circulation* 2006;113(6):e85–151.
2. Lloyd-Jones DM, Larson MG, Leip EP, Beiser A, D'Agostino RB, Kannel WB, *et al*. Lifetime risk for developing congestive heart failure: the Framingham Heart Study. *Circulation* 2002;106(24):3068–3072.
3. Lange RA, Hillis LD. Reperfusion therapy in acute myocardial infarction. *N Engl J Med* 2002;346(13):954–955.
4. Pfeffer MA, Braunwald E. Ventricular remodeling after myocardial infarction. Experimental observations and clinical implications. *Circulation* 1990;81(4):1161–1172.
5. Kherani AR, Garrido MJ, Cheema FH, Naka Y, Oz MC. Nontransplant surgical options for congestive heart failure. *Congest Heart Fail* 2003;9(1):17–24.
6. Westaby S, Narula J. Surgical options in heart failure. *Surg Clin North Am* 2004;84(1):xv–xix.
7. Nadal-Ginard B. [Generation of new cardiomyocytes in the adult heart: prospects of myocardial regeneration as an alternative to cardiac transplantation]. *Rev Esp Cardiol* 2001;54(5):543–550.
8. Gu W, Schneider JW, Condorelli G, Kaushal S, Mahdavi V, Nadal-Ginard B. Interaction of myogenic factors and the retinoblastoma protein mediates muscle cell commitment and differentiation. *Cell* 1993;72(3):309–324.
9. Endo T, Nadal-Ginard B. Reversal of myogenic terminal differentiation by SV40 large T antigen results in mitosis and apoptosis. *J Cell Sci* 1998;111(Pt 8):1081–1093.
10. Soonpaa MH, Field LJ. Survey of studies examining mammalian cardiomyocyte DNA synthesis. *Circ Res* 1998;83(1):15–26.
11. Kajstura J, Leri A, Finato N, Di Loreto C, Beltrami CA, Anversa P. Myocyte proliferation in end-stage cardiac failure in humans. *Proc Natl Acad Sci USA* 1998;95(15):8801–8805.
12. Cai L, Kang YJ. Cell death and diabetic cardiomyopathy. *Cardiovasc Toxicol* 2003;3(3):219–228.
13. Beltrami AP, Usrbanek K, Kajstura J, Yan SM, Finato N, Bussani R, *et al*. Evidence that human cardiac myocytes divide after myocardial infarction. *N Engl J Med* 2001;344(23):1750–1757.
14. Odorico JS, Zhang SC, Pedersen RA. *Human Embryonic Stem Cells*. Garland Science/BIOS Scientific Publishers, Abingdon, Oxon, 2005.
15. Bayes-Genis A, Muniz-Diaz E, Catasus L, Arilla M, Rodriguez C, Sierra J, *et al*. Cardiac chimerism in recipients of peripheral-blood and bone marrow stem cells. *Eur J Heart Fail* 2004;6(4):399–402.
16. Laflamme MA, Myerson D, Saffitz JE, Murry CE. Evidence for cardiomyocyte repopulation by extracardiac progenitors in transplanted human hearts. *Circ Res* 2002;90(6):634–640.
17. Quaini F, Urbanek K, Beltrami AP, Finato N, Beltrami CA, Nadal-Ginard B, *et al*. Chimerism of the transplanted heart. *N Engl J Med* 2002;346(1):5–15.
18. Kucia M, Dawn B, Hunt G, Guo Y, Wysoczynski M, Majka M, *et al*. Cells expressing early cardiac markers reside in the bone marrow and are mobilized into the peripheral blood after myocardial infarction. *Circ Res* 2004;95(12):1191–1199.

19. Gordon MY, Levicar N, Pai M, Bachellier P, Dimarakis I, Al-Allaf F, *et al.* Characterization and clinical application of human CD34+ stem/progenitor cell populations mobilized into the blood by granulocyte colony-stimulating factor. *Stem Cells* 2006;24(7):1822–1830.

20. Beltrami AP, Barlucchi L, Torella D, Baker M, Limana F, Chimenti S, *et al.* Adult cardiac stem cells are multipotent and support myocardial regeneration. *Cell* 2003;114(6):763–776.

21. Fazel S, Cimini M, Chen L, Li S, Angoulvant D, Fedak P, *et al.* Cardioprotective c-kit+ cells are from the bone marrow and regulate the myocardial balance of angiogenic cytokines. *J Clin Invest* 2006;116(7):1865–1877.

22. Oh H, Bradfute SB, Gallardo TD, Nakamura T, Gaussin V, Mishina Y, *et al.* Cardiac progenitor cells from adult myocardium: homing, differentiation, and fusion after infarction. *Proc Natl Acad Sci USA* 2003;100(21):12313–12318.

23. Pfister O, Mouquet F, Jain M, Summer R, Helmes M, Fine A, *et al.* CD31 − but not CD31+ cardiac side population cells exhibit functional cardiomyogenic differentiation. *Circ Res* 2005;97(1):52–61.

24. Messina E, De Angelis L, Frati G, Morrone S, Chimenti S, Fiordaliso F, *et al.* Isolation and expansion of adult cardiac stem cells from human and murine heart. *Circ Res* 2004;95(9): 911–921.

25. Tomita Y, Matsumura K, Wakamatsu Y, Matsuzaki Y, Shibuya I, Kawaguchi H, *et al.* Cardiac neural crest cells contribute to the dormant multipotent stem cell in the mammalian heart. *J Cell Biol* 2005;170(7):1135–1146.

26. Laugwitz KL, Moretti A, Lam J, Gruber P, Chen Y, Woodard S, *et al.* Post-natal isl1+ cardioblasts enter fully differentiated cardiomyocyte lineages. *Nature* 2005;433(7026):647–653.

27. Moretti A, Caron L, Nakano A, Lam JT, Bernshausen A, Chen Y, *et al.* Multipotent embryonic isl1+ progenitor cells lead to cardiac, smooth muscle, and endothelial cell diversification. *Cell* 2006;127(6):1151–1165.

28. Angelini P, Markwald RR. Stem cell treatment of the heart: a review of its current status on the brink of clinical experimentation. *Tex Heart Inst J* 2005;32(4):479–488.

29. Verfaillie CM. Adult stem cells: assessing the case for pluripotency. *Trends Cell Biol* 2002;12(11):502–508.

30. Krause DS, Theise ND, Collector MI, Henegariu O, Hwang S, Gardner R, *et al.* Multi-organ, multi-lineage engraftment by a single bone marrow-derived stem cell. *Cell* 2001;105(3): 369–379.

31. Osawa M, Hanada K, Hamada H, Nakauchi H. Long-term lymphohematopoietic reconstitution by a single CD34-low/negative hematopoietic stem cell. *Science* 1996;273(5272):242–445.

32. Orlic D, Kajstura J, Chimenti S, Jakoniuk I, Anderson SM, Li B, *et al.* Bone marrow cells regenerate infarcted myocardium. *Nature* 2001;410(6829):701–705.

33. Yeh ET, Zhang S, Wu HD, Korbling M, Willerson JT, Estrov Z. Transdifferentiation of human peripheral blood CD34+-enriched cell population into cardiomyocytes, endothelial cells, and smooth muscle cells *in vivo*. *Circulation* 28;108(17): 2070–2073.

34. Murry CE, Soonpaa MH, Reinecke H, Nakajima H, Nakajima HO, Rubart M, *et al.* Haematopoietic stem cells do not transdifferentiate into cardiac myocytes in myocardial infarcts. *Nature* 2004;428(6983):664–668.

35. Balsam LB, Wagers AJ, Christensen JL, Kofidis T, Weissman IL, Robbins RC. Haematopoietic stem cells adopt mature haematopoietic fates in ischaemic myocardium. *Nature* 2004;428(6983):668–673.

36. Deten A, Volz HC, Clamors S, Leiblein S, Briest W, Marx G, *et al.* Hematopoietic stem cells do not repair the infarcted mouse heart. *Cardiovasc Res* 2005;65(1):52–63.

37. Goodell MA, Brose K, Paradis G, Conner AS, Mulligan RC. Isolation and functional properties of murine hematopoietic stem cells that are replicating *in vivo*. *J Exp Med* 1996;183(4): 1797–1806.

38. McKinney-Freeman SL, Jackson KA, Camargo FD, Ferrari G, Mavilio F, Goodell MA. Muscle-derived hematopoietic stem cells are hematopoietic in origin. *Proc Natl Acad Sci USA* 2002;99(3):1341–1346.

39. Majka SM, Jackson KA, Kienstra KA, Majesky MW, Goodell MA, Hirschi KK. Distinct progenitor populations in skeletal muscle are bone marrow derived and exhibit different cell fates during vascular regeneration. *J Clin Invest* 2003;111(1):71–79.

40. Liadaki K, Kho AT, Sanoudou D, Schienda J, Flint A, Beggs AH, *et al.* Side population cells isolated from different tissues share transcriptome signatures and express tissue-specific markers. *Exp Cell Res* 2005;303(2):360–374.

41. Rochon C, Frouin V, Bortoli S, Giraud-Triboult K, Duverger V, Vaigot P, *et al.* Comparison of gene expression pattern in SP cell populations from four tissues to define common "stemness functions". *Exp Cell Res* 2006;312(11):2074–2082.

42. Choi K, Kennedy M, Kazarov A, Papadimitriou JC, Keller G. A common precursor for hematopoietic and endothelial cells. *Development* 1998;125(4):725–732.

43. Asahara T, Masuda H, Takahashi T, Kalka C, Pastore C, Silver M, *et al.* Bone marrow origin of endothelial progenitor cells responsible for postnatal vasculogenesis in physiological and pathological neovascularization. *Circ Res* 1999;85(3):221–228.

44. Kalka C, Masuda H, Takahashi T, Kalka-Moll WM, Silver M, Kearney M, *et al.* Transplantation of *ex vivo* expanded endothelial progenitor cells for therapeutic neovascularization. *Proc Natl Acad Sci USA* 2000;97(7):3422–3427.

45. Vasa M, Fichtlscherer S, Aicher A, Adler K, Urbich C, Martin H, *et al.* Number and migratory activity of circulating endothelial progenitor cells inversely correlate with risk factors for coronary artery disease. *Circ Res* 2001;89(1):E1–E7.

46. Hill JM, Zalos G, Halcox JP, Schenke WH, Waclawiw MA, Quyyumi AA, *et al.* Circulating endothelial progenitor cells, vascular function, and cardiovascular risk. *N Engl J Med* 2003;348(7):593–600.

47. Zimmet JM, Hare JM. Emerging role for bone marrow derived mesenchymal stem cells in myocardial regenerative therapy. *Basic Res Cardiol* 2005;100(6):471–481.

48. Pittenger MF, Mackay AM, Beck SC, Jaiswal RK, Douglas R, Mosca JD, *et al.* Multilineage potential of adult human mesenchymal stem cells. *Science* 1999;284(5411):143–147.

49. Fukuda K. Development of regenerative cardiomyocytes from mesenchymal stem cells for cardiovascular tissue engineering. *Artif Organs* 2001;25(3):187–193.

50. Wakitani S, Saito T, Caplan AI. Myogenic cells derived from rat bone marrow mesenchymal stem cells exposed to 5-azacytidine. *Muscle Nerve* 1995;18(12):1417–1426.

51. Xu W, Zhang X, Qian H, Zhu W, Sun X, Hu J, *et al.* Mesenchymal stem cells from adult human bone marrow differentiate into a cardiomyocyte phenotype *in vitro. Exp Biol Med (Maywood)* 2004;229(7):623–631.

52. Jiang XX, Zhang Y, Liu B, Zhang SX, Wu Y, Yu XD, *et al.* Human mesenchymal stem cells inhibit differentiation and function of monocyte-derived dendritic cells. *Blood* 2005;105(10): 4120–4126.

53. Amado LC, Saliaris AP, Schuleri KH, St John M, Xie JS, Cattaneo S, *et al.* Cardiac repair with intramyocardial injection of allogeneic mesenchymal stem cells after myocardial infarction. *Proc Natl Acad Sci USA* 2005;102(32):11474–11479.

54. Minguell JJ, Erices A. Mesenchymal stem cells and the treatment of cardiac disease. *Exp Biol Med (Maywood)* 2006;231(1):39–49.

55. Jiang Y, Jahagirdar BN, Reinhardt RL, Schwartz RE, Keene CD, Ortiz-Gonzalez XR, *et al.* Pluripotency of mesenchymal stem cells derived from adult marrow. *Nature* 2002;418(6893): 41–49.

56. Agbulut O, Mazo M, Bressolle C, Gutierrez M, Azarnoush K, Sabbah L, *et al.* Can bone marrow-derived multipotent adult progenitor cells regenerate infarcted myocardium? *Cardiovasc Res* 2006;72(1):175–183.
57. Yoon YS, Wecker A, Heyd L, Park JS, Tkebuchava T, Kusano K, *et al.* Clonally expanded novel multipotent stem cells from human bone marrow regenerate myocardium after myocardial infarction. *J Clin Invest* 2005;115(2):326–338.
58. Zuk PA, Zhu M, Ashjian P, De Ugarte DA, Huang JI, Mizuno H, *et al.* Human adipose tissue is a source of multipotent stem cells. *Mol Biol Cell* 2002;13(12):4279–4295.
59. Planat-Benard V, Menard C, Andre M, Puceat M, Perez A, Garcia-Verdugo JM, *et al.* Spontaneous cardiomyocyte differentiation from adipose tissue stroma cells. *Circ Res* 2004;94(2):223–229.
60. Agbulut O, Vandervelde S, Al Attar N, Larghero J, Ghostine S, Leobon B, *et al.* Comparison of human skeletal myoblasts and bone marrow-derived CD133+ progenitors for the repair of infarcted myocardium. *J Am Coll Cardiol* 2004;44(2):458–4563.
61. Leobon B, Garcin I, Menasche P, Vilquin JT, Audinat E, Charpak S. Myoblasts transplanted into rat infarcted myocardium are functionally isolated from their host. *Proc Natl Acad Sci USA* 2003;100(13):7808–7811.
62. Abraham MR, Henrikson CA, Tung L, Chang MG, Aon M, Xue T, *et al.* Antiarrhythmic engineering of skeletal myoblasts for cardiac transplantation. *Circ Res* 2005;97(2):159–167.
63. Itabashi Y, Miyoshi S, Yuasa S, Fujita J, Shimizu T, Okano T, *et al.* Analysis of the electrophysiological properties and arrhythmias in directly contacted skeletal and cardiac muscle cell sheets. *Cardiovasc Res* 2005;67(3):561–570.
64. Herreros J, Prosper F, Perez A, Gavira JJ, Garcia-Velloso MJ, Barba J, *et al.* Autologous intramyocardial injection of cultured skeletal muscle-derived stem cells in patients with non-acute myocardial infarction. *Eur Heart J* 2003;24(22):2012–2020.
65. Siminiak T, Kalawski R, Fiszer D, Jerzykowska O, Rzezniczak J, Rozwadowska N, *et al.* Autologous skeletal myoblast transplantation for the treatment of postinfarction myocardial injury: phase I clinical study with 12 months of follow-up. *Am Heart J* 2004 Sep;148(3):531–537.
66. Dib N, Michler RE, Pagani FD, Wright S, Kereiakes DJ, Lengerich R, *et al.* Safety and feasibility of autologous myoblast transplantation in patients with ischemic cardiomyopathy: four-year follow-up. *Circulation* 2005;112(12):1748–1755.
67. Hagege AA, Marolleau JP, Vilquin JT, Alheritiere A, Peyrard S, Duboc D, *et al.* Skeletal myoblast transplantation in ischemic heart failure: long-term follow-up of the first phase I cohort of patients. *Circulation* 2006;114(Suppl 1):I108–I113.
68. Steendijk P, Smits PC, Valgimigli M, van der Giessen WJ, Onderwater EE, Serruys PW. Intramyocardial injection of skeletal myoblasts: long-term follow-up with pressure-volume loops. *Nat Clin Pract Cardiovasc Med* 2006;3(Suppl 1):S94–S100.
69. Siminiak T, Fiszer D, Jerzykowska O, Grygielska B, Rozwadowska N, Kalmucki P, *et al.* Percutaneous trans-coronary-venous transplantation of autologous skeletal myoblasts in the treatment of post-infarction myocardial contractility impairment: the POZNAN trial. *Eur Heart J* 2005;26(12):1188–1195.
70. Kehat I, Kenyagin-Karsenti D, Snir M, Segev H, Amit M, Gepstein A, *et al.* Human embryonic stem cells can differentiate into myocytes with structural and functional properties of cardiomyocytes. *J Clin Invest* 2001;108(3):407–414.
71. Wei H, Juhasz O, Li J, Tarasova YS, Boheler KR. Embryonic stem cells and cardiomyocyte differentiation: phenotypic and molecular analyses. *J Cell Mol Med* 2005;9(4):804–817.
72. Min JY, Yang Y, Sullivan MF, Ke Q, Converso KL, Chen Y, *et al.* Long-term improvement of cardiac function in rats after infarction by transplantation of embryonic stem cells. *J Thorac Cardiovasc Surg* 2003;125(2):361–369.

73. Hodgson DM, Behfar A, Zingman LV, Kane GC, Perez-Terzic C, Alekseev AE, *et al.* Stable benefit of embryonic stem cell therapy in myocardial infarction. *Am J Physiol Heart Circ Physiol* 2004;287(2):H471–H479.
74. Min JY, Huang X, Xiang M, Meissner A, Chen Y, Ke Q, *et al.* Homing of intravenously infused embryonic stem cell-derived cells to injured hearts after myocardial infarction. *J Thorac Cardiovasc Surg* 2006;131(4):889–897.
75. Kehat I, Khimovich L, Caspi O, Gepstein A, Shofti R, Arbel G, *et al.* Electromechanical integration of cardiomyocytes derived from human embryonic stem cells. *Nat Biotechnol* 2004;22(10):1282–1289.
76. Zhang YM, Hartzell C, Narlow M, Dudley SC, Jr. Stem cell-derived cardiomyocytes demonstrate arrhythmic potential. *Circulation* 2002;106(10):1294–1299.
77. Henning RJ, Abu-Ali H, Balis JU, Morgan MB, Willing AE, Sanberg PR. Human umbilical cord blood mononuclear cells for the treatment of acute myocardial infarction. *Cell Transplant* 2004;13(7–8):729–739.
78. Ma N, Stamm C, Kaminski A, Li W, Kleine HD, Muller-Hilke B, *et al.* Human cord blood cells induce angiogenesis following myocardial infarction in NOD/scid-mice. *Cardiovasc Res* 2005;66(1):45–54.
79. Sakakibara Y, Tambara K, Lu F, Nishina T, Nagaya N, Nishimura K, *et al.* Cardiomyocyte transplantation does not reverse cardiac remodeling in rats with chronic myocardial infarction. *Ann Thorac Surg* 2002;74(1):25–30.
80. Gao LR, Wang ZG, Zhang NK, He S, Yao L, Ning HY, *et al.* [Transplantation of fetal cardiomyocyte regenerates infarcted myocardium in rats]. *Zhonghua Yi Xue Za Zhi* 2003;83(20):1818–1822.
81. Lanza R, Moore MA, Wakayama T, Perry AC, Shieh JH, Hendrikx J, *et al.* Regeneration of the infarcted heart with stem cells derived by nuclear transplantation. *Circ Res* 2004;94(6):820–827.
82. Koh GY, Klug MG, Soonpaa MH, Field LJ. Differentiation and long-term survival of C2C12 myoblast grafts in heart. *J Clin Invest* 1993;92(3):1548–1554.
83. Ma Q, Jones D, Borghesani PR, Segal RA, Nagasawa T, Kishimoto T, *et al.* Impaired B-lymphopoiesis, myelopoiesis, and derailed cerebellar neuron migration in CXCR4- and SDF-1-deficient mice. *Proc Natl Acad Sci USA* 1998;95(16):9448–9453.
84. Zou YR, Kottmann AH, Kuroda M, Taniuchi I, Littman DR. Function of the chemokine receptor CXCR4 in haematopoiesis and in cerebellar development. *Nature* 1998;393(6685):595–599.
85. Peled A, Petit I, Kollet O, Magid M, Ponomaryov T, Byk T, *et al.* Dependence of human stem cell engraftment and repopulation of NOD/SCID mice on CXCR4. *Science* 1999;283(5403):845–848.
86. Peled A, Kollet O, Ponomaryov T, Petit I, Franitza S, Grabovsky V, *et al.* The chemokine SDF-1 activates the integrins LFA-1, VLA-4, and VLA-5 on immature human CD34(+) cells: role in transendothelial/stromal migration and engraftment of NOD/SCID mice. *Blood* 2000;95(11):3289–3296.
87. Askari AT, Unzek S, Popovic ZB, Goldman CK, Forudi F, Kiedrowski M, *et al.* Effect of stromal-cell-derived factor 1 on stem-cell homing and tissue regeneration in ischaemic cardiomyopathy. *Lancet* 2003;362(9385):697–703.
88. Asahara T, Takahashi T, Masuda H, Kalka C, Chen D, Iwaguro H, *et al.* VEGF contributes to postnatal neovascularization by mobilizing bone marrow-derived endothelial progenitor cells. *EMBO J* 1999;18(14):3964–3972.
89. Ceradini DJ, Kulkarni AR, Callaghan MJ, Tepper OM, Bastidas N, Kleinman ME, *et al.* Progenitor cell trafficking is regulated by hypoxic gradients through HIF-1 induction of SDF-1. *Nat Med* 2004;10(8):858–864.

90. Heissig B, Hattori K, Dias S, Friedrich M, Ferris B, Hackett NR, *et al.* Recruitment of stem and progenitor cells from the bone marrow niche requires MMP-9 mediated release of kit-ligand. *Cell* 2002;109(5):625–637.

91. Abbott JD, Huang Y, Liu D, Hickey R, Krause DS, Giordano FJ. Stromal cell-derived factor-1 {alpha} plays a critical role in stem cell recruitment to the heart after myocardial infarction but is not sufficient to induce homing in the absence of injury. *Circulation* 2004;11:8.

92. De Falco E, Porcelli D, Torella AR, Straino S, Iachininoto MG, Orlandi A, *et al.* SDF-1 involvement in endothelial phenotype and ischemia-induced recruitment of bone marrow progenitor cells. *Blood* 2004;104(12):3472–3482.

93. Petit I, Szyper-Kravitz M, Nagler A, Lahav M, Peled A, Habler L, *et al.* G-CSF induces stem cell mobilization by decreasing bone marrow SDF-1 and up-regulating CXCR4. *Nat Immunol* 2002;3(7):687–694.

94. Yamaguchi J, Kusano KF, Masuo O, Kawamoto A, Silver M, Murasawa S, *et al.* Stromal cell-derived factor-1 effects on *ex vivo* expanded endothelial progenitor cell recruitment for ischemic neovascularization. *Circulation* 2003;107(9):1322–1328.

95. Banai S, Shweiki D, Pinson A, Chandra M, Lazarovici G, Keshet E. Upregulation of vascular endothelial growth factor expression induced by myocardial ischaemia: implications for coronary angiogenesis. *Cardiovasc Res* 1994;28(8):1176–1179.

96. Brogi E, Schatteman G, Wu T, Kim EA, Varticovski L, Keyt B, *et al.* Hypoxia-induced paracrine regulation of vascular endothelial growth factor receptor expression. *J Clin Invest* 1996;97(2):469–476.

97. Lee SH, Wolf PL, Escudero R, Deutsch R, Jamieson SW, Thistlethwaite PA. Early expression of angiogenesis factors in acute myocardial ischemia and infarction. *N Engl J Med* 2000;342(9):626–633.

98. Tuomisto TT, Rissanen TT, Vajanto I, Korkeela A, Rutanen J, Yla-Herttuala S. HIF-VEGF-VEGFR-2, TNF-alpha and IGF pathways are upregulated in critical human skeletal muscle ischemia as studied with DNA array. *Atherosclerosis* 2004;174(1):111–120.

99. Duelli D, Lazebnik Y. Cell fusion: a hidden enemy? *Cancer Cell* 2003;3(5):445–448.

100. Alvarez-Dolado M, Pardal R, Garcia-Verdugo JM, Fike JR, Lee HO, Pfeffer K, *et al.* Fusion of bone-marrow-derived cells with Purkinje neurons, cardiomyocytes and hepatocytes. *Nature* 2003;425(6961):968–973.

101. Nygren JM, Jovinge S, Breitbach M, Sawen P, Roll W, Hescheler J, *et al.* Bone marrow-derived hematopoietic cells generate cardiomyocytes at a low frequency through cell fusion, but not transdifferentiation. *Nat Med* 2004;10(5):494–501.

102. Andrade J, Lam JT, Zamora M, Huang C, Franco D, Sevilla N, *et al.* Predominant fusion of bone marrow-derived cardiomyocytes. *Cardiovasc Res* 2005;68(3):387–393.

103. Kajstura J, Rota M, Whang B, Cascapera S, Hosoda T, Bearzi C, *et al.* Bone marrow cells differentiate in cardiac cell lineages after infarction independently of cell fusion. *Circ Res* 2005;96(1):127–137.

104. Itescu S, Kocher AA, Schuster MD. Myocardial neovascularization by adult bone marrow-derived angioblasts: strategies for improvement of cardiomyocyte function. *Heart Fail Rev* 2003;8(3):253–258.

105. Isner JM, Kalka C, Kawamoto A, Asahara T. Bone marrow as a source of endothelial cells for natural and iatrogenic vascular repair. *Ann N Y Acad Sci* 2001;953:75–84.

106. Kinnaird T, Stabile E, Burnett MS, Lee CW, Barr S, Fuchs S, *et al.* Marrow-derived stromal cells express genes encoding a broad spectrum of arteriogenic cytokines and promote *in vitro* and *in vivo* arteriogenesis through paracrine mechanisms. *Circ Res* 2004;94(5):678–685.

107. Kinnaird T, Stabile E, Burnett MS, Shou M, Lee CW, Barr S, *et al.* Local delivery of marrow-derived stromal cells augments collateral perfusion through paracrine mechanisms. *Circulation* 2004;109(12):1543–1549.

108. Kamihata H, Matsubara H, Nishiue T, Fujiyama S, Tsutsumi Y, Ozono R, et al. Implantation of bone marrow mononuclear cells into ischemic myocardium enhances collateral perfusion and regional function via side supply of angioblasts, angiogenic ligands, and cytokines. *Circulation* 2001;104(9):1046–1052.

109. Thum T, Bauersachs J, Poole-Wilson PA, Volk HD, Anker SD. The dying stem cell hypothesis: immune modulation as a novel mechanism for progenitor cell therapy in cardiac muscle. *J Am Coll Cardiol* 2005;46(10):1799–1802.

110. Dimarakis I, Levicar N, Nihoyannopoulos P, Habib NA, Gordon MY. *In vitro* stem cell differentiation into cardiomyocytes: Part 1. Culture medium and growth factors. *J Cardiothorac Renal Res* 2006;1(2):107–114.

111. Dimarakis I, Levicar N, Nihoyannopoulos P, Gordon MY, Habib NA. *In vitro* stem cell differentiation into cardiomyocytes: Part 2: Chemicals, extracellular matrix, physical stimuli and coculture assays. *J Cardiothorac Renal Res* 2006;1(2):115–121.

112. Bittira B, Kuang JQ, Al-Khaldi A, Shum-Tim D, Chiu RC. *In vitro* preprogramming of marrow stromal cells for myocardial regeneration. *Ann Thorac Surg* 2002;74(4):1154–1159;9–60.

113. Lu L, Zhang JQ, Ramires FJ, Sun Y. Molecular and cellular events at the site of myocardial infarction: from the perspective of rebuilding myocardial tissue. *Biochem Biophys Res Commun* 2004;320(3):907–913.

114. Ciulla MM, Lazzari L, Pacchiana R, Esposito A, Bosari S, Ferrero S, et al. Homing of peripherally injected bone marrow cells in rat after experimental myocardial injury. *Haematologica* 2003;88(6):614–621.

115. Vulliet PR, Greeley M, Halloran SM, MacDonald KA, Kittleson MD. Intra-coronary arterial injection of mesenchymal stromal cells and microinfarction in dogs. *Lancet* 2004;363(9411):783–784.

116. Yoon YS, Park JS, Tkebuchava T, Luedeman C, Losordo DW. Unexpected severe calcification after transplantation of bone marrow cells in acute myocardial infarction. *Circulation* 2004;109(25):3154–3157.

117. Kang HJ, Kim HS, Zhang SY, Park KW, Cho HJ, Koo BK, et al. Effects of intracoronary infusion of peripheral blood stem-cells mobilised with granulocyte-colony stimulating factor on left ventricular systolic function and restenosis after coronary stenting in myocardial infarction: the MAGIC cell randomised clinical trial. *Lancet* 2004;363(9411):751–756.

118. Stenderup K, Justesen J, Clausen C, Kassem M. Aging is associated with decreased maximal life span and accelerated senescence of bone marrow stromal cells. *Bone* 2003;33(6):919–926.

119. Vacanti V, Kong E, Suzuki G, Sato K, Canty JM, Lee T. Phenotypic changes of adult porcine mesenchymal stem cells induced by prolonged passaging in culture. *J Cell Physiol* 2005;205(2):194–201.

120. Rubio D, Garcia-Castro J, Martin MC, de la Fuente R, Cigudosa JC, Lloyd AC, et al. Spontaneous human adult stem cell transformation. *Cancer Res* 2005;65(8):3035–3039.

121. Chang MG, Tung L, Sekar RB, Chang CY, Cysyk J, Dong P, et al. Proarrhythmic potential of mesenchymal stem cell transplantation revealed in an *in vitro* coculture model. *Circulation* 2006;113(15):1832–1841.

122. Botta R, Gao E, Stassi G, Bonci D, Pelosi E, Zwas D, et al. Heart infarct in NOD-SCID mice: therapeutic vasculogenesis by transplantation of human CD34+ cells and low dose CD34+KDR+ cells. *FASEB J* 2004;18(12):1392–1394.

123. Iwasaki H, Kawamoto A, Ishikawa M, Oyamada A, Nakamori S, Nishimura H, et al. Dose-dependent contribution of CD34-positive cell transplantation to concurrent vasculogenesis and cardiomyogenesis for functional regenerative recovery after myocardial infarction. *Circulation* 2006;113(10):1311–1325.

124. Retuerto MA, Schalch P, Patejunas G, Carbray J, Liu N, Esser K, *et al.* Angiogenic pretreatment improves the efficacy of cellular cardiomyoplasty performed with fetal cardiomyocyte implantation. *J Thorac Cardiovasc Surg* 2004;127(4):1041–1049; discussion 9–51.
125. Sakakibara Y, Nishimura K, Tambara K, Yamamoto M, Lu F, Tabata Y, *et al.* Prevascularization with gelatin microspheres containing basic fibroblast growth factor enhances the benefits of cardiomyocyte transplantation. *J Thorac Cardiovasc Surg* 2002;124(1):50–56.
126. Mangi AA, Noiseux N, Kong D, He H, Rezvani M, Ingwall JS, *et al.* Mesenchymal stem cells modified with Akt prevent remodeling and restore performance of infarcted hearts. *Nat Med* 2003;9(9):1195–1201.

3 Embryonic Stem Cells and Their Therapeutic Potential

Tomo Šarić, Shuhua Chen, Naidu Kamisetti,
Marcel Halbach, Michael Xavier Doss,
Johannes Winkler, Jürgen Hescheler and
Agapios Sachinidis

Embryonic Stem Cells

After the process of fertilization, the single-celled zygote is formed. The zygote is the first totipotent cell. Totipotency, i.e. the ability of a cell to give rise to all the tissues of the embryo proper as well as to extra-embryonic tissues, is retained until the eight-celled dividing structure, the morula. Then, the morula undergoes further cell divisions to form a blastocyst, which contains an undifferentiated inner cell mass (ICM), a fluid-filled cavity, and an outer layer of cells, called trophectoderm, which gives rise to extra-embryonic tissues. The cells of the ICM are pluripotent meaning that they can develop into all cell types of the embryo proper derived from endoderm, ectoderm, and mesoderm. Cells of the ICM can be harvested from blastocysts to produce embryonic stem (ES) cell lines that can be cultured *in vitro*. The first reports of mouse ES cells were in the early 1980s[1,2] and these cells have been instrumental in clarifying the function of many genes utilizing transgenic or knock out animals. However, broad interest in ES cells and their potential use for human therapy emerged when human ES cells were first described[3] and shown to have the capacity to differentiate into a wide range of mature cell types.[4]

Although murine and human ES cells share common properties, there are also important differences between them in regard to their growth characteristics, cell surface markers and gene expression profiles.[5] Whereas murine ES cells form tight, compact and rounded colonies, their human counterparts form flat and loose colonies. Undifferentiated murine and human ES cells have similarly high activity of alkaline phosphatase and telomerase enzymes but differ in the cell surface markers they express. Murine ES cells express stage-specific embryonic antigen 1 (SSEA-1)[6] and epiblast/germ cell-restricted transcription factor Oct3/4[7] whereas undifferentiated human ES cells are highly positive for stage-specific embryonic antigen-3 and -4 (SSEA-3, SSEA-4) and embryonal carcinoma markers TRA-1-60 and TRA-1-81,[8] which are absent on murine ES cells. However, SSEA-1 and Oct3/4 expression does not closely correlate with various functional assays for undifferentiated murine

ES cells and they are still highly expressed by early differentiating cells until up to five days after induction of differentiation.[9] Stem cell factor (SCF) receptor c-Kit appears to be more specific cell surface marker of undifferentiated murine ES cells, because its expression significantly decreases already 18 hours after leukemia inhibitory factor (LIF) removal and falls to about 12%–17% of its original level by day 3 of differentiation.[9]

ES cells derived from pre-implantation embryos were originally grown on mitotically inactivated mouse embryonic fibroblast (MEF) feeders, which allow their growth in undifferentiated state. Both murine and human ES cell lines can be generated to grow in the absence of feeder cells.[10] Mouse ES cells cultured without MEFs remain undifferentiated in the presence of LIF,[11] retain their unlimited self-renewal capacity almost indefinitely and also maintain their normal stable karyotype when passaged continuously. In contrary, human ES cells do not require LIF for maintaining pluripotency but rather depend on MEFs and serum supplemented with basic fibroblast growth factor (bFGF).[12] Human ES cells growing under feeder-free conditions[13–16] on human feeder cells[17,18] or on autogeneic human ES cell-derived fibroblasts[19] have an advantage over those lines maintained on animal feeder cells because the risk of contamination with infectious animal agents or with xenoantigens is eliminated. Pluripotency in murine ES cells is believed to be mediated via the signal transduction and activator of transcription-3 (STAT3) pathway.[20] LIF exerts its effects by binding to a heterodimeric receptor complex that consists of the LIF receptor and the gp130 receptor. Binding of LIF triggers the activation of the latent transcription factor STAT3, a necessary event *in vitro* for the continued proliferation of undifferentiated mouse ES cells. Nanog has also been implicated as a pluripotency-sustaining factor.[21] The role of Wnt signaling for self-renewal has recently been demonstrated whereby the canonical Wnt pathway is activated upon binding of Wnt to the Frizzled receptor. Activation of this pathway leads to inhibition of glycogen synthase kinase-3 (GSK-3), subsequent nuclear accumulation of β-catenin and the expression of target genes. With the help of a specific inhibitor of GSK-3, 6-bromoindirubin-3'-oxime (BIO), it was confirmed that that activation of the canonical Wnt pathway maintains the undifferentiated phenotype and sustains expression of the pluripotent state-specific transcription factors Oct-3/4, Rex-1 and Nanog.[22] This process is fully reversible after removal of BIO and which subsequently results in multidifferentiation of ES cells.[22]

ES Cell-Derived Cardiomyocytes

Generation of cardiomyocytes from ES cells

Differentiation of ES cells into cardiomyocytes *in vitro* is spontaneous.[23,24] In the absence of LIF, ES cells are differentiated into three-dimensional cell aggregates called embryoid bodies (EBs) in suspension cultures (Fig. 1). Eight days after differentiation, spontaneously beating cardiomyocyte foci emerge in murine EBs. Within

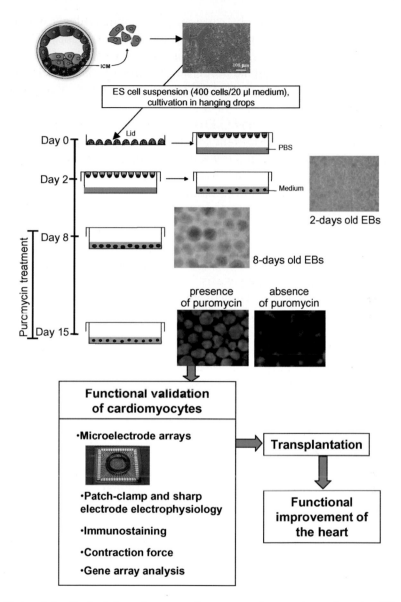

Fig. 1. Protocol for the isolation of pure cardiomyocytes from engineered murine ES cells by the lineage selection approach. ES cells are stably transfected with a bicistronic vector containing the puromycin resistance and EGFP genes under control of the α-myosin heavy chain promoter. Arrows indicate typical beating cardiac cell clusters expressing green fluorescent EGFP. Application of puromycin into the medium at day 8 of differentiation results in the killing of all the cells except cardiomyocytes. At day 15, a purity of cardiomyocytes of over 90% of viable cells can be achieved. After functional validation of ES cell-derived cardiomyocytes, they can be transplanted into cryoinfarcted hearts to improve the function of the injured heart.

the EB the cardiomyocytes form between an outer epithelial layer with characteristics of visceral endoderm and a basal layer of mesenchymal cells.[25] Depending on culture conditions, cardiomyocytes can be maintained viable for three months and more. The structural and functional properties of cardiomyocytes derived from ES cells have been studied in detail.[26] RT–PCR analyses have shown that cardiomyogenesis in EBs resembles that of the normal myocardial development. The appearance of transcription factors Nkx2.5 and GATA4 is followed by the expression of atrial natriuretic factor (ANF), myosin light chain (MLC2v) and then α- and β-myosin heavy chain (α-MHC and β-MHC) transcripts, which mimics the early myocardial development.[25] Early cardiomyocytes have sparse, irregularly organized myofibrils, but later develop into well-organized bundles of myofibrils and sarcomeres with clearly defined Z-disks and A- and I-bands, which are typical for terminally differentiated cardiomyocytes (observed in 36-day-old beating foci).[25] Several extrinsic growth factors such as bone morphogenetic proteins (BMPs), insulin-like growth factor-1 (IGF1), activin-A, members of the FGF2 and the Wnt family and their inhibitors are involved in cardiogenesis.[27–29] Intrinsic factors such as the transcription factor GATA4 and its co-activator Nkx2.5 are essential for cardiomyogenesis (for more comprehensive overview see Refs. 30 and 31).

Identification of extrinsic and intrinsic factors promoting cardiomyogenesis is critical for inducing cardiomyogenesis in ES cells. Gene expression profiling of ES cells at different stages of differentiation and functional validation of promising candidates may lead to identification of intrinsic factors required for specific developmental processes.[32] During differentiation of ES cells to cardiomyocytes, a number of other tissue specific cell types codifferentiate. Among those, the endoderm plays the most importnat role in inducing cardiomyocyte differentiation from ES cells. This has particularly evident in cocultures of human ES cells with visceral-endoderm-like cells from the mouse.[33,34] The molecular basis of this inductive effect of the endoderm has been recently elucidated by Behfar *et al.*[35] This group has demonstrated that tumor necrosis factor-alpha (TNFα) promotes cardiac differentiation of transplanted undifferentiated ES cells *in vivo* as well as in embryoid bodies *in vitro*. The procardiogenic action of this cytokine required an intact endoderm and was mediated by secreted endodermal cardio-inductive signals including TGF-β1, BMP-2, BMP-4, activin-A, VEGF-A, IL-6, FGF-2, FGF-4, IGF-1, IGF-2, and EGF. These growth factors applied as a recombinant cocktail to ES cells demonstrated a capacity to recruit from ES cells a monolayer of cells possessing an intermediate cardiopoietic progenitor cell phenotype. These cells completed the cardiac differentiation program with further cultivation in the presence of these factors. In contrast, removal of the cardiogenic cocktail after four days of recombinant stimulation resulted in continued engagement of cardiopoietic cells in the cell cycle without differentiation into cardiomyocytes and without withdrawal from cell cycle upon confluence. This study sets the basis for directed and scalable production of cardiomyocytes for heart repair under highly controlled culture conditions.

The use of serum, which is rich in growth factors and cytokines, makes it difficult to find out what factors support cardiomyogenesis and how signalling is functioning *in vitro*. This has led to the development of serum-free culture conditions. The use of such conditions could be a possible way to increase the number of cardiomyocytes derived from ES cells in the presence of specified growth factors.[36,37] Some of these factors capable of enhancing cardiogenic differentiation of ES cells are the platelet-derived growth factor (PDGF) and sphingosine-1-phosphate as extrinsic cardiomyogenic factors.[37] Retinoic acid, ascorbic acid and DMSO have been also reported to promote cardiac differentiation.[38–40] In one study, retinoic acid treated cultures contained approximately twice as many cardiac cells as compared to the non-treated cells.[41] There have been reports of other small molecules that promote cardiomyogenesis.[30,42] Screening of small molecule libraries for compounds capable of enhancing differentiation towards a specific cell type has a potential to discover new extrinsic factors. We have already identified by this approach several agents with potent pro- and anti-cardiomyogenic activity.[42]

In addition to the use of conditioned medium and defined factors, use of cell lineage specific promoters fused to selective markers could help in purification of the desired cells. The use of cardiac promoters such as α-MHC, cardiac actin, and MLC2v with a selection marker has been reported.[43–45] This methodology allows purification of cardiomyocytes by removing other cell types that also differentiate alongside the cardiac cells including remaining undifferentiated ES cells (Fig. 1) that have the potential of forming tumors. The various techniques for lineage selection aim to resolve this issue by exclusively providing cells that are already differentiated and thus are unlikely to form teratocarcinomas. Kolossov and coworkers reported a six- to ten-fold increase in yield of cardiomyocytes after puromycin selection as compared to cardiomyocyte numbers found in comparable amounts of unpurified embryoid bodies.[46] In a similar approach, the number of cardiomyocytes produced *in vitro* could be scaled up by using ES cells stably transfected with a fusion gene consisting of the αMHC promoter driving the aminoglycoside phosphotransferase (neomycin resistance) gene, wherein the authors have been able to obtain 99% pure cardiomyocytes from cardiac bodies (CB).[41]

Functional characteristics of ES cell-derived cardiomyocytes

The quality of ES-derived cardiomyocytes needs to be assessed by functional characterization. To avoid arrhythmias and to optimize the functional benefit of cardiomyoplasty, functional properties of ES cell-derived cardiomyocytes (ESCM) need to match those of adult host cardiomyocytes, including electrophysiology, hormonal regulation, Ca^{2+} handling and contraction force. In contrast to bone marrow-derived stem cells, which might promote myocardial regeneration by anti-apoptotic and angiogenic effects,[47] the potency for ES cells differentiating into cardiomyocytes has been proven undoubtedly.[26,34,48,49]

Electrophysiology of ESCM

Electrophysiology of cardiomyocytes is essential for cardiac function at the single cell (excitation-contraction coupling) as well as the whole heart (coordination of contraction) level. In native hearts, electrophysiological properties are diverse depending on the location of cardiomyocytes within the heart. Murine as well as human ESCM are capable of developing into the three major subtypes of cardiomyocytes, i.e. nodal-, atrial-, and ventricular-like cells.[26,48,49] Comparable to fetal and adult native cardiomyocytes, the three subtypes differ in resting membrane potential, upstroke velocity and action potential duration as well as in ion channel expression. All major cardiac ionic currents have been demonstrated by patch clamp measurements in murine ESCM,[26] including fast Na^+ current (I_{Na}), L-type and T-type Ca^{2+} currents (I_{Ca-L}, I_{Ca-T}), "funny" pacemaker current (I_f), inward rectifier K^+ current (I_{K1}), delayed rectifier K^+ current (I_K), transient outward K^+ current (I_{to}) as well as acetylcholine and ATP-dependent K^+ currents ($I_{K,Ach}$, $I_{K,ATP}$).

In human ESCMs, there is electrophysiological and pharmacological evidence for the presence of I_{Na}, I_{Ca-L}, I_{Kr} and I_f[50-52] and the absence of I_{K1}.[52] Despite the presence of I_f and I_{Ca-L} in human ESCM, their contribution to pacemaker activity is not clarified yet, since inconsistent effects of a blockage of I_f or I_{Ca-L} (decrease[53,54] or increase[52] of beating frequency) have been reported. Instead, I_{Na} might be important for the spontaneous electrical activity of human ESCM,[52] although it plays no significant role in native adult pacemaker cells. Changes of I_{Ca-L} in human ESCM at different stimulation frequencies were reported to be opposite to changes in adult cardiomyocytes.[55] While I_{Ca-L} increases at higher stimulation frequencies in the latter, I_{Ca-L} is diminished in human ESCM. Thus, basic studies in human ESCM revealed that these cells express characteristic cardiac ion channels and possess typical action potential morphologies, but raised questions regarding cardiac electrophysiology in detail. The reported inconsistencies might be due to differences in stem cell lines, differentiation protocols or developmental stages of cells under investigation and necessitate further electrophysiological and pharmacological studies on these variables. Despite clear evidence of a cardiac differentiation, it is noteworthy that basic electrophysiological properties of murine and human ESCM are immature as compared to adult native cardiomyocytes. Up to now, a terminal adult-like stage of ESCM has never been achieved.

Adrenergic and muscarinic regulation of chronotropy

Pacemaker activity in the native sinus node is predominantly regulated by adrenergic and muscarinic hormones. A physiological reaction of ESCM to hormonal regulation, which is essential for establishing cell-based pacemakers,[54] has been shown for murine[56] and human[34,53,54,57,58] ESCM. Isoprenaline, a β-adrenoceptor agonist, increased the spontaneous rate of electrical and contractile activity, while carbachol, a m-cholinoceptor agonist, provoked a decrease (Fig. 2). In murine ESCM at early stages (unpublished data) and in human ESCM,[58] the carbachol effect was

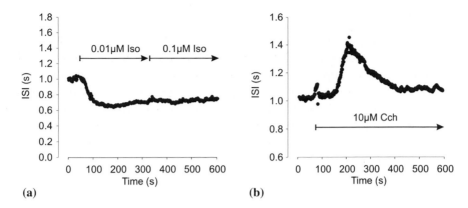

Fig. 2. Physiological adrenergic (**a**) and muscarinic (**b**) regulation of pacemaker activity in human ESCM. Application of isoprenaline (Iso) caused a stable decrease of the inter-spike interval (ISI), i.e. an increase of the frequency. In contrast, carbachol (Cch) provoked a transient prolongation of the ISI, i.e. a decrease of the frequency.

only transient, which might be due to a self-limiting NO signal cascade at early developmental stages described for murine ESCM.[59] In late stage murine ESCM, β-adrenergic and muscarinic signal cascades were comparable to those of adult cardiomyocytes,[59] as demonstrated by the ability of adreno- and cholinoceptors to control the adenyly cyclase activity and cAMP concentration via stimulatory and inhibitory G-proteins. Detailed data about signal cascades and their development in human ESCM at present are not available.

Ca^{2+} handling of ESCM

Periodic rises of the intracellular Ca^{2+} concentration are caused by transmembrane Ca^{2+} influx via voltage-gated channels. During the process of fetal development, there is furthermore an increasing contribution of Ca^{2+}-induced Ca^{2+} release via the ryanodine receptor from the sarcoplasmic reticulum in native cardiomyocytes. Besides Ca^{2+} influx, a sufficient extrusion of Ca^{2+} from the cytoplasm via the sarcolemmal Ca^{2+}-ATPase and Na^{2+}-Ca^{2+} exchanger and especially via the sarcoplasmic reticulum Ca^{2+}-ATPase (SERCA 2a) is crucial for cardiac contractile function. Functionality of ryanodine receptor and SERCA were shown in murine ESCM, even in early developmental stages $(7+2-4\,\text{days})$.[60] Ryanodine, a ryanodine receptor blocker, as well as thapsigargin, a SERCA blocker, induced a decline of the amplitude and upstroke velocity and a prolongation of the decay time of the Ca^{2+} transient. In contrast, ryanodine and thapsigargin had no effect on Ca^{2+} transients and contractions in human ESCM,[55] indicating that the sarcoplasmic reticulum did not contribute significantly to Ca^{2+} handling, but contractions depended on transmembrane Ca^{2+} influx. Thus, there is a marked difference between Ca^{2+} handling of human ESCM and human adult cardiomyocytes, which might be due to the immature developmental stage of human ESCM.

Contraction force

Although a passive stabilization of infarcted myocardium may yield beneficial effects on heart function,[61] active force development of transplanted ESCM is desired for an optimal improvement of heart function by transplanted ESCM. Measurements of contraction force clearly demonstrated an active force development of murine[62] and human[63] ESCM. Comparable to human adult cardiomyocytes, a positive length-tension relationship was found in human ESCM. However, absolute contraction force was low as compared to adult myocardium, which might be attributed to the irregular distribution of myofibrils at immature developmental stages. Measurements of cell shortening[55] revealed a negative force–frequency relation in human ESCM, while adult myocardium typically exhibits a positive force–frequency relation, which enforces the increase of cardiac output under exercise. The negative force–frequency relationship in human ESCM was associated with the above-mentioned decline of I_{Ca-L} at higher frequencies. Post-rest potentiation of contraction force, which is another characteristic feature of adult cardiomyocytes, was also absent in human ESCM, confirming the lack of a functional sarcoplasmic reticulum.

In conclusion, functional studies in murine ESCM demonstrated numerous cardiac-specific properties, including nodal-, atrial-, and ventricular-like action potentials, expression of numerous cardiac ion channels, vegetative regulation of chronotropy and function of the sarcoplasmic reticulum. Functional properties of late stage murine ESCM resemble those of late stage native embryonic cardiomyocytes. For human ESCM, a cardiac differentiation including typical action potential morphology, expression of major cardiac ion channels and vegetative regulation of chronotropy has been described, but there are remarkable functional differences as compared to human adult cardiomyocytes, especially regarding the electrophysiological basis of pacemaker activity, Ca^{2+} handling and characteristics of force development. A very immature developmental stage of human ESCM might be the main reason for these differences, but constitutional discrepancies between native cardiomyocyte and human ESCM differentiation using present protocols cannot be excluded. Therefore, future studies involving longer cultivation periods or improved differentiation protocols are demanded to further evaluate properties of human ESCM and to achieve a distinct compliance with native human cardiomyocytes.

Animal models of cardiomyocytes derived from ES cells

The first transplantation of α-MHC-neomycin-selected murine ESCMs into mouse hearts was described by Klug *et al.*[43] In this study, engrafted donor cells could be detected for up to seven weeks after implantation into uninjured hearts, whereas teratomas were not observed.[43] Since then, in a variety of studies both undifferentiated ES cells and ESCMs were transplanted into healthy or infarcted hearts, mostly in rodents. Yang *et al.* transplanted murine ES cell-derived cardiomyocytes into infarcted heart[64] and demonstrated that the myocardial contractility was improved in transplanted animals and was further improved by transfecting these cells to express the vascular endothelial growth factor (VEGF), presumably due to better neovascularization. Teratomas were not observed in this study. In another study, the survival of cells was improved by transplanting hESCM into SCID mice and applying of cytoprotective (allopurinol/uricase) and anti-inflammatory (ibuprofen) agents. No inflammatory reactions or tumors were observed and heart function was improved as measured by the ejection fraction.[65] Kolossov *et al.* used an antibiotic lineage selection approach based on the α-MHC-promoter to obtain highly purified murine cardiomyocyte populations. Engraftment of $3 \times 10^4 - 10^5$ of these cardiomyocytes into infarcted hearts of syngeneic mice resulted in poor survival of grafted cells but was significantly improved by cotransplanting equal numbers of syngeneic fibroblasts. Left ventricular function was significantly improved, and no teratoma formation was observed even after four to five months. Guo *et al.* enriched murine cardiomyocytes from embryoid bodies by Percoll-fractionation and generated *in vitro* artificial engineered cardiac tissue in a collagen–matrigel matrix capable of beating synchronously and responding to physical and chemical stimuli.[66] Although these constructs were not transplanted in this study, engineered tissue resembled neonatal mouse heart and markers for undifferentiated ES cells were absent. In some recent advances, hyaluronan linked to both butyric and retinoic acid (HBR) forced pluripotent ES cells into a cardiogenic pathway.[67] Therefore, artificial three-dimensional support for cells associated with defined chemical stimuli hold a promise of increasing the engraftment and function of cellular grafts.

While studies in rodents are useful and comparatively easy to perform, heart rates of 400–500 beats/min in rats and mice make arrythmias caused by grafted cardiomyocytes with pacemaker activity difficult to detect. Indeed, it has been shown that the pacemaker activity of hESCMs transplanted into pigs with a complete atrioventricular block can replace an electronic pacemaker and pace the pig hearts.[68] Similar results were reported for the transplantation of hESCMs into guinea pigs. Here, hESCMs integrated electrophysiologically and could work as pacemakers.[54] While these results are promising with regards to the use of such cells as biological pacemakers, this also highlights the risk of inadvertently causing arrythmias by grafting ESCMs.

Several studies examined if transplantation of undifferentiated ES cells directly into hearts could lead to their predominant differentiation into cardiomyocytes and improve the function of injured heart without formation of teratomas. Behfar *et al.* reported transplantation of undifferentiated mESCs into a rat model with heart failure after myocardial infarction. They observed that stem cells differentiate into functional cardiomyocytes within infarcted hearts, resulting in improved function. They further argued that the proliferative and differentiation capacity of undifferentiated cells as opposed to fully differentiated, non-proliferative cardiomyocytes would be required to efficiently repopulate ischemic areas weeks after an infarction took place. Teratomas were only observed in TGF-β-signaling-deficient ESCs.[69]

In an another study, undifferentiated CGR8 mESCs transplanted into infarcted rat myocardium improved heart function without signs of graft rejection or tumor formation.[70] Kofidis *et al.* transplanted the undifferentiated mESCs supplemented with insulin-like growth factor-1 (IGF1) as an adjuvant to promote cardiomyocyte differentiation into ischemic areas of mouse hearts after left anterior descending (LAD) artery ligation. While functional improvements were detectable, all recipient animals developed teratomas after two weeks.[71] To overcome the high mortality of cells grafted by injection as a cell suspension, tissue engineering might offer a solution. When a suspension of undifferentiated, GFP-transgenic mESCs was engrafted in a matrigel solution that solidified upon injection into ischemic hearts of rats and mice the injected cells expressed cardiomyocyte markers and, compared to cells injected without matrigel, heart function was better restored. Teratoma formation was not observed, but animals were only monitored for two weeks (rat) or four (mice) weeks.[72,73]

To monitor the fate of implanted cells *in vivo*, a variety of *in vivo* imaging techniques are being explored. They have the advantage of allowing non-invasive, quantitative, and repetitive visualization of grafts within living subjects. In order to follow the fate of transplanted cells, Himes *et al.* have used magnetic resonance imaging (MRI) of mESCs loaded with superparamagnetic iron (SPIO) and injected into mouse hearts.[74] Grafted cells could be tracked for a minimum of five weeks, although dilution effects due to cell division and uptake of SPIO particles by, e.g. macrophages complicate interpretation of results. Cao *et al.* transfected undifferentiated mESCs with reporter constructs allowing tracking of transfected ES cells for several weeks both by optical bioluminescence imaging as well as positron emission tomography (PET).[75] The quick development on imaging techniques enables us to control closely of the cell commitment after transplanted. As a return, the feedback points the direction for further improvement.

Other Animal Models for ES Cell-Based Therapeutics

Animal models have also been used successfully to study the feasibility of transplanting ES cell-derived cells to treat other degenerative diseases. Collecting of information from animal models useful for an ES-based therapy is fairly advanced. Moreover, significant progresses in translating those information to clinical applications has been made.

Models for neural diseases

Neural developmental and degenerative diseases such as Parkinson's disease, retinal degeneration, brain injury, and spinal cord injury represent prime targets for ES cell-based therapy. Animal models for neural diseases include mouse ones such as the myelin basic protein (MBP)-deficient shiverer mice, an experimental model for multiple sclerosis (MS),[76] 6-hydroxydopamin (6-OHDA)-treatment of mouse striatum as an acute model of Parkinson's disease,[77] cryogenic brain injury as a stroke model[75] and central nervous system (CNS) injury and demyelination.[78] Rat models include dysmyelinating mutants, chemically induced foci of myelin loss, antibody/complement-induced demyelination models[79] and spinal cord injury.[80]

Hara *et al.* transplanted undifferentiated ES cells into normal adult mouse retinas to observe how the physiological micro-environment influences ES cell differentiation *in situ*. The ES cells grew along the retinal surface, developed fine neuronal cell processes around cell nuclei and generated neuronal networks into the retinal inner plexiform layer 30 days after transplantation. The implanted ES cells further differentiated *in vivo* and expressed retinal and neuronal markers.[81] The results demonstrate on the one hand that ES cells can directly differentiate into functional neurons, on the other hand, this immediate application of ES cells without *in vitro* differentiation stage caused apoptosis of most transplanted ES cells. In this study, there is no report on tumor formation due to persistence of undifferentiated ES cells.

To avoid the risk of teratoma formation from undifferentiated ES cell transplantation, more groups are trying with transplanting *in vitro* differentiated ES cells. Neural precursors and neurons can be induced *in vitro* from ES cells by supplementing Noggin, a transforming growth factor-β (TGFβ) antagonist[82,83] or by using SDIA.[84–86] A system which uses soluble interleukin-6 (IL6) receptor (IL6R) fused to IL6 in the culture enhanced the differentiation of ES cells to neurosphere cells, which further differentiated into oligodendrocytes. Furthermore, IL6R-IL6 fusion protein stimulated the myelinating function of oligodendrocyte precursors when IL6R-IL6-pretreated cells were transplanted into brain slices of MBP-deficient shiverer mice.[76]

The first step for early animal experiments was the survival of implanted cells. A stromal cell-derived inducing activity (SDIA) is a well-defined factor used for inducing differentiation of neural cells from mouse ES cells and efficiently generating midbrain tyrosine hydroxylase (TH$^+$)-positive dopaminergic neurons. These neurons survive well when implanted in the 6-OHDA-treated mouse striatum. No functional test was performed.[77] When ES cells were differentiated into motoneurons *in vitro* and transplanted into the spinal cords of adult rats with motoneuron injury, 25% of input motoneurons survived for more than a month in the spinal cord. In combination with dbcAMP administration, approximately 80 ES cell-derived motor axons were observed within the ventral roots.[80] No further functional improvement was detected. However, the survival of transplanted cells may be considered a promising first step.

Once the survival point is resolved, the second step was to see whether those transplanted cells can perform their functions or not. The outputs are positive. Neural cells derived from ES cells which express neurotrophin-3 (NT-3) were transplanted underneath the injured motor cortex of mice damaged by cryogenic brain injury, ajacent to the paraventricular region. Some of these transplanted cells further differentiated and migrated into the damaged region, possibly in response to signals from the lesion and contributed to the functional recovery of the recipient mice.[87] Mouse

ES cell-derived glial precursors (ESGPs) implanted into brain efficiently survive and further differentiate *in vivo* into both oligodendrocytes and astrocytes. These cells restore myelin function by forming new myelin sheaths in acutely demyelinated lesion in mice.[79] Roy *et al.* worked in a similar direction using a rat model using human ES cells were differentiated to express the A9 nigrostriatal phenotype. Upon transplantation into the neostriata of 6-OHDA-lesioned Parkinsonian rats, the implanted cells yielded a significant, substantial and long-lasting restitution of motor function. However, there are still concerns on teratoma formation and phenotype instability.[88]

Loss of hearing is traditionally treated with allografts and xenografts to regenerate auditory hair cells, spiral ganglion neurons or their axons. The biggest challenge is to establish connections from the inner ear to the central auditory system. This challenge was addressed by cell implantation of undifferentiated ES cells along the auditory nerve in an adult host. When implanted into the injured vestibulocochlear nerve of adult rats and guinea pigs, ES cells differentiated at the site of injection as well as within the brain stem.[89] Similarly, when transplanted to adult rats, mouse ES cells could migrate into the brain stem close to the ventral cochlear nucleus.[90] Although no teratoma formation was described in these two papers, it is very likely the case. Experimental data however suggest that ES cells have the potential of recreating a neuronal conduit following neuronal injury between the inner ear and the central nervous system (CNS).

Models for therapy for type I diabetes mellitus

The use of ES cell-derived β-cells for the therapy of Type 1 diabetes mellitus may overcome the shortcomings of injecting insulin by adjusting the release of insulin according to the level of blood sugar. However, despite numerous efforts by many groups (for review, see Ref. 91), the efficiency of differentiation of ES cells into islets is still very low and fully differentiated true β-cells are not yet available.

Streptozotocin-treated mice represent a model of chemically induced pancreatic damage that causes hyperglycemia, and this model is often used for *in vivo* investigations of diabetes. However, the efficiency of differentiation of ES cells into islets is still very low. This low efficiency reflects a lack of understanding of the intrinsic and extrinsic signals which regulate the developmental processes of the pancreas.[92] Insulin production is the terminal marker for pancreatic differentiation. In this process, several differentiation stages are involved, namely from pluripotent ES cells to primitive ectoderm, then to mesendoderm, further towards definitive endoderm, followed by pancreatic commitment and finally to insulin producing cells (IPC). Considering the complexity of this process, the frequency of insulin-positive cells in ES cell differentiation cultures remains very low, and increasing the efficiency of IPC differentiation is an important issue for an ES cell-based therapeutic approach.

Animal models for liver and kidney regeneration

Undifferentiated or differentiated ES cells have been transplanted into animal models with acute liver failure and were able to recover liver function to a certain extent.[93,94] Kai *et al.* transfected HNF4-overexpressing mouse ES cells to Sprague–Dawley rats were subjected to 90% hepatectomy and given 5% oral dextrose; 33% of treated rats survived more than one month whereas all rats in control groups died in three days.[93] Similarly, treatment of 90% hepatectomized mice with a subcutaneously implanted improved by bioartificial liver (BAL) seeded with mouse ES cell–derived hepatocytes improved liver function and prolonged survival of more than 12 days and had regenerated their remnant liver, whereas treatment with a BAL seeded with control cells did not within four days.[94] Encouraging results are also reported from other groups.[95–97]

In order to obtain higher yields of hepatocyte-like cells from mouse ES cells, improved cell culturing protocols have been developed based on a three-dimensional culture system.[98] In this system, the EBs were cultured on collagen scaffolds and stimulated with exogenous growth factors and hormones to induce hepatocyte-like cells. The scaffolds including EB-derived hepatocyte-like cells were transplanted into the median lobes of partially hepatectomized nude mice. Interestingly, transplanted cells continued to differentiate and expressed both albumin and cytokeratin-18.[98] Recently there is an alternative method to differentiated mouse ES cells into hepatocytes by coculture with a combination of human liver non-parenchymal cell lines and fibroblast growth factor-2, human activin-A and hepatocyte growth factor. Those cocultured cells provided necessary atmosphere for hepatotic commitment of ES cells. Those derived hepatocytes expressed liver-specific genes, secreted albumin and metabolized ammonia, lidocaine and diazepam. When cocultured ES cell-derived hepatocytes transplanted to the mouse model, animal survival rate is more than two folds than those from transplantation with normal ES cells.[94]

There are three cell compartments that physiologically contribute to vertebrate liver parenchymal maintenance and regeneration after injury:[99] the mature liver cells (hepatocytes, cholangiocytes), intraorgan stem/progenitor cells (cells of the proximal biliary tree, periductal cells) and extraorgan stem cells (from the circulation and the bone marrow). Considering these favorable conditions *in vivo*, ES cell-based liver regeneration is expected to move forward faster than other endoderm derived organ tissues such as pancreas, insulin β-cells, lung, and kidney. However, the quality and purity of differentiated hepatocyte-like cells as well as other ES cell derivatives need to be improved.

Differentiated mouse ES cells, which express kidney early developmental genes (Pax-2, Lim-1, c-Ret, Emx2, Sall1, WT-1, Eya-1, GDNF, and Wnt-4) were transplanted into kidneys of adult mice. Those transplanted differentiated ES cells continued to differentiate *in vivo*. The collecting ducts are positive for Pax-2, endo A cytokeratin, kidney-specific cadherin, and Dolichos biflorus agglutinin, which suggests that the these renal primordial ducts started to branch to mesonephric ducts and uretric buds.[100] Later these continued to differentiate further *in vivo* and showed different gene expression patterns and phenotypes. This animal model provides not only an insight to kidney regeneration, better understanding of differentiation dynamic from ES cells to kidney precursors as well.

Cell replacement therapy for other organ systems

Recently, intensive research of the application of ES cells to treat lung diseases has started focusing on differentiation to lung specific surfactant protein C (SPC)-expressing lung epithelial cells (type II pneumocytes) in serum-free small airway growth medium (SAGM).[101] With slight modification of the SAGM culture system (removal of retinoic acid and triiodothryonine), a threefold increase of SPC expression was achieved.[102] Samadikuchaksaraei's group used another approach and first cultured cells in traditional DMEM medium for 20 days without splitting, followed by culture in SAGM medium for three days. This sequential culture increased the mRNA expression level of SPC eightfold compared to cells cultured in DMEM only.[103] When cocultured with embryonic pulmonary mesenchyme, murine ES cells differentiated to epithelium lining cells expressing cytokeratin-1, thyroid transcription factor-1 and SPC. Those lining cells formed channels in cocultured EBs.[104] Type II pneumocytes differentiated from murine ES cells have been tested composed of basal, ciliated, intermediate, and Clara cells, similar to those of native tracheobronchial airway epithelium. These Type II pneumocytes showed similar properties as natural airway epithelial tissue regarding their ultrastructure and secretory function.[105] As lung specific markers are known and culture systems have been developed, *in vivo* studies are soon expected once suitable animal models are established.

Endothelial growth factor receptor-2 (Flk1)-expressing ES cell derivatives can be further differentiated to endothelial, mural (vascular smooth muscle cells and pericytes) and blood cells. Therefore Flk-1$^+$ cells are called vascular progenitor cells (VPCs) (for review, see Ref. 106). When implanted into mice, these VPCs were often detected as non-vascular cells. Only in the presence of vascular endothelial growth factor (VEGF), cultured VPCs are capable of differentiation into a mixture of endothelial cells (ECs) and mural cells (MCs). Among them the ECs were more specifically incorporated into developing vasculature and the MCs in vascular walls.[107] More advanced research is done with ES cells from non-human primate. The differentiation features of monkey ES cells to vascular cells requires more selection steps based on the expression of VEGF-R2, platelet endothelial cell adhesion molecule-1 (PECAM1), VE-cadherin, endothelial nitric oxide synthase (eNOS), and smooth muscle actin (SMA). It has been shown that the differentiated VEGF-R2$^+$ cells can form tube-like structures in a three-dimensional culture.[108] More promising results are expected once a primate animal model is established.

Although blood transplantation is currently performed with adult bone marrow-derived hematopoietic stem cells, efforts are also made to apply ES cells to replace adult stem cells. Until now, ES cell-derived, high-proliferative potential colony-forming cells (HPP-CFC) give rise to secondary granulocytes, erythrocytes, macrophages, and mast cells. The regeneration capacity of HPP-CFCs are similar to those from yolk sac, which are superior to those from adult bone marrow. However, their *in vivo* reconstitution ability has not been detected yet.[109]

Barriers to Successful Use of ES Cell-Based Therapies

Ethical concerns

Since derivation of human ES cells eight years ago serious ethical objections to use of ES cells have been raised. It has been argued that derivation of ES cell lines from human blastocyst is morally objectionable because this procedure destroys potential human life. To obviate the use of ES cells, adult stem cells have been proposed as an alternative source of cells for regenerative medicine. However, adult stem cells are difficult to culture for extended periods of time and they seem to possess only limited ability to differentiate into a variety of tissue-specific cells. Therefore, adult stem cells can hardly surpass the pluripotential nature of ES cells so that the derivation of many clinically useful cell types may still be dependent on the availability of well characterized ES cell lines. But, is it possible to develop ethically acceptable ES cell lines? Hurlbut suggested that one way of getting around the ethical concerns might be the derivation of pluripotent ES cell lines from blastocysts generated by a so-called altered nuclear transfer.[110] This technology involves temporal inactivation before nuclear transfer of a gene required for normal development, such as *Cdx2*. This gene is essential for trophectoderm formation in mouse. Blastocysts with disabled *Cdx2* lack trophectoderm and cannot implant but can serve as a source of normal ES cells after removal of a transgene producing the *Cdx2*-interfering RNA.[111] According to proponents of this concept, blastocysts which lack the full developmental potential should be ideologically acceptable. Another suggested procedure for generation of human ES cell lines that might be ethically acceptable is by using a single-cell biopsy similar to that used in preimplantation genetic diagnosis. In this concept, a single blastomere is removed from an eight-cell embryo to derive a new ES cell line and the remaining seven-cell embryos are transferred into surrogate mothers, in which they develop into normal mice.[112,113] However, the validity of both approaches for production of ideologically acceptable human ES cell lines has been strongly questioned.[114,115] Critics of the conditional *Cdx2*-knockdown strategy argue that reversible inactivation of *Cdx2* is ethically not distinct from destroying the embryo by the immunosurgical method that is routinely used to derive human ES cells. Moreover, it is not clear if the *Cdx2* would have the same indispensable function in human development as it has in a mouse and testing this by using human embryos would not be allowed for ethical reasons. Even if this could be verified, the method for knocking down the gene by RNA interference would never be in the position to insure that absolutely every embryo produced by this technology is incapable of normal development. One of the critics' strongest arguments against the second proposal by Klimanskaya et al. is the lack of evidence that a single blastomere from a human embryo at the eight-cell stage has no capacity to develop into a normal human being.[113] If this cell would be capable of normal development, as is the case with single blastomeres isolated from a rabbit or sheep, the ES cells derived from such blastomeres would also not be ethically acceptable according to the opponents of this approach. The difficulty of this concept is that the developmental potential of human blastomeres, similarly as the developmental potential of Cdx2-deficient embryos, can never be tested in humans for ethical reasons since it would require the transfer of a single blastomere into the uterus. Heated debate instigated by these proposals illustrates that it is very difficult, if not impossible, to solve ethical questions by scientific means. One possible solution to this problem would be to completely abandon the idea of deriving cells for replacement therapy from blastocysts but to derive the patient's own ES cells in vitro by nuclear reprogramming of her or his own adult somatic cells to a pluripotent state, from which other cell types required for therapeutic purposes can be generated.[116] However, even if this goal can be achieved, insuring and standardizing the quality of each individual patient's cell line will be an economical and logistical challenge.

Safety concerns and purity

Besides ethical issues, safety concerns related to teratoma development and transmission of infection pose another obstacle to therapeutic use of ES cell derivatives. Undifferentiated ES cells form teratomas when injected in syngeneic, immunodeficient and even allogeneic mice,[117,118] and this property is frequently used as a proof of their pluripotency. Under appropriate conditions, the injection of as few as two transplanted ES cells can form teratoma,[119] which indicates that cells used for replacement therapy will have to be absolutely avoid of undifferentiated ES cells. In addition, undifferentiated cells capable of developing into teratomas persist in cultured embryoid bodies even at late stages of differentiation.[120] Therefore, the achievement of optimal engraftment and function of ES cell-derived transplants without teratoma development represents a major challenge. The only way of achieving this goal is by using only those cells for therapeutic transplantation which have lost their teratogenic potential such as ES cell-derived somatic precursors or mature cells. There are two strategies for producing pure cell populations having a defined somatic phenotype. In one strategy, differentiation of ES cells into a desired cell type is conducted in a stepwise manner in strictly controlled culture conditions containing defined differentiation and survival factors. The advantage of this method is that no genetic modification of donor cells is required. However, this approach is complicated because exact factors and culture conditions must be known in order for it to be successful and for many required cell types the maximally attainable purity may not be satisfactory. Another strategy is a so-called lineage selection. This concept involves genetic modification of ES cells to express a selectable marker, which is expressed in differentiated cells in a cell type-specific manner. Approaches that build on this strategy utilize selection techniques based on antibiotic resistance, immunoselection or fluorescence activated cell sorting and had already been successfully applied in generation of purified cardiomyocytes,[43,46] neurons,[121,122] pancreatic β-cells[123] and endothelial cells.[124] Besides enrichment for a specific cell population these strategies may lead to elimination of teratoma formation upon their transplantation as has been reported for antibiotic-selected cardiomyocytes.[46] Genetic engineering strategies can also be used for negative selection of unwanted undifferentiated ES cells, which may be present as contaminants in purified differentiated cells of interest. In this approach, expression of a "suicide" gene, such as the herpes simplex virus thymidine kinase gene, specifically in undifferentiated ES cells would render them sensitive to a drug, such as ganciclovir in case of thymidine kinase, which could be administered as a second-line safety measure in case teratoma occurs upon transplantation.[75,125]

By combining positive and negative selection strategies with targeted differentiation protocols to favor generation of a desired cell type it will be possible to improve the safety to the level where transplantations involving ES cell-derived tissues can be performed without the risk of teratoma formation. In addition, adoption of standardized procedures based on good manufacturing practice guidelines and stringent quality control protocols will help eliminate risk for transmission of infectious agents to prospective patients.

Rejection of allogeneic cells

Even if all other obstacles discussed above were removed one important barrier that could limit the clinical use of ES cell-derived replacement therapy will be the immune rejection of allogeneic cells (reviewed by Ref. 126). The most important, although not only, immunological barrier to transplantation of ES cell derivatives will be presumably the polymorphic major histocompatibility (MHC) class I antigens because MHC class II antigens are normally present only on macrophages, B-cells and dendritic cells. In contrast, MHC class I molecules are constitutively expressed at various levels on all adult nucleated cells and are strongly upregulated by inflammatory cytokines such as interferon-γ and tumor necrosis factor-α. Transcripts encoding murine MHC and associated molecules have been detected already in one- or two-cell preimplantation stage embryos.[127] Recent work by several groups has established that MHC class I molecules are expressed at low levels on the surface of human ES cells and their differentiated derivatives.[128–130] We and others have demonstrated that MHC class I molecules are not constitutively detectable on undifferentiated murine ES cells and their differentiated derivatives by standard flow cytometry[131,132] (Fig. 3), but only with highly sensitive methods such as T-cell hybridoma assay.[131] With the exception of murine undifferentiated ES cells, which are not responsive to interferon γ,[131,132] this cytokine strongly induces MHC class I molecules on human ES cells and on both human and murine EB-derived cells[129,131]

(Fig. 3). These data indicate that tissues developed from ES cells for therapeutic purposes will most likely express MHC class I molecules and that they may be rejected upon transplantation.

However, several studies have reported that human ES cells and their derivatives may possess immune-privileged properties.[129,130,133] Li *et al.* demonstrated that undifferentiated and differentiated ES cells failed to directly stimulate proliferation of alloreactive human T-cells *in vitro* and inhibited third-party allogeneic dendritic cell-mediated T-cell proliferation.[130] Murine ES cells also fail to directly stimulate allogeneic responder leukocytes but are able to completely abrogate an ongoing allostimulation in a mixed lymphocyte reaction in a process involving a direct cell-cell contact.[134] Upon intramuscular injection of undifferentiated human ES cells into immunocompetent mice there was no induction of an immune response after a two-day follow-up.[130] Unfortunately, authors did not monitor the immune response at later time points, so the role of the adaptive immune system in this xenogeneic model could not be evaluated. However, two recent reports demonstrated that xenotransplanted human ES cells under the kidney capsule were readily rejected as demonstrated by the absence of teratomas one month after transplantation.[133,135] These experiments also showed that xenorejection was mediated by murine T-cells, as deduced from the observation that teratomas developed only in T-cell-deficient but not B-cell- or NK-cell-deficient animals. However, transplantation of undifferentiated or differentiated human ES cells into mice reconstituted with human peripheral blood mononuclear cells resulted in only a weak response of human leukocytes toward transplanted cells.[133] Even preimmunization of the reconstituted mice with irradiated human ES cells did not induce subsequent rejection of ES cell grafts. Interestingly, when cytotoxic T-cells were stimulated *in vitro* by influenza-derived peptide and cultured with antigen-loaded and interferon γ-stimulated human ES cells, weak lysis occured. It has been suggested that transplanted cells evaded immune destruction due to their low immunostimulatory potential and immunological immaturity. Indeed, human ES cells do not express costimulatory molecules and many other immune-related genes as demonstrated by flow cytometry or by comparison of gene expression profiles in different human ES cell lines, leukocytes and various adult organs.[133,135]

Similarly to human ES cells, immune privileged properties have also been described for mouse ES cells. Fändrich *et al.* reported the induction of immunologic tolerance after intraportal injection of undifferentiated rat ES cell-like cells into fully MHC-mismatched rats.[136] These cells engrafted permanently without supplementary host conditioning and allowed for the subsequent long-term acceptance of second-set transplanted cardiac allografts. In a related study, intravenously administered undifferentiated allogeneic murine ES cells populated the thymus, spleen and liver of sublethaly irradiated host mice and induced mixed hematopoietic chimerism without forming teratomas.[134] In these animals the mixed chimerism was no longer detectable after four weeks, suggesting that cells were rejected or lacked specific factors required for long-term renewal. In contrast to immuno-compromised mice,

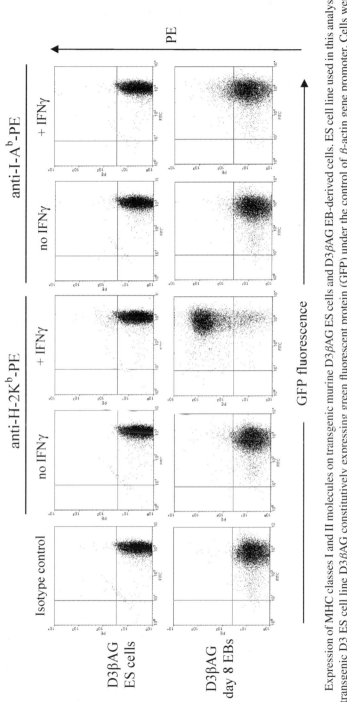

Fig. 3. Expression of MHC classes I and II molecules on transgenic murine D3βAG ES cells and D3βAG EB-derived cells. ES cell line used in this analysis was a transgenic D3 ES cell line D3βAG constitutively expressing green fluorescent protein (GFP) under the control of β-actin gene promoter. Cells were treated with 20 ng/ml interferon-γ for two days or were left untreated. Monocellular suspensions of undifferentiated ES cells or EBs taken at day 8 after induction of differentiation were prepared by Trypsin/EDTA treatment. Cells were stained with phycoerythrin (PE)-conjugated H-2K^b antibodies (clone AF6-88.5) or PE-conjugated isotype control antibodies and analyzed with FACScan and the CellQuest software (T. Šarić, unpublished data).

only very modest degree of mixed chimerism in only 30%–40% of animals could be observed when the ES cells were intravenously injected to immunocompetent animals. The authors reasoned that ES cells were unable to engraft in this setting probably because the niches required for ES cell engraftment were occupied by the recipient's cells and not because the ES cells were rejected by allogeneic immune system, since only irradiation but not cyclosporine-A or rapamycin enhanced the engraftment. Noteworthy, engrafted ES cells were shown to induce apoptosis to neighboring T-cells in lymphoid organs by the process involving engagement of Fas on T-cells and FasL on ES cells.[134] However, it is unclear if this protective mechanism is active in all ES cell lines because we were unable to detect FasL in murine CGR8 ES cells both at the transcript and protein level (*Šarić T., unpublished observation*). FasL also does not seem to be expressed in human undifferentiated ES cells and their derivatives.[133,135] Although these studies demonstrate that undifferentiated ES cells are non-immunogenic and may be capable of inhibiting immune responses, the clinical use of undifferentiated ES cells still carries the serious risk of teratocarcinoma formation and it is highly unlikely that it will be used for induction of immune tolerance or for tissue regeneration in humans.

While the above studies suggest a critical role for Fas–FasL engagement in ES cell engraftment, recently published data reveal yet another mechanism utilized by ES cells for protection against T-cell-mediated immune responses. Abdullah *et al.* demonstrated that ES cells infected with lymphocytic choriomeningitis virus are readily recognized by virus-specific cytotoxic T-cells in a process which leads to a secretion of interferon-γ and cytotoxic mediators by T-cells at the immunological synapse.[131] This recognition could be demonstrated even with undifferentiated ES cells expressing only very low levels of MHC class I molecules. However, despite this recognition, no or only very weak killing of ES cells and their derivatives could be observed, corroborating the results obtained with alloreactive T-cells.[137] This was true even when MHC class I molecules were highly upregulated on differentiated ES cell derivatives by interferon-γ treatment. This resistance to T-cell-mediated lysis was shown to be conferred by a high level expression of granzyme-B inhibitor SPI-6 (serpin protease inhibitor-6) in undifferentiated and differentiated ES cells.[131] These data indicate that ES cells possess cell-autonomous protection mechanisms against cellular immune responses, which may be in part responsible for the successful engraftment of allotransplanted human and murine ES cells as observed in studies discussed above. Notably, embryonic tissues from early gestational stages are also known to be less immunogenic compared with their adult counterparts and thus resemble immunological properties of ES cells.[138] As ES cells further differentiate and mature in host tissues upon transplantation they may acquire immunological maturity and be subject to rejection.

Above results are contrasted by recently published observations showing activation of the recipient's humoral and cellular adaptive immune system upon allotransplantation of undifferentiated murine ES cells into infarcted hearts of immunocompetent mice.[117,118] Teratomas formed in the hearts shortly after transplantation of ES cells both in syngeneic and allogeneic animals and four to eight weeks later the allogeneic grafts were infiltrated by immune cells and rejection ensued. Reasons for different outcome of these transplantations as compared to those described above are unclear but may lie in different ES cell lines used (D3 *versus* R1), different amount of cells implanted, or different routes of implantation (intramyocardial *versus* intra-venous).

Despite the apparent immune privilege of ES cells, pure organotypic cells derived from ES cells may still have different immunological properties and be more prone to rejection. Therefore, strategies for preventing rejection will most likely be needed and they may include either induction of immune tolerance in the recipient or gen-eration of ES cell-derivatives having a property to evade rejection. For example, the generation of major histocompatibility complex (MHC)-deficient ES cell line has been suggested but such ES cell-based grafts would likely still encounter rejection just as it has been described for tissue grafts from MHC$^{-/-}$ mice.[139] There is also a suggestion of a possible knockin of the patient's MHC locus in the ES cell line that would be used for cell therapy, but is not feasible in normal cell replacement therapy. Another possible way to circumvent the immune rejection problem could be therapeutic cloning. Somatic cell nuclear transfer from a patient's adult somatic cell into an enucleated oocyte would reprogram the somatic nucleus. ES cells derived from such a cloned blastocyst are then autologous in nature and hence there would be no histoincompatibility. The cells derived by this strategy may therefore be used as a patient-tailored source of replacement cells.[140]

Conclusions

Transplantation of differentiated cells derived from ES cells opens new possibilities for therapy of a number of diseases characterized by irreparable tissue injury. Several promising animal studies have already demonstrated the functional improvement of injured organs after transplantation of tissue-specific cells derived from embryonic stem cells. In this context, significant progress has been made in developing method-ologies for generation of large amounts of safe and pure tissue specific cells. Exploitation and practical application of human ES cells in cell replacement therapy are still at the beginning and further exploration and refinement of differentiation conditions, purification protocols and delivery techniques are needed. Although results from transplantation experiments in mouse models hold promise that ES-cell based replacement therapy may one day be used for human therapy, several barriers must be resolved before clinical studies in this direction can be initiated. In addition to their potential for therapy, the immense value of ES cells is in serving as an irreplaceable research tool for exploring the nature of pluripotency, self-renewal, and elucidating the mechanisms of differentiation into specialized cell types. The development of other strategies for obtaining pluripotent cell-for-cell replacement therapy in humans, such as reprogramming of somatic cells to a pluripotent state or therapeutic cloning by nuclear transfer, will greatly if not exclusively depend on the knowledge generated in the research with ES cells.

References

1. Evans MJ, Kaufman MH. Establishment in culture of pluripotential cells from mouse embryos. *Nature* 1981;292(5819):154–156.
2. Axelrod HR. Embryonic stem-cell lines derived from blastocysts by a simplified technique. *Dev Biol* 1984;101(1):225–228.
3. Thomson JA, Itskovitz-Eldor J, Shapiro SS, Waknitz MA, Swiergiel JJ, Marshall VS *et al.* Embryonic stem cell lines derived from human blastocysts. *Science* 1998;282(5391):1145–1147.
4. Odorico JS, Kaufman DS, Thomson JA. Multilineage differentiation from human embryonic stem cell lines. *Stem Cells* 2001;19(3):193–204.
5. Ginis I, Luo Y, Miura T, Thies S, Brandenberger R, Gerecht-Nir S, *et al.* Differences between human and mouse embryonic stem cells. *Dev Biol* 2004;269(2):360–380.
6. Solter D, Knowles BB. Monoclonal antibody defining a stage-specific mouse embryonic antigen (Ssea-1). *Proc Natl Acad Sci USA* 1978;75(11):5565–5569.
7. Pesce M, Anastassiadis K, Scholer HR. Oct-4: Lessons of totipotency from embryonic stem cells. *Cells Tissues Organs* 1999;165(3–4):144–152.
8. Henderson JK, Draper JS, Baillie HS, Fishel S, Thomson JA, Moore H, *et al.* Preimplantation human embryos and embryonic stem cells show comparable expression of stage-specific embryonic antigens. *Stem Cells* 2002;20(4):329–337.
9. Palmqvist L, Glover CH, Hsu L, Lu M, Bossen B, Piret JM, *et al.* Correlation of murine embryonic stem cell gene expression profiles with functional measures of pluripotency. *Stem Cells* 2005;23(5):663–680.
10. Austin GS. Culture and differentiation of embryonic stem cells. *Methods Cell Sci* 1991;V13(2):89–94.
11. Williams RL, Hilton DJ, Pease S, Willson TA, Stewart CL, Gearing DP, *et al.* Myeloid leukaemia inhibitory factor maintains the developmental potential of embryonic stem cells. *Nature* 1988;336(6200):684–687.
12. Wang G, Zhang H, Zhao Y, Li J, Cai J, Wang P, *et al.* Noggin and bFGF cooperate to maintain the pluripotency of human embryonic stem cells in the absence of feeder layers. *Biochem Biophys Res Commun* 2005;330(3):934–942.
13. Xu C, Inokuma MS, Denham J, Golds K, Kundu P, Gold JD, *et al.* Feeder-free growth of undifferentiated human embryonic stem cells. *Nat Biotechnol* 2001;19(10): 971–974.
14. Mallon BS, Park KY, Chen KG, Hamilton RS, McKay RD. Toward xeno-free culture of human embryonic stem cells. *Int J Biochem Cell Biol* 2006;38(7):1063–1075.
15. Amit M, Shariki C, Margulets V, Itskovitz-Eldor J. Feeder layer- and serum-free culture of human embryonic stem cells. *Biol Reprod* 2004;70(3):837–845.
16. Ludwig TE, Bergendahl V, Levenstein ME, Yu J, Probasco MD, Thomson JA. Feeder-independent culture of human embryonic stem cells. *Nat Methods* 2006;3(8):637–646.
17. Richards M, Fong CY, Chan WK, Wong PC, Bongso A. Human feeders support prolonged undifferentiated growth of human inner cell masses and embryonic stem cells. *Nat Biotechnol* 2002;20(9):933–936.
18. Genbacev O, Krtolica A, Zdravkovic T, Brunette E, Powell S, Nath A, *et al.* Serum-free derivation of human embryonic stem cell lines on human placental fibroblast feeders. *Fertil Steril* 2005;83(5):1517–1529.

19. Stojkovic P, Lako M, Stewart R, Przyborski S, Armstrong L, Evans J, *et al.* An autogeneic feeder cell system that efficiently supports growth of undifferentiated human embryonic stem cells. *Stem Cells* 2005;23(3):306–314.
20. Niwa H, Burdon T, Chambers I, Smith A. Self-renewal of pluripotent embryonic stem cells is mediated via activation of STAT3. *Genes Dev* 1998;12(13):2048–2060.
21. Mitsui K, Tokuzawa Y, Itoh H, Segawa K, Murakami M, Takahashi K *et al.* The homeoprotein nanog is required for maintenance of pluripotency in mouse epiblast and ES cells. *Cell* 2003;113(5):631–642.
22. Sato N, Meijer L, Skaltsounis L, Greengard P, Brivanlou AH. Maintenance of pluripotency in human and mouse embryonic stem cells through activation of Wnt signaling by a pharmacological GSK-3-specific inhibitor. *Nat Med* 2004;10(1):55–63.
23. Doetschman TC, Eistetter H, Katz M, Schmidt W, Kemler R. The *in vitro* development of blastocyst-derived embryonic stem cell lines: formation of visceral yolk sac, blood islands and myocardium. *J Embryol Exp Morphol* 1985;87:27–45.
24. Wobus AM, Wallukat G, Hescheler J. Pluripotent mouse embryonic stem-cells are able to differentiate into cardiomyocytes expressing chronotropic responses to adrenergic and cholinergic agents and Ca2+ channel blockers. *Differentiation* 1991;48(3): 173–182.
25. Boheler KR, Czyz J, Tweedie D, Yang HT, Anisimov SV, Wobus AM. Differentiation of pluripotent embryonic stem cells into cardiomyocytes. *Circ Res* 2002;91(3):189–201.
26. Maltsev VA, Wobus AM, Rohwedel J, Bader M, Hescheler J. Cardiomyocytes differentiated *in vitro* from embryonic stem cells developmentally express cardiac-specific genes and ionic currents. *Circ Res* 1994;75(2):233–244.
27. Lough J, Barron M, Brogley M, Sugi Y, Bolender DL, Zhu X. Combined BMP-2 and FGF-4, but neither factor alone, induces cardiogenesis in non-precardiac embryonic mesoderm. *Dev Biol* 1996;178(1):198–202.
28. Kajstura J, Cheng W, Reiss K, Anversa P. The IGF-1-IGF-1 receptor system modulates myocyte proliferation but not myocyte cellular hypertrophy *in vitro*. *Exp Cell Res* 1994;215(2):273–283.
29. Pandur P, Lasche M, Eisenberg LM, Kuhl M. Wnt-11 activation of a non-canonical Wnt signalling pathway is required for cardiogenesis. *Nature* 2002;418(6898):636–641.
30. Sachinidis A, Fleischmann BK, Kolossov E, Wartenberg M, Sauer H, Hescheler J. Cardiac specific differentiation of mouse embryonic stem cells. *Cardiovasc Res* 2003;58(2):278–291.
31. Winkler J, Hescheler J, Sachinidis A. Embryonic stem cells for basic research and potential clinical applications in cardiology. *Biochim Biophys Acta* 2005;1740(2): 240–248.
32. Hescheler J, Sachinidis A, Chen S, Winkler J. The FunGenES consortium: functional genomics in engineered embryonic stem cells. *Stem Cell Rev* 2006;2(1):1–4.
33. Bin Z, Sheng LG, Gang ZC, Hong J, Jun C, Bo Y, *et al.* Efficient cardiomyocyte differentiation of embryonic stem cells by bone morphogenetic protein-2 combined with visceral endoderm-like cells. *Cell Biol Int* 2006;30(10):769–776.
34. Mummery C, Ward-van Oostwaard D, Doevendans P, Spijker R, van den BS, Hassink R, *et al.* Differentiation of human embryonic stem cells to cardiomyocytes: role of coculture with visceral endoderm-like cells. *Circulation* 2003;107(21): 2733–2740.
35. Behfar A, Perez–Terzic C, Faustino RS, Arrell DK, Hodgson DM, Yamada S, *et al.* Cardiopoietic programming of embryonic stem cells for tumor-free heart repair. *J Exp Med* 2007;204(2):405–420.
36. Gissel C, Voolstra C, Doss MX, Koehler CI, Winkler J, Hescheler J, *et al.* An optimized embryonic stem cell model for consistent gene expression and developmental studies: a fundamental study. *Thromb Haemost* 2005;94(4):719–727.
37. Sachinidis A, Gissel C, Nierhoff D, Hippler-Altenburg R, Sauer H, Wartenberg M, *et al.* Identification of platelet-derived growth factor-BB as cardiogenesis-inducing factor in mouse embryonic stem cells under serum-free conditions. *Cell Physiol Biochem* 2003;13(6):423–429.

38. Wobus AM, Kaomei G, Shan J, Wellner MC, Rohwedel J, Ji G, *et al.* Retinoic acid accelerates embryonic stem cell-derived cardiac differentiation and enhances development of ventricular cardiomyocytes. *J Mol Cell Cardiol* 1997;29(6):1525–1539.
39. Takahashi T, Lord B, Schulze PC, Fryer RM, Sarang SS, Gullans SR, *et al.* Ascorbic acid enhances differentiation of embryonic stem cells into cardiac myocytes. *Circulation* 2003;107(14):1912–1916.
40. Hidaka K, Lee JK, Kim HS, Ihm CH, Iio A, Ogawa M, *et al.* Chamber-specific differentiation of Nkx2.5-positive cardiac precursor cells from murine embryonic stem cells. *FASEB J* 2003;17(2):740–742.
41. Zandstra PW, Bauwens C, Yin T, Liu Q, Schiller H, Zweigerdt R, *et al.* Scalable production of embryonic stem cell-derived cardiomyocytes. *Tissue Eng* 2003;9(4): 767–778.
42. Sachinidis A, Schwengberg S, Hippler-Altenburg R, Mariappan D, Kamisetti N, Seelig B, *et al.* Identification of small signalling molecules promoting cardiac-specific differentiation of mouse embryonic stem cells. *Cell Physiol Biochem* 2006;18(6):303–314.
43. Klug MG, Soonpaa MH, Koh GY, Field LJ. Genetically selected cardiomyocytes from differentiating embronic stem cells form stable intracardiac grafts. *J Clin Invest* 1996;98(1):216–224.
44. Kolossov E, Fleischmann BK, Liu Q, Bloch W, Viatchenko-Karpinski S, Manzke O, *et al.* Functional characteristics of ES cell-derived cardiac precursor cells identified by tissue-specific expression of the green fluorescent protein. *J Cell Biol* 1998;143(7):2045–2056.
45. Muller M, Fleischmann BK, Selbert S, Ji GJ, Endl E, Middeler G, *et al.* Selection of ventricular-like cardiomyocytes from ES cells *in vitro. FASEB J* 2000;14(15): 2540–2548.
46. Kolossov E, Bostani T, Roell W, Breitbach M, Pillekamp F, Nygren JM, *et al.* Engraftment of engineered ES cell-derived cardiomyocytes but not BM cells restores contractile function to the infarcted myocardium. *J Exp Med* 2006;203(10):2315–2327.
47. Vandervelde S, van Luyn MJ, Tio RA, Harmsen MC. Signaling factors in stem cell-mediated repair of infarcted myocardium. *J Mol Cell Cardiol* 2005;39(2):363–376.
48. Hui JH, Ouyang HW, Hutmacher DW, Goh JC, Lee EH. Mesenchymal stem cells in musculoskeletal tissue engineering: a review of recent advances in National University of Singapore. *Ann Acad Med Singapore* 2005;34(2):206–212.
49. Maltsev VA, Rohwedel J, Hescheler J, Wobus AM. Embryonic stem cells differentiate *in vitro* into cardiomyocytes representing sinusnodal, atrial and ventricular cell types. *Mech Dev* 1993;44(1):41–50.
50. He JQ, Ma Y, Lee Y, Thomson JA, Kamp TJ. Human embryonic stem cells develop into multiple types of cardiac myocytes: action potential characterization. *Circ Res* 2003;93(1):32–39.
51. Reppel M, Pillekamp F, Brockmeier K, Matzkies M, Bekcioglu A, Lipke T, *et al.* The electrocardiogram of human embryonic stem cell-derived cardiomyocytes. *J Electrocardiol* 2005;38(Suppl 4):166–170.
52. Satin J, Kehat I, Caspi O, Huber I, Arbel G, Itzhaki I, *et al.* Mechanism of spontaneous excitability in human embryonic stem cell derived cardiomyocytes. *J Physiol* 2004;559(Pt 2):479–496.
53. Xu C, Police S, Rao N, Carpenter MK. Characterization and enrichment of cardiomyocytes derived from human embryonic stem cells. *Circ Res* 2002;91(6):501–508.
54. Xue T, Cho HC, Akar FG, Tsang SY, Jones SP, Marban E, *et al.* Functional integration of electrically active cardiac derivatives from genetically engineered human embryonic stem cells with quiescent recipient ventricular cardiomyocytes: insights into the development of cell-based pacemakers. *Circulation* 2005;111(1):11–20.
55. Dolnikov K, Shilkrut M, Zeevi-Levin N, Gerecht-Nir S, Amit M, Danon A, *et al.* Functional properties of human embryonic stem cell-derived cardiomyocytes: intracellular Ca^{2+} handling and the role of sarcoplasmic reticulum in the contraction. *Stem Cells* 2006;24(2):236–245.

56. Banach K, Halbach MD, Hu P, Hescheler J, Egert U. Development of electrical activity in cardiac myocyte aggregates derived from mouse embryonic stem cells. *Am J Physiol Heart Circ Physiol* 2003;284(6):H2114–H2123.

57. Kehat I, Kenyagin-Karsenti D, Snir M, Segev H, Amit M, Gepstein A, *et al.* Human embryonic stem cells can differentiate into myocytes with structural and functional properties of cardiomyocytes. *J Clin Invest* 2001;108(3):407–414.

58. Reppel M, Boettinger C, Hescheler J. Beta-adrenergic and muscarinic modulation of human embryonic stem cell-derived cardiomyocytes. *Cell Physiol Biochem* 2004;14 (4–6):187–196.

59. Ji GJ, Fleischmann BK, Bloch W, Feelisch M, Andressen C, Addicks K, *et al.* Regulation of the L-type Ca^{2+} channel during cardiomyogenesis: switch from NO to adenylyl cyclase-mediated inhibition. *FASEB J* 1999;13(2):313–324.

60. Fu JD, Li J, Tweedie D, Yu HM, Chen L, Wang R, *et al.* Crucial role of the sarcoplasmic reticulum in the developmental regulation of Ca^{2+} transients and contraction in cardiomyocytes derived from embryonic stem cells. *FASEB J* 2006;20(1): 181–183.

61. Muller-Ehmsen J, Kedes LH, Schwinger RH, Kloner RA. Cellular cardiomyoplasty — a novel approach to treat heart disease. *Congest Heart Fail* 2002;8(4):220–227.

62. Metzger JM, Lin WI, Samuelson LC. Transition in cardiac contractile sensitivity to calcium during the *in vitro* differentiation of mouse embryonic stem cells. *J Cell Biol* 1994;126(3):701–711.

63. Pillekamp F, Reppel M, Rubenchyk O, Pfannkuche K, Matzkies M, Bloch W, *et al.* Force measurements of human embryonic stem cell-derived cardiomyocytes in an *in vitro* transplantation model. *Stem Cells* 2007;25(1):174–180.

64. Yang Y, Min JY, Rana JS, Ke Q, Cai J, Chen Y, *et al.* VEGF enhances functional improvement of postinfarcted hearts by transplantation of ESC-differentiated cells. *J Appl Physiol* 2002;93(3):1140–1151.

65. Kofidis T, Lebl DR, Swijnenburg RJ, Greeve JM, Klima U, Gold J, *et al.* Allopurinol/uricase and ibuprofen enhance engraftment of cardiomyocyte-enriched human embryonic stem cells and improve cardiac function following myocardial injury. *Eur J Cardiothorac Surg* 2006;29(1):50–55.

66. Guo XM, Zhao YS, Chang HX, Wang CY, Ling L, Zhang XA, *et al.* Creation of engineered cardiac tissue *in vitro* from mouse embryonic stem cells. *Circulation* 2006;113(18):2229–2237.

67. Ventura C, Maioli M, Asara Y, Santoni D, Scarlata I, Cantoni S, *et al.* Butyric and retinoic mixed ester of hyaluronan. A novel differentiating glycoconjugate affording a high throughput of cardiogenesis in embryonic stem cells. *J Biol Chem* 2004;279(22):23574–23579.

68. Kehat I, Khimovich L, Caspi O, Gepstein A, Shofti R, Arbel G, *et al.* Electromechanical integration of cardiomyocytes derived from human embryonic stem cells. *Nat Biotechnol* 2004;22(10):1282–1289.

69. Behfar A, Zingman LV, Hodgson DM, Rauzier JM, Kane GC, Terzic A, *et al.* Stem cell differentiation requires a paracrine pathway in the heart. *FASEB J* 2002;16(12):1558–1566.

70. Hodgson DM, Behfar A, Zingman LV, Kane GC, Perez-Terzic C, Alekseev AE, *et al.* Stable benefit of embryonic stem cell therapy in myocardial infarction. *Am J Physiol Heart Circ Physiol* 2004;287(2):H471–H479.

71. Kofidis T, de Bruin JL, Yamane T, Balsam LB, Lebl DR, Swijnenburg RJ, *et al.* Insulin-like growth factor promotes engraftment, differentiation, and functional improvement after transfer of embryonic stem cells for myocardial restoration. *Stem Cells* 2004;22(7):1239–1245.

72. Kofidis T, de Bruin JL, Hoyt G, Lebl DR, Tanaka M, Yamane T, *et al.* Injectable bioartificial myocardial tissue for large-scale intramural cell transfer and functional recovery of injured heart muscle. *J Thorac Cardiovasc Surg* 2004;128(4):571–578.

73. Kofidis T, Lebl DR, Martinez EC, Hoyt G, Tanaka M, Robbins RC. Novel injectable bioartificial tissue facilitates targeted, less invasive, large-scale tissue restoration on the beating heart after myocardial injury. *Circulation* 2005;112(Suppl 9):I173–I177.

74. Himes N, Min JY, Lee R, Brown C, Shea J, Huang X, et al. *In vivo* MRI of embryonic stem cells in a mouse model of myocardial infarction. *Magn Reson Med* 2004;52(5):1214–1219.

75. Cao F, Lin S, Xie X, Ray P, Patel M, Zhang X, et al. *In vivo* visualization of embryonic stem cell survival, proliferation, and migration after cardiac delivery. *Circulation* 2006;113(7):1005–1014.

76. Zhang PL, Izrael M, Ainbinder E, Ben Simchon L, Chebath J, Revel M. Increased myelinating capacity of embryonic stem cell derived oligodendrocyte precursors after treatment by interleukin-6/soluble interleukin-6 receptor fusion protein. *Mol Cell Neurosci* 2006;31(3):387–398.

77. Takagi Y, Takahashi J, Saiki H, Morizane A, Hayashi T, Kishi Y, et al. Dopaminergic neurons generated from monkey embryonic stem cells function in a Parkinson primate model. *J Clin Invest* 2005;115(1):102–109.

78. McDonald JW, Howard MJ. Repairing the damaged spinal cord: a summary of our early success with embryonic stem cell transplantation and remyelination. *Prog Brain Res* 2002;137:299–309.

79. Perez-Bouza A, Glaser T, Brustle O. ES cell-derived glial precursors contribute to remyelination in acutely demyelinated spinal cord lesions. *Brain Pathol* 2005;15(3):208–216.

80. Harper JM, Krishnan C, Darman JS, Deshpande DM, Peck S, Shats I, et al. Axonal growth of embryonic stem cell-derived motoneurons *in vitro* and in motoneuron-injured adult rats. *Proc Natl Acad Sci USA* 2004;101(18):7123–7128.

81. Hara A, Niwa M, Kunisada T, Yoshimura N, Katayama M, Kozawa O, et al. Embryonic stem cells are capable of generating a neuronal network in the adult mouse retina. *Brain Res* 2004;999(2):216–221.

82. Gerrard L, Rodgers L, Cui W. Differentiation of human embryonic stem cells to neural lineages in adherent culture by blocking bone morphogenetic protein signaling. *Stem Cells* 2005;23(9):1234–1241.

83. Itsykson P, Ilouz N, Turetsky T, Goldstein RS, Pera MF, Fishbein I, et al. Derivation of neural precursors from human embryonic stem cells in the presence of noggin. *Mol Cell Neurosci* 2005;30(1):24–36.

84. Perrier AL, Tabar V, Barberi T, Rubio ME, Bruses J, Topf N, et al. Derivation of midbrain dopamine neurons from human embryonic stem cells. *Proc Natl Acad Sci USA* 2004;101(34):12543–12548.

85. Brederlau A, Correia AS, Anisimov SV, Elmi M, Paul G, Roybon L, et al. Transplantation of human embryonic stem cell-derived cells to a rat model of Parkinson's disease: effect of *in vitro* differentiation on graft survival and teratoma formation. *Stem Cells* 2006;24(6):1433–1440.

86. Kawasaki H, Mizuseki K, Nishikawa S, Kaneko S, Kuwana Y, Nakanishi S, et al. Induction of midbrain dopaminergic neurons from ES cells by stromal cell-derived inducing activity. *Neuron* 2000;28(1):31–40.

87. Chiba S, Iwasaki Y, Sekino H, Suzuki N. Transplantation of motoneuron-enriched neural cells derived from mouse embryonic stem cells improves motor function of hemiplegic mice. *Cell Transplant* 2003;12(5):457–468.

88. Roy NS, Cleren C, Singh SK, Yang L, Beal MF, Goldman SA. Functional engraftment of human ES cell-derived dopaminergic neurons enriched by coculture with telomerase-immortalized midbrain astrocytes. *Nat Med* 2006;12(11):1259–1268.

89. Regala C, Duan M, Zou J, Salminen M, Olivius P. Xenografted fetal dorsal root ganglion, embryonic stem cell and adult neural stem cell survival following implantation into the adult vestibulocochlear nerve. *Exp Neurol* 2005;193(2):326–333.

90. Hu Z, Ulfendahl M, Olivius NP. Central migration of neuronal tissue and embryonic stem cells following transplantation along the adult auditory nerve. *Brain Res* 2004;1026(1):68–73.

91. Bonner-Weir S, Weir GC. New sources of pancreatic beta-cells. *Nat Biotechnol* 2005;23(7):857–861.

92. Kume S. Stem-cell-based approaches for regenerative medicine. *Dev Growth Differ* 2005;47(6):393–402.

93. Kuai XL, Cong XQ, Du ZW, Bian YH, Xiao SD. Treatment of surgically induced acute liver failure by transplantation of HNF4-overexpressing embryonic stem cells. *Chin J Dig Dis* 2006;7(2):109–116.

94. Soto-Gutierrez A, Kobayashi N, Rivas-Carrillo JD, Navarro-Alvarez N, Zhao D, Okitsu T, *et al.* Reversal of mouse hepatic failure using an implanted liver-assist device containing ES cell-derived hepatocytes. *Nat Biotechnol* 2006;24(11):1412–1419.

95. Kobayashi N, Ando M, Kosaka Y, Yong C, Okitsu T, Arata T, *et al.* Partial hepatectomy and subsequent radiation facilitates engraftment of mouse embryonic stem cells in the liver. *Transplant Proc* 2004;36(8):2352–2354.

96. Teramoto K, Asahina K, Kumashiro Y, Kakinuma S, Chinzei R, Shimizu-Saito K, *et al.* Hepatocyte differentiation from embryonic stem cells and umbilical cord blood cells. *J Hepatobiliary Pancreat Surg* 2005;12(3):196–202.

97. Yamamoto H, Quinn G, Asari A, Yamanokuchi H, Teratani T, Terada M, *et al.* Differentiation of embryonic stem cells into hepatocytes: biological functions and therapeutic application. *Hepatology* 2003;37(5):983–993.

98. Imamura T, Cui L, Teng R, Johkura K, Okouchi Y, Asanuma K, *et al.* Embryonic stem cell-derived embryoid bodies in three-dimensional culture system form hepatocyte-like cells *in vitro* and *in vivo. Tissue Eng* 2004;10(11–12):1716–1724.

99. Theise ND. Liver stem cells: prospects for treatment of inherited and acquired liver diseases. *Expert Opin Biol Ther* 2003;3(3):403–408.

100. Yamamoto M, Cui L, Johkura K, Asanuma K, Okouchi Y, Ogiwara N, *et al.* Branching ducts similar to mesonephric ducts or ureteric buds in teratomas originating from mouse embryonic stem cells. *Am J Physiol Renal Physiol* 2006;290(1):F52–F60.

101. Ali NN, Edgar AJ, Samadikuchaksaraei A, Timson CM, Romanska HM, Polak JM *et al.* Derivation of type II alveolar epithelial cells from murine embryonic stem cells. *Tissue Eng* 2002;8(4):541–550.

102. Rippon HJ, Ali NN, Polak JM, Bishop AE. Initial observations on the effect of medium composition on the differentiation of murine embryonic stem cells to alveolar type II cells. *Cloning Stem Cells* 2004;6(2):49–56.

103. Samadikuchaksaraei A, Bishop AE. Derivation and characterization of alveolar epithelial cells from murine embryonic stem cells *in vitro. Methods Mol Biol* 2006;330:233–248.

104. Van Vranken BE, Romanska HM, Polak JM, Rippon HJ, Shannon JM, Bishop AE. Coculture of embryonic stem cells with pulmonary mesenchyme: a microenvironment that promotes differentiation of pulmonary epithelium. *Tissue Eng* 2005;11(7–8): 1177–1187.

105. Coraux C, Nawrocki-Raby B, Hinnrasky J, Kileztky C, Gaillard D, Dani C, *et al.* Embryonic stem cells generate airway epithelial tissue. *Am J Respir Cell Mol Biol* 2005;32(2):87–92.

106. Yamashita JK. Differentiation and diversification of vascular cells from embryonic stem cells. *Int J Hematol* 2004;80(1):1–6.

107. Yurugi-Kobayashi T, Itoh H, Yamashita J, Yamahara K, Hirai H, Kobayashi T, *et al.* Effective contribution of transplanted vascular progenitor cells derived from embryonic stem cells to adult neovascularization in proper differentiation stage. *Blood* 2003;101(7):2675–2678.

108. Sone M, Itoh H, Yamashita J, Yurugi-Kobayashi T, Suzuki Y, Kondo Y, *et al.* Different differentiation kinetics of vascular progenitor cells in primate and mouse embryonic stem cells. *Circulation* 2003;107(16):2085–2088.

109. Liu B, Hou CM, Wu Y, Zhang SX, Mao N. [The investigation of hematopoietic capacity of HPP-CFC derived from murine embryonic stem cells *in vitro* and *in vivo*]. *Sheng Wu Gong Cheng Xue Bao* 2003;19(3):312–316.

110. Hurlbut WB. Altered nuclear transfer as a morally acceptable means for the procurement of human embryonic stem cells. *Perspect Biol Med* 2005;48(2):211–228.

111. Meissner A, Jaenisch R. Generation of nuclear transfer-derived pluripotent ES cells from cloned Cdx2-deficient blastocysts. *Nature* 2006;439(7073):212–215.

112. Chung Y, Klimanskaya I, Becker S, Marh J, Lu SJ, Johnson J, *et al.* Embryonic and extraembryonic stem cell lines derived from single mouse blastomeres. *Nature* 2006;439(7073):216–219.

113. Klimanskaya I, Chung Y, Becker S, Lu SJ, Lanza R. Human embryonic stem cell lines derived from single blastomeres. *Nature* 2006;444(7118):481–485.

114. Melton DA, Daley GQ, Jennings CG. Altered nuclear transfer in stem-cell research — a flawed proposal. *N Engl J Med* 2004;351(27):2791–2792.

115. Solter D. Politically correct human embryonic stem cells? *N Engl J Med* 2005; 353(22):2321–2323.

116. Alberio R, Campbell KH, Johnson AD. Reprogramming somatic cells into stem cells. *Reproduction* 2006;132(5):709–720.

117. Kofidis T, deBruin JL, Tanaka M, Zwierzchoniewska M, Weissman I, Fedoseyeva E, *et al.* They are not stealthy in the heart: embryonic stem cells trigger cell infiltration, humoral and T-lymphocyte-based host immune response. *Eur J Cardiothorac Surg* 2005;28(3):461–466.

118. Swijnenburg RJ, Tanaka M, Vogel H, Baker J, Kofidis T, Gunawan F, *et al.* Embryonic stem cell immunogenicity increases upon differentiation after transplantation into ischemic myocardium. *Circulation* 2005;112(Suppl 9):166–172.

119. Lawrenz B, Schiller H, Willbold E, Ruediger M, Muhs A, Esser S. Highly sensitive biosafety model for stem-cell-derived grafts. *Cytotherapy* 2004;6(3):212–222.

120. Teramoto K, Hara Y, Kumashiro Y, Chinzei R, Tanaka Y, Shimizu-Saito K, *et al.* Teratoma formation and hepatocyte differentiation in mouse liver transplanted with mouse embryonic stem cell-derived embryoid bodies. *Transplant Proc* 2005;37(1):285–286.

121. Li M, Pevny L, Lovell-Badge R, Smith A. Generation of purified neural precursors from embryonic stem cells by lineage selection. *Curr Biol* 1998;8(17):971–974.

122. Wernig M, Tucker KL, Gornik V, Schneiders A, Buschwald R, Wiestler OD, *et al.* Tau EGFP embryonic stem cells: an efficient tool for neuronal lineage selection and transplantation. *J Neurosci Res* 2002;69(6):918–924.

123. Soria B, Roche E, Berna G, Leon-Quinto T, Reig JA, Martin F. Insulin-secreting cells derived from embryonic stem cells normalize glycemia in streptozotocin-induced diabetic mice. *Diabetes* 2000;49(2):157–162.

124. Gimond C, Marchetti S, Pages G. Differentiation of mouse embryonic stem cells into endothelial cells: genetic selection and potential use *in vivo*. *Methods Mol Biol* 2006;330:303–329.

125. Schuldiner M, Itskovitz-Eldor J, Benvenisty N. Selective ablation of human embryonic stem cells expressing a "suicide" gene. *Stem Cells* 2003;21(3):257–265.

126. Bradley JA, Bolton EM, Pedersen RA. Stem cell medicine encounters the immune system. *Nat Rev Immunol* 2002;2(11):859–871.

127. Cooper JC, Fernandez N, Joly E, Dealtry GB. Regulation of major histocompatibility complex and TAP gene products in preimplantation mouse stage embryos. *Am J Reprod Immunol* 1998;40(3):165–171.

128. Draper JS, Pigott C, Thomson JA, Andrews PW. Surface antigens of human embryonic stem cells: changes upon differentiation in culture. *J Anat* 2002;200(Pt 3):249–258.

129. Drukker M, Katz G, Urbach A, Schuldiner M, Markel G, Itskovitz-Eldor J, *et al.* Characterization of the expression of MHC proteins in human embryonic stem cells. *Proc Natl Acad Sci USA* 2002;99(15):9864–9869.

130. Li L, Baroja ML, Majumdar A, Chadwick K, Rouleau A, Gallacher L, *et al.* Human embryonic stem cells possess immune-privileged properties. *Stem Cells* 2004;22(4):448–456.

131. Abdullah Z, Saric T, Kashkar H, Baschuk N, Yazdanpanah B, Fleischmann BK, *et al.* Serpin-6 expression protects embryonic stem cells from lysis by antigen-specific cytotoxic T lymphocytes. *J Immunol* 2007;178(6):3390–3399.

132. Tian L, Catt JW, O'Neill C, King NJ. Expression of immunoglobulin superfamily cell adhesion molecules on murine embryonic stem cells. *Biol Reprod* 1997;57(3): 561–568.

133. Drukker M, Katchman H, Katz G, Even-Tov Friedman S, Shezen E, Hornstein E, *et al.* Human embryonic stem cells and their differentiated derivatives are less susceptible to immune rejection than adult cells. *Stem Cells* 2006;24(2):221–229.

134. Bonde S, Zavazava N. Immunogenicity and engraftment of mouse embryonic stem cells in allogeneic recipients. *Stem Cells* 2006;24(10):2192–2201.

135. Grinnemo KH, Kumagai-Braesch M, Mansson-Broberg A, Skottman H, Hao X, Siddiqui A, *et al.* Human embryonic stem cells are immunogenic in allogeneic and xenogeneic settings. *Reprod Biomed Online* 2006;13(5):712–724.

136. Fandrich F, Lin X, Chai GX, Schulze M, Ganten D, Bader M, *et al.* Preimplantation-stage stem cells induce long-term allogeneic graft acceptance without supplementary host conditioning. *Nat Med* 2002;8(2):171–178.

137. Fabricius D, Bonde S, Zavazava N. Induction of stable mixed chimerism by embryonic stem cells requires functional Fas/FasL engagement. *Transplantation* 2005;79(9):1040–1044.

138. Dekel B, Burakova T, Arditti FD, Reich-Zeliger S, Milstein O, Aviel-Ronen S, *et al.* Human and porcine early kidney precursors as a new source for transplantation. *Nat Med* 2003;9(1):53–60.

139. Grusby MJ, Auchincloss H, Jr., Lee R, Johnson RS, Spencer JP, Zijlstra M, *et al.* Mice lacking major histocompatibility complex class I and class II molecules. *Proc Natl Acad Sci USA* 1993;90(9):3913–3917.

140. Munsie MJ, Michalska AE, O'Brien CM, Trounson AO, Pera MF, Mountford PS. Isolation of pluripotent embryonic stem cells from reprogrammed adult mouse somatic cell nuclei. *Curr Biol* 2000;10(16):989–992.

4   Umbilical Cord Blood Cells for Cardiac Repair

Elad Maor, Arnon Nagler and Jonathan Leor

Introduction

In 1989, human umbilical cord blood (UCB) cells were transplanted for the first time in a child with Fanconi Anemia.[1] Since then thousands of cases of cord blood transplantations have taken place worldwide for a variety of diseases.[2,3] The availability of human UCB units has improved since the first public cord blood bank was created in New York in 1991. Currently, there are between 175,000 and 200,000 UCB units in storage in public cord blood banks worldwide, in addition to those in the private sector. UCB and blood from the mother are typed for HLA antigens and blood group and screened for infectious diseases, while information is collected about the medical history of the mother and the family.[4,5]

With the progress in the field of stem cell research and cell-based therapies, as well as widespread clinical experience in bone marrow transplantation, new applications for human UCB-derived cells in the field of regenerative medicine are being studied. This chapter will focus on the potential of UCB cell population in the field of cardiac repair.

Myocardial Infarction and Heart Failure

The heart is unable to regenerate significantly after ischemic injury. The resultant irreversible loss of cardiomyocytes leads to the formation of fibrous scar tissue, thereby creating a major clinical problem. Beyond the contraction and fibrosis of the myocardial scar tissue, the progressive remodeling process of the non-ischemic myocardium further reduces cardiac function in the weeks and months following myocardial infarction. This progressive process involves molecular, cellular and physiological responses that can lead to life-threatening arrhythmias, sudden cardiac death, cardiac aneurysms, and heart failure.[6] Despite the rapid restoration of blood flow, post-infarction heart failure remains a major clinical problem. Current therapeutic options for heart failure patients are limited and often cardiac transplantation remains the only available treatment for dealing with end stage heart failure.

Myocardial Regeneration and Cell-Based Therapy

Myocardial regeneration aims to replace the scar tissue of the infarcted myocardium with viable new tissue, consisting of functional cardiomyocytes, blood vessels, and connective tissue. There are different approaches to regenerate the infarcted myocardium, and the use of stem cell-based therapy is one of the most exciting and most studied approaches.

In recent years, there has been substantial progress in the field of cell-based therapy, and today it is one of the leading fields in myocardial regeneration. The cell-based approach for myocardial regeneration originally focused on the repopulation of the cardiac scar with new transplanted healthy cells. It has been shown in animal models that stem cell-based therapy can induce angiogenesis, partially regenerate the infarcted myocardium and improve cardiac function after myocardial infarction.[7,8]

Preliminary results in humans have demonstrated the feasibility and safety of adult bone marrow mononuclear cell transplantation through intracoronary injection to the infarcted heart.[9–13] However, the results of clinical trials are inconsistent and conflicting.[9–14] Furthermore, these trials enrolled low-risk patients (younger age and higher ejection fraction than average MI patient).

Problems with the Available Cell Sources

Clinical trials to date have focused mainly on adult progenitor cell therapy for cardiac repair. Table 1 shows the main cell sources investigated for cardiac repair. The three main categories for stem cell-based therapy are embryonic, adult tissues or blood, and cord blood.

Table 1. Other sources for stem/progenitor cells.

Source	Advantage	Disadvantage
Embryonic stem cells	• Multipotent • Expandable	• Teratogenic • Allogeneic • Ethical concerns • No clinical experience
Adult stem cells*	• Autologous • Expandable • Clinical experience in humans	• Rare • Limited regenerative capacity • Affected by age and co-morbidites • Invasive collection • Time between collection and implantation

*Suggested adult stem cells include: bone marrow-derived cells, endothelial progenitor cells, skeletal myoblasts, and cardiac stem cells.

The main problems regarding embryonic stem cell therapy are ethical research restrictions, the allogeneic nature of the cells and their potential teratogenicity.[15,16] Adult stem cells are autologous and have already been evaluated in clinical trials. These cells demonstrate numerous limitations, such as reduced potential with age, obesity and other co-morbidities.[17–21] In a typical target population of elderly patients with ischemic heart disease, the regenerative capacity of autologous adult stem cells is probably reduced.

Heeschen *et al.*[22] demonstrated that bone marrow mononuclear cells from patients with chronic ischemic cardiomyopathy have a lower colony forming capacity and a lower migratory response to stromal cell-derived factor-1 and vascular endothelial growth factor (VEGF). Moreover, they showed that the angiogenic effect of these cells is reduced *in vivo*, in a hindlimb ischemia model.

Choi *et al.*[23] demonstrated that endothelial progenitor-cell colony forming capacity and migratory response were reduced in patients with chronic renal failure. Their group showed that the number of circulating endothelial progenitor cells was reduced in patients with chronic renal failure, compared with normal patients, even among those with the same burden of Framingham's risk factors.

A further limitation of adult stem cell-based therapy is the limited number of cells that can be collected. Expanding and preparing these cells for transplantation is a process requiring a lengthy time period. An additional limitation is the collection procedure. Although adult stem cells can be collected from peripheral blood in a non-invasive approach, numerous clinical trials have used bone marrow aspiration for collection of adult stem cells from the bone marrow.[12,13] Bone marrow mononuclear cells collection is an invasive procedure and can be problematic in elderly patients with chronic ischemic cardiomyopathy or patients treated with anti-coagulant medications.

The Rationale for Using UCB Progenitor Cells for Myocardial Repair

As discussed earlier, both embryonic and adult stem cells suffer from specific limitations. Human UCB progenitor cells are a source for cell-based therapy that can overcome many of these limitations. Human UCB mononuclear cells are rich in both stem and progenitor cells and have improved proliferative characteristics.[24–28] These cells have the potential to transdifferentiate to endothelial cells, they can induce angiogenesis and help in the regeneration of the infarcted myocardium.[8] It appears that there is a reduced risk of rejection by the host's immune response with human UBC cells compared to adult HLA mismatched cells.[29,30] In addition, significant research has provided the clinical experience required to facilitate the collection, expansion, storage, and transplantation of human UCB.[5] There are many public and private UCB banks worldwide with cells readily available for transplantation at the most optimal time. Furthermore, there are no ethical limitations

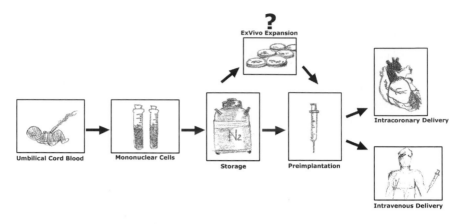

Fig. 1. The clinical approach for UCB cell therapy.

regarding the use of these cells, and no risks to the mother or to the newborn in collecting the cells after labor.

UCB cell therapy will take advantage of the existing clinical approach for UCB collection and storage. The desired cell population (CD34, CD133, mesenchymal stem cells) will be separated and expanded *ex vivo*. The cells will be transplanted directly to the myocardium through the coronary artery in the coronary catheterization lab, or indirectly by an intravenous injection (Fig. 1).

Data from Animal Models (Table 2)

The UCB cells contain a relatively high concentration (0.1–1.1 CD34$^+$ cells per 100 nucleated cells) of progenitor cells (CD34$^+$38$^-$, CD34$^+$DR$^-$, CD34$^+$Thy$^+$).[31–33] The cord blood progenitor cell population possesses homing, differentiation and angiogenic potential relevant for myocardial repair.[34,35] In recent years, several experiments demonstrated the angiogenic and regenerative potential of human UCB cells in the ischemic hindlimb model, as well as in the myocardial infarction model (Table 2).

Murohara *et al.*[28] demonstrated that transplanted human UCB endothelial precursor cells augment postnatal neovascularization in an ischemic hindlimb model. The experiment was performed on immunodeficient nude rats, and 14 days after cell transplantation neovascularization was evident both by laser Doppler and histology.

Yang *et al.*[36] demonstrated similar results on a similar model. They expanded human CD133$^+$ UCB cells *ex vivo*, and injected 5 × 10^5 to the tail vein of nude mice one day after femoral artery ligation. The UCB cells successfully incorporated into the nude mice capillary network, augmented neovascularization and improved ischemic hindlimb salvage.

Table 2. UCB cells to repair the ischemic tssues: data from different animal models.

Ref.	Cell type	Model	*Ex vivo* expansion	Follow-up	Results	Immunosuppression
43	1×10^8 unrestrictive somatic stem cells (Kourion Therapeutics)	Pigs, 1 week post-MI, intracoronary injection	Yes	4 weeks	No difference in global or regional LV function. Larger infarct size in the cell therapy group. Intracoronary injection of UCB cells caused micro-infarctions. Cells survived in the border zone but did express cardiomyocytes or endothelial markers	Prednisolone and cyclosporine-A
8	$1.2-2 \times 10^6$ CD133$^+$ human UCB	Nude rats, intravenous injection to the femoral vein on day 7	No	1 month	Improved LV function and scar thickness, a few cells survived and incorporated into vessel walls	Nude rat
41	5×10^5, CD133 UCB vs. bone marrow	SCID mice, cryoinjury model, intramyocardial injection on day 0	No	4 weeks	Improved angiogenesis, reduced apoptosis and reduced post-MI mortality	SCID mice
37	1×10^7 human mononuclear	Rats, day 0, directly to the scar	No	4 weeks	Improved angiogenesis, reduced collagen content, improved LV function	Non (survival not checked)
42	1×10^8 unrestrictive somatic stem cells (Kourion Therapeutics)	Pigs, 4 weeks post-MI, directly to the scar	Yes	4 weeks	Improved regional perfusion, improved LV function, and scar thickness	Cyclosporine-A (cells survived 4 weeks)
44	CD133$^+$ UCB [99m] Tc-labeled cells	A single pig infarcted myocardium	No	4 weeks	Colonization of cells at pig's infarcted myocardium	None

(Continued)

Table 2. *(Continued)*

Ref.	Cell type	Model	*Ex vivo* expansion	Follow-up	**Results**	Immunosuppression
32	6×10^6, mononuclear	SCID, intravenous injection to the tail	No	24 hours, 1 and 3 weeks	UCB cells survived after 3 weeks, reduced infarct size	SCID mice
40	1×10^6, cd34+	SCID, day 0 directly to the scar	Yes, with transfection	4 weeks	Improved angiogenesis, reduced scar size, improved LV function	SCID mice
60	10^6 human mononuclear	Rats, 1 hour after MI	No	4 months	Improved LV function, reduced scar size, improved scar thickness, improved dp/dt	None
36	5×10^5, CD133+	Hindlimb ischemia in nude mice. Injection to the tail vein on day 1	Yes	Up to 3 weeks	Cells incorporated into capillary networks, improved angiogenesis, improved ischemic limb salvage	Nude mice
28	3×10^5, CD34+	Hindlimb ischemia in nude rats. Muscular injection on day 0	Yes	2 weeks	Cells survived and participated in capillary networks. Enhanced angiogenesis and blood flow	Nude rat

LV — Left ventricle; MI — myocardial infarction; UCB — umbilical cord blood; SCID — severe combined immunodeficiency.

In recent years, many groups have focused on the potential of UCB cells for cardiac repair, and most of them successfully demonstrated that UCB cell transplantation to the infarcted myocardium reduces scar size, enhances angiogenesis, and improves left ventricular function. Hu *et al.*[37] demonstrated that the injection of UCB mononuclear cells without *in vitro* expansion and without immunosuppression, reduced end-diastolic left ventricular, increased $+dp/dt_{max}$, induced angiogenesis and reduced scar size and thickness. In their experiment, 1×10^7 human UCB mononuclear cells were injected to a rat heart immediately after inducing myocardial infarction. Although cell survival was not determined in this experiment, functional improvement was evident, without the use of immunosuppression in an immunocompetent animal.

Henning *et al.*[38] injected 1×10^6 human UCB mononuclear cells immediately after myocardial infarction in a rat model. Immunosuppression was not given, but compared with controls, UCB cells reduced infarct size and left ventricular dilatation, and improved ejection fraction.

Hirata *et al.*[39] injected 2×10^5 human UCB CD34$^+$ cells immediately after ligation of the coronary artery in a rat model. Rats were treated with immunosuppression (FK506). Four weeks after myocardial infarction, myocardial contractility had improved in the UCB-treated rats. Injected cells survived in the scar tissue, improved vessel density and about 1% incorporated into the vessel walls.

Two groups demonstrated improved cardiac function and scar size, and angiogenesis was enhanced in severe combined immunodeficient (NOD/SCID) mice. Ma *et al.*[32] demonstrated that injection of 6×10^6 human UCB mononuclear cells to the mouse tail vein was effective, and that the UCB cells reached the infarcted myocardium through the circulation. Chen *et al.*[40] demonstrated that forced expression of angiogenic genes — angiopoietin-1 (Ang1) cDNA and VEGF165 cDNA, in combination or alone — in UCB cells prior to injection, reduces infarct size, attenuates progression of cardiac dysfunction and increases capillary density in acute ischemic myocardium.

In another study, Ma *et al.*[41] showed that human CD133$^+$ UCB cells and human CD133$^+$ adult bone marrow cells have a similar effect on post-myocardial infarction mortality in a SCID mice cryoinjury model. In the experiment, 5×10^5 CD133$^+$ cells were injected intramyocardially immediately following the cryoinjury. Cells could not be detected on histology slides 48 hours after injection, although human DNA PCR was positive after four weeks. There was no effect on scar size, but left ventricular contractility was superior in the bone marrow group compared with the UCB group. One possible explanation for the superiority of the bone marrow cells over the UCB cells, was a higher percentage of CD117$^+$ (c-Kit) cells in the bone marrow cells used.

Recently, our group has demonstrated the positive effect of intravenous injection of CD133$^+$ human UCB cells.[8] In our experiment, $1.2–2 \times 10^6$ cells were injected to the femoral vein of nude rats, seven days after a myocardial infarction. One month later, left ventricular fractional shortening improved in the treated animal group — compared with control. Moreover, anterior wall thinning was decreased in the treated animal group. Human cells were identified near blood vessels and near the left ventricular cavity, and were occasionally incorporated into vessel walls. However, there was no protective effect on LV dilatation.

There are only two studies in large animals that have addressed the issue of UCB cell therapy for myocardial infarction. Kim et al.[42] injected 1×10^8 specific UCB sub-population that was CD45 negative and was expanded *ex vivo* at Kourion Therapeutics AG Laboratories before transplantation. The cells were injected directly into the scar of pigs four weeks post-myocardial infarction. The pigs in this experiment received immunosuppression and cell survival and incorporation was determined four weeks after cell transplantation by immunohistochemistry and by *in situ* hybridization. Animals treated with UCB cell therapy in this study showed improved regional perfusion and wall motion of the infarcted region, compared with the control group. Scar thickness of the transplant group was higher than the control. The grafted cells were detected four weeks after transplantation by both immunohistochemistry and *in situ* hybridization.

Moelker et al.[43] had disappointing results with the same cell line of Kourion Therapeutics AG. In their study, the same type and amount of cells were injected intracoronary one week post-MI to pigs. Four weeks after injection, cells survived at the border zone of the infarction, but did not express any cardiomyocyte or endothelial markers. Magnetic resonance imaging failed to show any improvement in global or regional left ventricular function. Moreover, infarct size was larger in the cell therapy group compared with the control group. One explanation for the larger infarct size in the cell therapy group was micro-infarctions caused by intracoronary injections of the UCB cells. Histologic analysis of normal porcine hearts after intracoronary injection of the same cells confirmed the presence of these micro-infarctions.

In a different experiment, we tested the feasibility and safety of intracoronary injection of CD133[+] UCB cells to an infarcted myocardium in a pig model.[44] Cells were labeled immediately prior to infusion with [99m]Tc Exametazime. Whole body distribution two hours after the infusion clearly showed the ^{99m}Tc–labeled cells in the infarcted myocardium. Histology and immunohistochemistry four weeks after the infusion showed clusters of human cell survival in pig's myocardium. In this experiment, we successfully showed the feasibility and safety of intracoronary injection of UCB cells, and the ability of the cells to survive for at least one month without immunosuppression.

How Do Cord Blood Cells Repair Ischemic Tissue?

To date, our understanding of the exact mechanism by which UCB cell-based therapy helps cardiac regeneration and repair is limited. In many studies the mechanism by which cell transplantation improves cardiac function remains speculative.

First, UCB cells can transdifferentiate in the myocardium and help regenerate both myocardium and blood vessels.[45] This mechanism may be relevant to the UCB cell population, which has a high concentration of progenitor cells.[25] However, this notion of stem cell transdifferentiation is controversial. Second, stem cells are able to secrete cytokines and have a paracrine effect on the remaining viable tissue. Pomyje *et al.* demonstrated that the expression of angiopoietin-1, angiopoietin-2 and vascular growth factors, as well as their mRNA receptors in human CD34$^+$/CD133$^+$ human UCB cells *in vitro*, supporting a paracrine angiogenic effect of UCB cells.[46] This mechanism is supported by the fact that most of the cells injected to the infarcted myocardium disappear in the first days after transplant, but their positive effect lasts for months.[7,41,47−50] Third, it has been postulated that apoptosis of transplanted cells modulates the local immune response in the infarcted myocardium and enhance healing and repair.[51]

Ex vivo Expansion

The number of human UCB cells is limited by the low-cell dose available in each UCB unit. Although the number of cells for cardiac cell-based therapy has not yet been established, the small number of cells in each UCB unit will necessitate the use of multiple doses from different UCB units, or *in vitro* expansion. One concern is that the *in vitro* expansion process produces "lower" quality progenitor cells, at the expense of "higher" quality progenitor cells.[52] Although discussed and designed specifically with respect to UCB hematopoeitic stem cells (HSC), the *ex vivo* expansion techniques are relevant to other cell populations. The data regarding the functional quality of *ex vivo* expanded HSC are contradictory. Some groups have demonstrated that cell quality is preserved,[53,54] while others have shown a reduced quality of *ex vivo* expanded HSC.[55−57]

In a preliminary study, we demonstrated that UCB CD34$^+$ cells can be successfully expanded *ex vivo* and improve cardiac function following myocardial infarction. *Ex vivo* expansion in the presence of a copper chelator and four early hematopoietic acting cytokines (SCF, TPO, FLt−3 and IL6) increased the number of CD34$^+$ cells by 31 ± 6 fold and the early progenitor (CD34$^+$CD38$^-$) cells by 409.2 ± 148.9 fold. These expanded cells engrafted and colonized the infarcted myocardium in a rat model of myocardial infarction. Moreover, cell therapy with these *ex vivo* expanded cord blood-derived cells resulted in new vessel formation and prevention of LV dilatation and myocardial dysfunction.[58]

Kim *et al.*[42] demonstrated that *ex vivo* expanded human UCB cells improve cardiac function in pigs four weeks following myocardial infarction. They successfully transferred *ex vivo* expanded UCB cells from laboratories in Germany to the experimental site in Toronto on the day of the experiment. Cells were detected four weeks after transplantation using immunohistochemistry and *in situ* hybridization and improved both reginal and ejection fractions. They did not compare their results with a non-expanded human UCB cell population.

Immunosuppression and Graft versus Host Disease

UCB cells are considered less immunogenic than adult cells,[29,30] but their allogeneic nature raises the question of immunosuppressive drugs during transplantation. Immunosuppressive drugs have a high-toxicity profile, and their use might hamper the prospect of clinical use of UCB for cardiac repair. The subject of immunosuppression and UCB cells transplantation was discussed thoroughly in a recent review by Riordan et al.,[31] although the issues of cardiac regeneration and repair were not addressed.

Although most animal studies have been conducted on immunodeficient animals, two groups did prove that transplantation of human UCB cells to immunocompetent animals improves cardiac function after myocardial infarction without the need for immunosupression.[37,38] These two groups, however, did not check cell survival in the infarcted myocardium after transplant, although we do know from previous reports that immunosuppression improves progenitor cell engraftment and survival.[59]

The success of transplantation without immunosuppression can be partially explained by the fact that even without long-term survival, transplanted cells can have a positive effect on cardiac remodeling.[7,47,51] This positive effect could be related to a paracrine effect, to an influence on resident cardiac stem cells, or to an immunomodulation of the local inflammatory response. Another explanation for the success of UCB transplantation without immunosuppression could be the fact that UCB cells are partially immune-privileged, and less immunogenic compared with adult cells.[29,30] It is possible that UCB cells survive in the infarcted myocardium and transdifferentiate to muscle and blood vessels.

A better understanding of the mechanism by which UCB cells effect cardiac regeneration will help to solve the dilemma of the need for immunosuppression in UCB cell transplantation for cardiac repair.

Summary

Umbilical cord blood cells are emerging as a potential option for cardiac repair. Several groups in recent years have contributed towards our understanding of the therapeutic options for UCB cells, and the data collected to date have been encouraging. However, there are still a few unanswered questions that need to be resolved. The first is the mechanism by which UCB cells repair the infarcted heart. The second is how the allogenic nature of UCB cells will affect the safety and therapeutic potential of this approach. Do we need immunosuppression? If the answer is – yes – for how long and in which protocol? Finally, what effect will cell expansion have on regeneration and homing capacity?

Advances in cell expansion technology and immunology, as well as additional clinical trials focusing on cell-based therapy, will increase the prospects of utilizing UCB cells to treat myocardial infarction patients, particularly elderly patients with a decreased number and function of autologous progenitor cells.

References

1. Gluckman E, Broxmeyer HA, Auerbach AD. Hematopoietic reconstitution in a patient with Fanconi's anemia by means of umbilical cord blood from an HLA-identical sibling. *N Engl J Med* 1989;321:1174–1178.
2. Cohen Y, Nagler A. Umbilical cord blood transplantation — how, when and for whom? *Blood Rev* 2004;18(3):167–179.
3. Goldstein G, Toren A, Nagler A. Human umbilical cord blood biology, transplantation and plasticity. *Curr Med Chem* 2006;13:1249–1259.
4. Kurtzberg J, Lyerly AD, Sugarman J. Untying the Gordian knot: policies, practices, and ethical issues related to banking of umbilical cord blood. *J Clin Invest* 2005;115(10):2592–2597.
5. Steinbrook R. The cord-blood bank controversies. *N Engl J Med* 2004;22:2255–2257.
6. Mann DL. Mechanisms and models in heart failure: a combinatorial approach. *Circulation* 1999;100(9):999–1008.
7. Reinecke H, Zhang M, Bartosek T, Murry CE. Survival, integration, and differentiation of cardiomyocyte grafts: a study in normal and injured rat hearts. *Circulation* 1999;100(2):193–202.
8. Leor J, Guetta E, Feinberg MS, Galski H, Bar I, Holbova R, *et al.* Human umbilical cord blood-derived CD133$^+$ cells enhance function and repair of the infarcted myocardium. *Stem Cells* 2006;24(3):772–780.
9. Assmus B, Honold J, Schachinger V, Britten MB, Fischer-Rasokat U, Lehmann R, *et al.* Transcoronary transplantation of progenitor cells after myocardial infarction. *N Engl J Med* 2006;355(12):1222–1232.
10. Meyer GP, Wollert KC, Lotz J, Steffens J, Lippolt P, Fichtner S, *et al.* Intracoronary bone marrow cell transfer after myocardial infarction: eighteen months' follow-up data from the randomized, controlled boost (Bone marrow transfer to enhance ST-elevation infarct regeneration) trial. *Circulation* 2006;113(10):1287–1294.
11. Schaefer A, Meyer GP, Fuchs M, Klein G, Kaplan M, Wollert KC, *et al.* Impact of intracoronary bone marrow cell transfer on diastolic function in patients after acute myocardial infarction: results from the BOOST trial. *Eur Heart J* 2006;27(8):929–935.
12. Schachinger V, Erbs S, Elsasser A, Haberbosch W, Hambrecht R, Holschermann H, *et al.* Intracoronary bone marrow-derived progenitor cells in acute myocardial infarction. *N Engl J Med* 2006;355(12):1210–1221.
13. Lunde K, Solheim S, Aakhus S, Arnesen H, Abdelnoor M, Egeland T, *et al.* Intracoronary injection of mononuclear bone marrow cells in acute myocardial infarction. *N Engl J Med* 2006;355(12):1199–1209.
14. Janssens S, Dubois C, Bogaert J, Theunissen K, Deroose C, Desmet W, *et al.* Autologous bone marrow-derived stem-cell transfer in patients with ST-segment elevation myocardial infarction: double-blind, randomised controlled trial. *Lancet* 2006;367(9505):113–121.
15. Thomson JA, Itskovitz-Eldor J, Shapiro SS, Waknitz MA, Swiergiel JJ, Marshall VS, *et al.* Embryonic stem cell lines derived from human blastocysts. *Science* 1998;282(5391):1145–1147.
16. Thomson JA, Kalishman J, Golos TG, Durning M, Harris CP, Becker RA, *et al.* Isolation of a primate embryonic stem cell line. *Proc. Natl Acad Sci USA* 1995;92(17): 7844–7848.
17. Scheubel RJ, Zorn H, Silber R-E, Kuss O, Morawietz H, Holtz J, *et al.* Age-dependent depression in circulating endothelial progenitor cells inpatients undergoing coronary artery bypass grafting. *J Am Coll Cardiol* 2003;42(12):2073–2080.

18. Vasa M, Fichtlscherer S, Aicher A, Adler K, Urbich C, Martin H, et al. Number and migratory activity of circulating endothelial progenitor cells inversely correlate with risk factors for coronary artery disease. Circ Res 2001;89:E1–E7.

19. Eizawa T, Ikeda U, Murakami Y, Matsui K, Yoshioka T, Takahashi M, et al. Decrease in circulating endothelial progenitor cells in patients with stable coronary artery disease. Heart 2004;90(6):685–686.

20. Hao Zhang SF, Tian H, Mickle DAG, Weisel RD, Fujii T, Li R-K. Increasing donor age adversely impacts beneficial effects of bone marrow but not smooth muscle myocardial cell therapy. Am J Physiol Heart Circ Physiol 2005;289:H2089–H2096.

21. Li T-S, Furutani A, Takahashi M, Ohshima M, Qin S-L, Kobayashi T, et al. Impaired potency of bone marrow mononuclear cells for inducing therapeutic angiogenesis in obese diabetic rats. Am J Physiol Heart Circ Physiol 2006;290(4):H1362–1369.

22. Heeschen C, Lehmann R, Honold J, Assmus B, Aicher A, Walter DH, et al. Profoundly reduced neovascularization capacity of bone marrow mononuclear cells derived from patients with chronic ischemic heart disease. Circulation 2004;109(13):1615–1622.

23. Choi J-H, Kim KL, Huh W, Kim B, Byun J, Suh W, et al. Decreased number and impaired angiogenic function of endothelial progenitor cells in patients with chronic renal failure. Arterioscler Thromb Vasc Biol 2004;24(7):1246–1252.

24. Lewis ID, Verfaillie CM. Multi-lineage expansion potential of primitive hematopoietic progenitors: superiority of umbilical cord blood compared to mobilized peripheral blood. Exp Hematol 2000;28(9):1087–1095.

25. Broxmeyer HE. Biology of cord blood cells and future prospects for enhanced clinical benefit. Cytotherapy 2005;7(3):209–218.

26. Nieda M, Nicol A, Denning-Kendall P, Sweetenham J, Bradley B, Hows J. Endothelial cell precursors are normal components of human umbilical cord blood. Br J Haematol 1997;98(3):775–777.

27. Hristov M, Erl W, Weber PC. Endothelial progenitor cells: isolation and characterization. Trends Cardiovasc Med 2003;13(5):201–206.

28. Murohara T, Ikeda H, Duan J, Shintani S, Sasaki K, Eguchi H, et al. Transplanted cord blood-derived endothelial precursor cells augment postnatal neovascularization. J Clin Invest 2000;105(11):1527–1536.

29. Bracho F, van de Ven C, Areman E, Hughes RM, Davenport V, Bradley MB, et al. A comparison of ex vivo expanded DCs derived from cord blood and mobilized adult peripheral blood plastic-adherent mononuclear cells: decreased alloreactivity of cord blood DCs. Cytotherapy 2003;5(5):349–361.

30. Tsafrir A, Brautbar C, Nagler A, Elchalal U, Miller K, Bishara A. Alloreactivity of umbilical cord blood mononuclear cells: specific hyporesponse to noninherited maternal antigens. Hum Immunol 2000;61(6):548–554.

31. Riordan N, Chan K, Marleau A, Ichim T. Cord blood in regenerative medicine: do we need immune suppression? J Trans Med 2007;5(1):8.

32. Ma N, Stamm C, Kaminski A, Li W, Kleine H-D, Muller-Hilke B, et al. Human cord blood cells induce angiogenesis following myocardial infarction in NOD/scid-mice. Cardiovasc Res 2005;66(1):45–54.

33. Nagler A, Peacock M, Tantoco M, Lamons D, Okarma TB, Okrongly DA. Red blood cell depletion and enrichment of CD34+ hematopoietic progenitor cells from human umbilical cord blood using soybean agglutinin and CD34 immunoselection. Exp Hematol 1994;22(12):1134–1140.

34. Hristov M, Erl W, Weber PC. Endothelial progenitor cells: mobilization, differentiation, and homing. Arterioscler Thromb Vasc Biol 2003;23(7):1185–1189.

35. Kocher AA, Schuster MD, Szabolcs MJ, Takuma S, Burkhoff D, Wang J, *et al.* Neovascularization of ischemic myocardium by human bone-marrow-derived angioblasts prevents cardiomyocyte apoptosis, reduces remodeling and improves cardiac function. *Nat Med* 2001;7(4):430–436.

36. Yang C, Zhi HZ, Zong JL, Ren CY, Guan QQ, Zhong CH. Enhancement of neovascularization with cord blood CD133+ cell-derived endothelial progenitor cell transplantation. *Thromb Haemost* 2004;91(6):1202–1212.

37. Hu CH, Wu GF, Wang XQ, Yang YH, Du ZM, He XH, *et al.* Transplanted human umbilical cord blood mononuclear cells improve left ventricular function through angiogenesis in myocardial infarction. *Chin Med J (Engl)* 2006;119(18):1499–1506.

38. Henning RJ, Abu-Ali H, Balis JU, Morgan MB, Willing AE, Sanberg PR. Human umbilical cord blood mononuclear cells for the treatment of acute myocardial infarction. *Cell Transplant* 2004;13(7–8):729–739.

39. Hirata Y, Sata M, Motomura N, Takanashi M, Suematsu Y, Ono M, *et al.* Human umbilical cord blood cells improve cardiac function after myocardial infarction. *Biochem Biophys Res Commun* 2005;327(2):609–614.

40. Chen HK, Hung HF, Shyu KG, Wang BW, Sheu JR, Liang YJ, *et al.* Combined cord blood stem cells and gene therapy enhances angiogenesis and improves cardiac performance in mouse after acute myocardial infarction. *Eur J Clin Invest* 2005;35(11): 677–686.

41. Ma N, Ladilov Y, Moebius JM, Ong L, Piechaczek C, David A, *et al.* Intramyocardial delivery of human CD133+ cells in a SCID mouse cryoinjury model: bone marrow vs. cord blood-derived cells. *Cardiovasc Res* 2006;71(1):158–169.

42. Kim B-O, Tian H, Prasongsukarn K, Wu J, Angoulvant D, Wnendt S, *et al.* Cell transplantation improves ventricular function after a myocardial infarction: a preclinical study of human unrestricted somatic stem cells in a porcine model. *Circulation* 2005;112(Suppl 9):I-96–104.

43. Moelker AD, Baks T, Wever KM, Spitskovsky D, Wielopolski PA, van Beusekom HM, *et al.* Intracoronary delivery of umbilical cord blood derived unrestricted somatic stem cells is not suitable to improve LV function after myocardial infarction in swine. *J Mol Cell Cardiol* 2007; 42(4):735–745.

44. Leor J, Guetta E, Chouraqui P, Guetta V, Nagler A. Human umbilical cord blood cells: a new alternative for myocardial repair? *Cytotherapy* 2005;7(3):251–257.

45. Orlic D, Kajstura J, Chimenti S, Jakoniuk I, Anderson SM, Li B, *et al.* Bone marrow cells regenerate infarcted myocardium. *Nature* 2001;410(6829):701–705.

46. Pomyje J, Zivny J, Sefc L, Plasilova M, Pytlik R, Necas E. Expression of genes regulating angiogenesis in human circulating hematopoietic cord blood CD34+/CD133+ cells. *Eur J Haematol* 2003;70(3):143–150.

47. Wu JC, Chen IY, Sundaresan G, Min J-J, De A, Qiao J-H, *et al.* Molecular imaging of cardiac cell transplantation in living animals using optical bioluminescence and positron emission tomography. *Circulation* 2003;108(11):1302–1305.

48. Agbulut O, Vandervelde S, Al Attar N, Larghero J, Ghostine S, Leobon B, *et al.* Comparison of human skeletal myoblasts and bone marrow-derived CD133+ progenitors for the repair of infarcted myocardium. *J Am Coll Cardiol* 2004;44(2):458–463.

49. Barbash IM, Chouraqui P, Baron J, Feinberg MS, Etzion S, Tessone A, *et al.* Systemic delivery of bone marrow-derived mesenchymal stem cells to the infarcted myocardium: feasibility, cell migration, and body distribution. *Circulation* 2003;108(7):863–868.

50. Balsam LB, Wagers AJ, Christensen JL, Kofidis T, Weissman IL, Robbins RC. Haematopoietic stem cells adopt mature haematopoietic fates in ischaemic myocardium. *Nature* 2004;428(6983):668–673.

51. Thum T, Bauersachs J, Poole-Wilson PA, Volk H-D, Anker SD. The dying stem cell hypothesis: immune modulation as a novel mechanism for progenitor cell therapy in cardiac muscle. *J Am Coll Cardiol* 2005;46(10):1799–1802.

52. Robinson S, Niu T, de Lima M, Ng J, Yang H, McMannis J, *et al. Ex vivo* expansion of umbilical cord blood. *Cytotherapy* 2005;7(3):243–250.

53. Piacibello W, Sanavio F, Severino A, Dane A, Gammaitoni L, Fagioli F, *et al.* Engraftment in nonobese diabetic severe combined immunodeficient mice of human cd34$^+$ cord blood cells after *ex vivo* expansion: evidence for the amplification and self-renewal of repopulating stem cells. *Blood* 1999;93(11):3736–3749.

54. Lewis ID, Almeida-Porada G, Du J, Lemischka IR, Moore KA, Zanjani ED, *et al.* Umbilical cord blood cells capable of engrafting in primary, secondary, and tertiary xenogeneic hosts are preserved after *ex vivo* culture in a noncontact system. *Blood* 2001;97(11):3441–3449.

55. Von Drygalski A, Alespeiti G, Ren L, Adamson JW. Murine bone marrow cells cultured *ex vivo* in the presence of multiple cytokine combinations lose radioprotective and long-term engraftment potential. *Stem Cells Dev* 2004;13(1):101–111.

56. McNiece IK, Almeida-Porada G, Shpall EJ, Zanjani E. *Ex vivo* expanded cord blood cells provide rapid engraftment in fetal sheep but lack long-term engrafting potential. *Exp Hematol* 2002;30(6):612–616.

57. Holyoake TL, Alcorn MJ, Richmond L, Farrell E, Pearson C, Green R, *et al.* CD34 positive PBPC expanded *ex vivo* may not provide durable engraftment following myeloablative chemoradiotherapy regimens. *Bone Marrow Transplant* 1997;19(11):1095–101.

58. Nagler A, Grynspan F, Peled T, Mandel J, Guetta E, Holbova R, *et al.* Expansion of human umbilical cord blood-derived cd34$^+$ stem/progenitor cells to treat myocardial infarction. *Blood* 2003;102(11): (abstract).

59. Yan J, Xu L, Welsh AM, Chen D, Hazel T, Johe K, *et al.* Combined immunosuppressive agents or cd4 antibodies prolong survival of human neural stem cell grafts and improve disease outcomes in amyotrophic lateral sclerosis transgenic mice. *Stem Cells* 2006;24(8):1976–1985.

60. Henning RJ, Abu-Ali H, Balis JU, Morgan MB, Willing AE, Sanberg PR. Human umbilical cord blood mononuclear cells for the treatment of acute myocardial infarction. *Cell Transplant* 2004;13(7–8):729–739.

5   Amniotic Stem Cells

Paolo De Coppi, Anthony Atala and Saverio Sartore

Introduction

In the last few years our as well as other groups have described the presence of stem cell with various differentiative and proliferative potential in the amniotic fluid. Amniocentesis is a widely accepted method for prenatal diagnosis. Minimal or no ethical concerns would be present if embryonic and fetal stem cells would be taken from amniotic fluid before or at birth. In this chapter, we will briefly describe the techniques in use for amniocentesis and examine the different progenitor cells described so far.

Amniocentesis

One of the primary uses of amniocentesis is a safe method of isolating cells from the fetus that can be karyotyped and examined for chromosomal abnormalities.[1−4] The protocol generally consists of acquiring 10–20 ml of fluid using a transabdominal approach. Amniotic fluid samples are centrifuged, and cell supernatant is resuspended in culture medium. Approximately 10^4 cells are seeded on $22 \times 22 \, mm^2$ cover slips. Cultures are grown to confluence for 3–4 weeks in 5% CO_2 at 37°C, and the chromosomes are characterized from mitotic phase cells.[5] Amniocentesis is performed typically around 16 weeks of gestation, although in some cases it may be performed as early as 14 weeks when the amnion fuses with the chorion and the risk of bursting the amniotic sac by needle puncture is minimized. Amniocentesis can be performed as late as term. The amniotic sac is usually noticed first by ultrasound around the ten-week gestational time point.

Differentiated Cells from Amniotic Fluid

Amniotic fluid cell culture consists of a heterogeneous cell population displaying a range of morphologies and behaviors. Studies on these cells have characterized them into many shapes and sizes varying from 6 to $50 \, \mu m$ in diameter and from round to squamous in shape. Most cells in the fluid are terminally differentiated along epithelial lineages and have limited proliferative and differentiation capabilities.

73

Previous studies have noted an interesting composition of the fluid consisting of a heterogeneous cell population expressing markers from all three germ layers.[6–9] The source of these cells and of the fluid itself underwent a great deal of research. Current theories suggest that the fluid is largely derived from urine and pulmonary secretion from the fetus as well as from some ultra-filtrate from the plasma of the mother entering though the placenta. The cells in the fluid have been shown to be overwhelmingly from the fetus and are thought to be the mostly cells sloughed off the epithelium and digestive and urinary tract of the fetus as well as the amnion.[10,11]

Mesenchymal Stem Cells from Amniotic Fluid

Several studies have been published in the last few years describing very simple protocols for the isolation of mesenchymal stem cells similar to the one described to be present in various adult tissues such as bone marrow from amniotic fluid.[12–14] These cells were able to proliferate *in vitro*, to be engineered in a three-dimensional structure and used *in vivo* to repair tissue defects.[13] A few years later In't Anker *et al.* was able to prove for the first time that both amniotic fluid and placenta were abundant sources of fetal mesenchymal stem cells (MSCs) that exhibit a phenotype and multilineage differentiation potential similar to that of postnatal bone marrow (BM)-derived MSCs.[15] Isolation and expansion protocols are similar to what has been used to isolate MSCs from other sources. Briefly, amniotic fluid samples were centrifuged for ten minutes at 1283 rpm. Pellets were resuspended in Iscove's modified Dulbecco's medium containing 2% fetal calf serum and antibiotics (defined as washing medium). Similarly, for the placenta, approximately 1 cm^3 was washed in phosphate-buffered saline (PBS) and single-cell suspensions were made by mincing and flushing the tissue parts through a 100 μm nylon filter with washing medium. Single cell suspensions of amniotic fluid and placenta were plated in six-well plates and cultured in M199 supplemented with 10% FCS, 20 μg/ml endothelial cell growth factor, heparin (8 U/ml) and antibiotics. After seven days, non-adherent cells were removed and the medium was refreshed. When grown to confluence, adherent cells were detached with trypsin/EDTA and expanded in culture flasks pre-coated with 1% gelatin and kept in a humidified atmosphere at 37°C. The expansion potency of fetal MSCs was higher compared with adult bone marrow-derived MSCs. As a result, they were able to expand amniotic fluid MSCs to about 180 × 106 cells within four weeks (three passages). The phenotype of the culture-expanded amniotic fluid-derived cells was similar to that reported for MSCs derived from second-trimester fetal tissues and adult BM. They were able to show that amniotic fluid-derived MSCs showed multilineage differentiation potential into fibroblasts, adipocytes, and osteocytes.[16] Furthermore, amniotic fluid-derived MSCs were successfully isolated, cultured, and enriched without interfering with the routine process of fetal karyotyping. Flow cytometry analyses showed that they were positive for SH2, SH3, SH4, CD29 and CD44, low positive for CD90 and CD105, but negative for CD10, CD11b, CD14, CD34, CD117, and EMA.[17] Most importantly, immuno-phenotypic analyses demonstrated that these cells expressed HLA-ABC, class I major histocompatibility complex (MHC-I), but they did not express HLA-DR, DP, DQ (MHC-II molecules).[18] Li *et al.* have extensively investigated their immuno-logical role and described that when mononucleated cells recovered from placentas by density gradient fractionation were added to umbilical cord blood (UCB) lympho-cytes stimulated by human adult lymphocytes or potent T-cell mitogen phytohemag-glutinin, a significant reduction in lymphocyte proliferation was observed. This immunoregulatory feature strongly implies that they may have potential application in allograft transplantation. Since, it is possible to obtain placenta and UCB from the same donor, they suggested the placenta as an attractive source of MSCs for co-transplantation in conjunction with UCB-derived hematopoietic stem cells (HSCs) to reduce the potential graft-versus-host disease (GVHD) in recipients.[19] Finally, ovine mesenchymal amniocytes have also been cultured and engineered into a col-lagen hydrogel in order to replace partial diaphragmatic lost or absences. The authors showed that diaphragmatic repair with an autologous tendon engineered from mes-enchymal amniocytes leads to improved mechanical and functional outcomes when compared with an equivalent acellular bioprosthetic repair, depending on scaffold composition.[20] Different groups have shown that MSCs from placenta and amniotic fluid may have more plasticity than what initially thought. Phenotypic and gene expression studies indicated mesenchymal stem cell-like profiles in both amnion and chorion cells that were positive for neuronal, pulmonary, adhesion, and migration markers. In addition, transplantation in neonatal swine and rats resulted in human microchimerism in various organs and tissues suggesting that amnion and chorion cells may represent an advantageous source of progenitor cells with potential appli-cations in a variety of cell therapy and transplantation procedures.[21] Similarly, Zhao *et al.* have reported that human amniotic mesenchymal cells (hAMC) may also be a suitable cell source for cardiomyocytes.[22] They showed that freshly isolated hAMC expressed cardiac-specific transcription factor GATA4, cardiac-specific genes, such as myosin light chain (MLC)-2a, MLC-2v, cTnI, and cTnT, and the alpha-subunits of the cardiac-specific L-type calcium channel (alpha1c). After stimulation with basic fibroblast growth factor (bFGF) or activin A, hAMC expressed Nkx2.5, a spe-cific transcription factor for the cardiomyocyte and cardiac-specific marker atrial natriuretic peptide. In addition, the cardiac-specific gene alpha-myosin heavy chain was detected after treatment with activin A. Co-culture experiments confirmed that hAMC were able to both integrate into cardiac tissues and differentiate into cardiomyocyte-like cells. After transplantation into the myocardial infarcts (AMI) in rat hearts, hAMC survived in the scar tissue for at least two months and differentiated into cardiomyocyte-like cells.[22] However, we have recently shown that this potential does not belong to mesenchymal progenitor cells in bigger animals such as pigs. Amniotic fluid-derived mesenchymal cells autotransplanted in a porcine model of AMI were able to transdifferentiate to cells of vascular cell lineages but failed to give origin to cardiomyocytes.[23,24] Neuron regeneration has been also described using rat amniotic epithelial (RAE) cells. In particular they expressed *in vitro* both neuronal and neural stem cell markers, neurofilament microtubule-associated protein-2 and nestin. RT–PCR revealed that these cells expressed nestin mRNA. The RAE cells were also transplanted into the hippocampus of adult gerbils that were subjected to temporal occlusion of bilateral carotid arteries. Five weeks after transplantation, grafted cells migrated into the CA1 pyramidal layer that showed selective neuronal death, and survived in a manner similar to CA1 pyramidal neurons.[25] Different reports suggest that also human amniotic epithelial cells (HAEC) possess certain properties similar to that of neural and glial cells.[26] When transplanted into the transection cavities in the spinal cord of bonnet monkeys, HAEC were able to survive, support the growth of host axons through them, prevent the formation of glial scar at the cut ends and may prevent death in axotomized cells or attract the growth of new collateral sprouting.[25] Amniotic epithelial cells isolated from human term placenta express surface markers normally present on embryonic stem and germ cells. In addition, they express the pluripotent stem cell-specific transcription factors octamer-binding protein-4 (Oct-4) and nanog. Under certain culture conditions, amniotic epithelial cells form spheroid structures that retain stem cell characteristics. Amniotic epithelial cells did not require other cell-derived feeder layers to maintain Oct-4 expression, did not express telomerase, and are non-tumorigenic upon transplantation. Based on immunohistochemical and genetic analysis, amniotic epithelial cells had the potential to differentiate to all three germ layers–endoderm (liver, pancreas), mesoderm (cardiomyocyte), and ectoderm (neural cells) *in vitro*.[27] Sarkar *et al.* have also shown that human amniotic epithelial cells obtained from human placenta were able to survive into the transection cavities in the spinal cord of bonnet monkeys, to support the growth of host axons through them, to prevent the formation of glial scar at the cut ends and may prevent death in axotomized cells or attract the growth of new collateral sprouting. They speculated that HAEC may be having certain properties equal to the beneficial effects of neural tissue in repairing spinal cord injury. Apart from this speculation, there are two more reasons for why HAEC transplantation studies are warranted to understand the long-term effects of such transplantations. First, there was no evidence of immunological rejection probably due to the non-antigenic nature of the HAEC. Second, unlike neural tissue, procurement of HAEC does not involve many legal or ethical problems.[28–31]

Amniotic Fluid Stem Cells

We have recently described a pluripotent population of cells derived from amniotic fluid defined as amniotic fluid stem (AFS) cells. Similar cells can be derived from chorionic villi (placenta) samples. We will describe in the following sections in details their isolation, characterization and differentiation *in vitro* into different lineages.[32]

Cell preparation and culture methods

Chorionic villi samples and human amniotic fluid was obtained under informed consent at 12–18 weeks of pregnancy from a total of 300 women between 23 and 42 years of age. In all cases, the karyotype evaluated from the cultured chorionic villi and amniotic fluid cells was normal. Samples were seeded in a $22 \times 22 \, mm^2$ cover slip in a volume of 2 ml and grown to confluence for three to four weeks at 95% humidity and 37°C. Fresh medium was applied after five days of culture and every third day thereafter. The culture medium consisted of (MEM (GIBCO/BRL, Grand Island, NY), 18% Chang Medium B (Irvine Scientific, Santa Ana, CA), 2% Chang Medium C (Irvine Scientific, Santa Ana, CA) with 15% embryonic stem cell-certified fetal bovine serum (ES- FBS, GIBCO/BRL, Grand Island, NY), 1% antibiotics (GIBCO/BRL, Grand Island, NY), and L-glutamine (Sigma-Aldrich, St. Louis, MO). The cells were subcultured using 0.25% trypsin containing 1-mM EDTA for five minutes at 37°C.

Cell cloning

In order to test the hypothesis that placenta and amniotic fluid could contain stem cells that would be able to differentiate into multiple lineages, cell colonies derived from single cells were expanded. Isolation and characterization of chorionic villi and amniotic-derived stem cells. The cells were successfully isolated from 300 fetuses and maintained in culture in Chang medium. The presence of cells of maternal origin in placenta and amniotic fluid is extremely low. In order to evaluate for the presence of maternal cells, the studies were performed using cells from male fetuses. Karyotypic analyses of the ckit^{pos} cells showed an *xy* phenotype in all the cells. Female fetuses were used as controls and they did not show any difference in their pluripotential ability. Cytofluorimetric analysis and immunocytochemistry showed that most of the amniotic cells were epithelial and stained positive for cytokeratins. Most of the stromal cells stained for α-actin, and only a few cells were positive for desmin or myosin. FACS analyses showed that between 18% and 21% of the cells expressed CD105, while approximately the same proportion of cells (between 0.8% and 3%) expressed ckit and CD34. The ckit^{pos} cells were successfully isolated and maintained in culture in Chang medium. The ckit^{pos} cells were shown to be pluripotent. They maintained a round shape when cultured in bacterial plates for almost a week while they had a very low-proliferative capability. After the first week the cells started to adhere to the plate and they changed their morphology, becoming more elongated, and they started to proliferate. The medium was changed every three days and they were passed whenever they reached confluence. If the cells were not passaged, they aggregated, forming embryoid shaped tissue-like structures measuring $1 \times 5 \, mm^2$. Serial sections of these structures showed specific markers for the three embryonic germ layers immunohistochemically. The embryo-shaped tissue, if disaggregated, was still able to differentiated into different lineages under appropriate growth conditions. The CD105, CD90, and CD34 immunoseparated cells, and the remaining non-immunoseparated cells did not show any pluripotential ability. No feeder layers were required, either for maintenance, or expansion.[33,34]

Telomerase activity

Telomerase activity was evaluated using the Telomerase Repeat Amplification Protocol (TRAP) assay and the presence of telomerase was analyzed immunocytochemically. No telomerase activity could be detected with the TRAP-assay, either in the *ckit^pos* cells (lane 1), or in the control bone marrow stem cells (lane 2). In contrast, the prostate cancer cell line PC3 and an epithelial tumor cell lines (HeLa), as a control, showed high telomerase activity (lanes 4 and 5). Anti-telomerase antibodies positively stained the amniotic *ckit^pos* cells, suggesting that the cells may express telomerase protein, but the levels were too low to be detected by the TRAP-assay.

Differentiation potential induction of osteogenic phenotype

Light microscopy analysis showed that *ckit^pos* cells, within four days in osteogenic medium, developed an osteoblastic-like appearance with fingerlike excavations into the cells cytoplasm (Fig. 1A).[35,36] At 16 days the cells aggregated in the typical lamellar bone-like structures and increased their expression of alkaline phosphatase. Calcium accumulation was evident after one week and increased over time. To confirm the cytochemical findings, AP activity was measured using a quantitative assay, which measured p-nitrophenol, equivalent to AP production. The *ckit^pos* cells showed more than a 200-time increase in AP production in the osteogenic-inducing medium compared to cells grown in control medium at days 16 and 24. After that time the levels of AP decreased. No AP production was detected in *ckit^neg* amniotic cells cultured in osteogenic medium at any time point. AP expression was confirmed at the RNA level. No activation of the AP gene was detected at eight, 16, 24, and 32 days in the *ckit^pos* cells grown in control medium. In contrast, *ckit^pos* cells grown in osteogenic medium showed an activation of the AP gene at each time point. Expression of cbfa1, a transcription factor specifically expressed in osteoblasts and hypertrophic chondrocytes, was highest in cells grown in osteogenic inducing medium at day 8, and decreased slightly at days 16, 24, and 32. The expression of cbfa1 in the controls was significantly lower at each time point. Osteocalcin was expressed only in the *ckit^pos* cells in osteogenic conditions at eight days. No expression of osteocalcin was detectedable in the *ckit^pos* cells in the control medium and in the *ckit^neg* cells in the osteogenic conditions at any time point. A major feature of osteogenic differentiation is the ability of the cells to precipitate calcium. Cell-associated mineralization can be analyzed using von Kossa staining and by measuring the calcium content of cells in culture. Von Kossa staining of cells grown in the osteogenic medium showed enhanced silver nitrate precipitations by day 16, indicating high levels of calcium. Calcium precipitation continued to increase exponentially at 24 and 32 days. In contrast, cells in the control medium did not form silver nitrate precipitations after 32 days. Microscopic examination of stained cells showed no calcification in the osteogenic-treated cells at days 4 or 8, but strong black silver nitrate precipitates were noticed in osteogenic induced cells after 16, 24, and 32 days in culture. In cells cultured in control medium, no precipitates were noticed over the 32-day time period. Calcium deposition by the cells was also measured with a quantitative chemical assay, which measures calcium-cresolophthalein complexes. Cells undergoing osteogenic induction showed a significant increase in

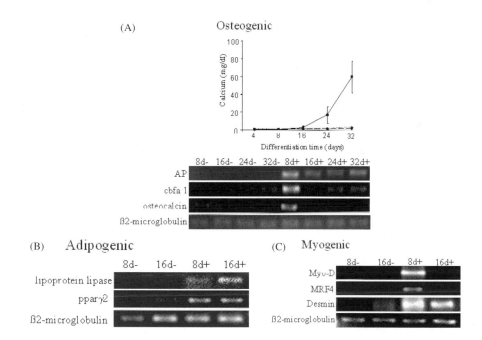

Fig. 1. The differentiated cell types expressed functional and biochemical characteristics of the target tissue. (**A**) Myogenic-induced cells showed a strong expression of *desmin* expression at day 16 (lane 4). *MyoD* and *MRF4* were induced with myogenic treatment at day 8 (lane 3). Specific PCR-amplified DNA fragments of *MyoD*, *MRF4*, and *desmin* could not be detected in the control cells at days 8 and 16 (lanes 1 and 2). (**B**) Gene expression of *ppary* and *lipoprotein lipase* in cells grown in adipogenic-inducing medium was noted at days 8 and 16 (lanes 3 and 4). (**C**) RT-PCR revealed an upregulation of albumin gene expression. Western blot analyses of cell lysate showed the presence of the hepatic lineage-related proteins HNF-4α, c-met, MDR, albumin, and α-fetoprotein. Undifferentiated cells were used as negative control. (**D**) Osteogenic-induced progenitor cells showed a significant increase of calcium deposition starting at day 16 (solid line). No calcium deposition was detected in progenitor cells grown in control medium or the negative control cells grown in osteogenic conditions (dashed line). RT-PCR showed presence of *cbfa1* and *osteocalcin* at day 8 and confirmed the expression of *AP* in the osteogenic-induced cells. (**E**) Only the progenitor cells cultured under neurogenic conditions showed the secretion of glutamic acid in the collected medium. The secretion of glutamic acid could be induced (20 minutes in 50-mM KCl buffer). (**F**) RT-PCR of progenitor cells induced in endothelial medium (lane 2) showed the expression of *CD31* and *VCAM*.

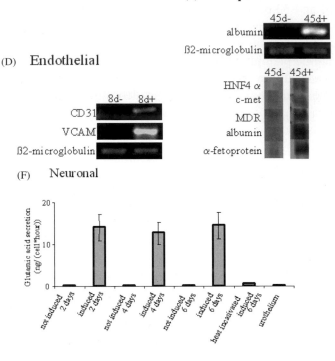

Fig. 1. (*Continued*)

calcium precipitation after 16 days (up to 4 mg/dl). The precipitation of calcium increased up to 70 mg/dl at 32 days. In contrast, cells grown in control medium did not show any increase in calcium precipitation (1.6 mg/dl) by day 32.

Induction of adipogenic lineage

Ckit^pos a cells cultured in a medium containing dexametasone, insulin, indomethacin, and 3-isobutyl-1-methylxanthine, within eight days changed their morphology from elongated to round (Fig. 1B).[37] This coincided with the accumulation of intracellular triglyceride droplets. After 16 days in culture, more than 95% of the cells had their cytoplasm completely filled with lipid-rich vacuoles, which stained positively with Oil-O-Red. The amniotic *ckit^neg* cells that were induced with the same medium and the *ckit^pos* cells cultured in control medium did not show any phenotypic change of adipogenic differentiation and did not stain with Oil-O-Red after 16 days of culture. Adipogenic differentiation was confirmed by RT–PCR analysis. The expression of peroxisome proliferation-activated receptor (2 (ppar(2), a transcription factor that regulates adipogenesis and of lipoprotein lipase was analyzed. Expression of these genes was up regulated in the *ckit^pos* cells under adipogenic conditions. *Ckit^{pos}* cells cultured under control conditions and *ckit^{neg}* cells in adipogenic medium did not express either gene at any time point.

Induction of myogenic phenotype

Ckit^{pos} cells were cultured with myogenic medium on Matrigel coated dishes (Fig. 1C).[38,39] Induction with 5-azacytidine for 24 hours promoted the formation of multinucleated cells over a 24- to 48-hour period. After 16 days, the cells grown with myogenic medium formed myofiber-like structures that stained immunocytochemically with Desmin and Sarcomeric Tropomyosin. *Ckit^{pos}* cells grown in control medium and *ckit^{neg}* cells cultured in myogenic medium did not lead to cell fusion or multinucleated cells. Only a few Desmin cells were present in the *ckit^{neg}* amniotic cells cultured in myogenic medium at 16 days. Expression of *MyoD*, *Myf5*, *Myf 6* (MRF4) and *Desmin*, were analyzed using RT–PCR. *MyoD* and MRF4 were expressed by the *ckit^{pos}* cells in culture at eight days, and suppressed at 16 days. Both these genes were not expressed either at eight or 16 days in the controls. *Desmin* expression was induced at eight days and increased by 16 days in the *ckit^{pos}* cells cultured in myogenic medium. In contrast, there was no activation of *Desmin* in the control cells at eight and 16 days. *Myf5* was present at eight days and increased at 16 days in the *ckit^{pos}* cells. Lower levels of the *myf5* gene were detected in the cells maintained in culture with the control medium at 16 days.

Induction of endothelial phenotype

Ckit^{pos} cells were cultured with endothelial medium in PBS-gelatin coated dishes (Fig. 1D). After one week in culture the cells started to change their morphology, and by the second week, were mostly tubular. The cells stained positively for FVIII, KDR, and P1H12. *Ckit^{neg}* cells cultured in the same conditions and *Ckit^{pos}* cells cultured in Chang medium for the same period were not able to form tubular structures and did not stain for endothelial specific markers. The cells, once differentiated, were able to grow in culture for more than one month.

Induction of hepatocytes phenotype

When cultured in hepatic conditions cells exhibited morphological changes after seven days showing a changing in the morphology from an elongated to a cobblestone appearance (Fig. 1E).[40,41] The cells showed positive staining for albumin at day 45 post-differentiation, and were also found to express the transcription factor HNF4α, the c-met receptor, the MDR membrane transporter, albumin, and α-fetoprotein. RT–PCR analysis further provided evidence of albumin production. The maximum rate of urea production for hepatic differentiation induced cells was 1.21×10^3 ng urea/h/cell as opposed to 5.0×10^1 ng urea/h/cell for control progenitor cell populations.

Induction of neurogenic phenotype

$Ckit^{pos}$ cells cultured in neurogenic conditions changed their morphology within the first 24 hours (Fig. 1F).[42,43] Responsive cells progressively assumed neuronal morphological characteristics; initially the cytoplasm retracted towards the nucleus, forming contracted multipolar structures. Over the subsequent hours, the cells displayed primary and secondary branches, and cone-like terminal expansions. Induced $Ckit^{pos}$ cells stained positively for β III Tubulin and Nestin. $Ckit^{neg}$ cells cultured in the same conditions and $Ckit^{pos}$ cells cultured in Chang medium for the same period were not able to form tubular structures and did not stain for endothelial specific markers. The cells, once differentiated, were able to grow in culture for more than one month.

Clonal and proliferative analyses

$Ckit^{pos}$ cells were able to be expanded clonally. After serial dilution, we observed that most of the wells contained no cells, and only a few of the 96 wells contained a single cell. Cells from numerous clones showed similar morphology and growth behavior. Clonal lineages from different patients were tested. All the cells underwent osteogenic, adipogenic, myogenic, neurogenic and endothelial differentiation. Amniotic stem cells did not show any decrease in their growth ability after more than 80 cell divisions, and they maintained their ability to differentiate into different lineages.

Conclusion

Fetal tissue has been used in the past for transplantation and tissue engineering research because of its pluripotency and proliferative ability. Fetal cells maintain a higher capacity to proliferate than adult cells and may preserve their pluripotency longer in culture. However, fetal cell transplants are plagued by problems that are very difficult to overcome. Beyond the ethical concerns regarding the use of cells from aborted fetuses or living fetuses, there are other issues which remain a challenge. Previous studies have shown that it takes almost six fetuses to provide enough material to treat one patient with Parkinson's disease. In the last few years it has been shown that placenta and amniotic fluid contain a large variety of cells. The vast majority of the cells in the placenta and in the amniotic fluid are already differentiated, and therefore, have a limited proliferative ability. In this chapter, we have summarized the different progenitors, described in the last few years, present in both the amniotic fluid and placenta. In particular, we have focused our attention on mesenchymal stem cells and AFS cells, isolated using *ckit* cells, and their different proliferative and differentiative potential. In conclusion, placenta and amniotic fluid could be an excellent cell source for therapeutic applications. When compared with ES cells, fetal stem cells are easily differentiated into specific cell lineages, they do not need any feeder layer to grow, and avoid the current controversies associated with the use of human embryonic stem cells.

References

1. Crane JP, Cheung SW. An embryogenic model to explain cytogenetic inconsistencies observed in chorionic villus versus fetal tissue. *Prenat Diagn* 1988;8:119–129.
2. Milunsky A. Amniotic fluid cell culture. In: Milunsky A (ed.) *Genetic Disorder of the Fetus.* Plenum Press, New York, 1979, pp. 75–84.
3. Hoehn H, Salk D. Morphological and biochemical heterogeneity of amniotic fluid cells in culture. *Methods Cell Biol* 1982;26:11–34.
4. Gosden CM. Amniotic fluid cell types and culture. *Br Med Bull* 1983;39:348–354.
5. Brace RA, Resnik R. Dynamics and disorders of amniotic fluid. In: Creasy RK, Renik R (eds.) *Maternal Fetal Medicine.* Saunders, Philadelphia, 1999, pp. 623–643.
6. Medina-Gomez P, Johnston TH. Cell morphology in long-term cultures of normal and abnormal amniotic fluids. *Hum Genet* 1982;60(4):310–313.
7. Sarkar S, Chang HC, Porreco RP, Jones OW. Neural origin of cells in amniotic fluid. *Am J Obstet Gynecol* 1980;136(1):67–72.
8. von Koskull H, Aula P, Trejdosiewicz LK, Virtanen I. Identification of cells from fetal bladder epithelium in human amniotic fluid. *Hum Genet* 1984;65(3):262–267.
9. Cousineau J, Potier M, Dallaire L, Melancon SB. Separation of amniotic fluid cell types in primary culture by Percoll density gradient centrifugation. *Prenat Diagn* 1982;2(4):241–249.
10. Lotgering FK, Wallenburg HC. Mechanisms of production and clearance of amniotic fluid. *Semin Perinatol* 1986;10(2):94–102.
11. Underwood MA, Gilbert WM, Sherman MP. Amniotic fluid: not just fetal urine anymore. *J Perinatol* 2005;25(5):341–348.
12. Kaviani A, Perry TE, Dzakovic A, *et al.* The amniotic fluid as a source of cells for fetal tissue engineering. *J Pediatr Surg* 2001;36:1662–1665.
13. Kaviani A, Guleserian K, Perry TE, *et al.* Tissue engineering from amniotic fluid. *J Am Coll Surg* 2003;196:592–597.
14. Haigh T, Chen C, Jones CJ, Aplin JD. Studies of mesenchymal cells from 1st trimester human placenta: expression of cytokeratin outside the trophoblast lineage. *Placenta* 1999;20:615–625.
15. In't Anker PS, Scherjon SA, Kleijburg-van der Keur C, Noort WA, Claas FH, Willemze R, Fibbe WE, Kanhai HH. Amniotic fluid as a novel source of mesenchymal stem cells for therapeutic transplantation. *Blood* 2003;102(4):1548–1549.
16. In't Anker PS, Scherjon SA, Kleijburg-van der Keur C, de Groot-Swings GM, Claas FH, Fibbe WE, Kanhai HH. Isolation of mesenchymal stem cells of fetal or maternal origin from human placenta. *Stem Cells* 2004;22(7):1338–1345.
17. Tsai MS, Lee JL, Chang YJ, Hwang SM. Isolation of human multipotent mesenchymal stem cells from second-trimester amniotic fluid using a novel two-stage culture protocol. *Hum Reprod* 2004;19(6):1450–1456.
18. Li CD, Zhang WY, Li HL, Jiang XX, Zhang Y, Tang P, Mao N. Isolation and identification of a multilineage potential mesenchymal cell from human placenta. *Placenta* 2005.
19. Li CD, Zhang WY, Li HL, Jiang XX, Zhang Y, Tang PH, Mao N. Mesenchymal stem cells derived from human placenta suppress allogeneic umbilical cord blood lymphocyte proliferation. *Cell Res* 2005;15(7):539–547.
20. Fuchs JR, Kaviani A, Oh JT, LaVan D, Udagawa T, Jennings RW, Wilson JM, Fauza DO. Diaphragmatic reconstruction with autologous tendon engineered from mesenchymal amniocytes. *J Pediatr Surg* 2004;39(6):834–838.

21. Bailo M, Soncini M, Vertua E, Signoroni PB, Sanzone S, Lombardi G, Arienti D, Calamani F, Zatti D, Paul P, Albertini A, Zorzi F, Cavagnini A, Candotti F, Wengler GS, Parolini O. Engraftment potential of human amnion and chorion cells derived from term placenta. *Transplantation* 2004;78(10):1439–1448.

22. Zhao P, Ise H, Hongo M, Ota M, Konishi I, Nikaido T. Human amniotic mesenchymal cells have some characteristics of cardiomyocytes. *Transplantation* 2005;79(5):528–535.

23. Sartore S, Lenzi M, Angelini A, Chiavegato A, Gasparotto L, De Coppi P, Bianco R, Gerosa G. Amniotic mesenchymal cells autotransplanted in a porcine model of cardiac ischemia do not differentiate to cardiogenic phenotypes. *Eur J Cardiothorac Surg* 2005;28(5):677–684.

24. Sankar V, Muthusamy R. Role of human amniotic epithelial cell transplantation in spinal cord injury repair research. *Neuroscience* 2003;118(1):11–17.

25. Okawa H, Okuda O, Arai H, Sakuragawa N, Sato K. Amniotic epithelial cells transform into neuron-like cells in the ischemic brain. *Neuroreport* 2001;12(18):4003–4007.

26. Tsai MS, Hwang SM, Tsai YL, Cheng FC, Lee JL, Chang YJ. Clonal amniotic fluid-derived stem cells express characteristics of both mesenchymal and neural stem cells. *Biol Reprod* 2006; 74(3):545–551.

27. Miki T, Lehmann T, Cai H, Stolz DB, Strom SC. Stem cell characteristics of amniotic epithelial cells. *Stem Cells* 2005;23(10):1549–1559.

28. Sakuragawa N, Thangavel R, Mizuguchi M, *et al.* Expression of markers for both neuronal and glial cells in human amniotic epithelial cells. *Neurosci Lett* 1996;209:9–12.

29. Elwan MA, Sakuragawa N. Evidence for synthesis and release of catecholamines by human amniotic epithelial cells. *Neuroreport* 1997;8:3435–3438.

30. Sakuragawa N, Enosawa S, Ishii T, *et al.* Human amniotic epithelial cells are promising transgene carriers for allogeneic cell transplantation into liver. *J Hum Genet* 2000;45:171–176.

31. Takahashi N, Enosawa S, Mitani T, *et al.* Transplantation of amniotic epithelial cells into fetal rat liver by *in utero* manipulation. *Cell Transplant* 2002;11:443–449.

32. De Coppi P, Bartsch G Jr, Siddiqui MM, Xu T, Santos CC, Perin L, Mostoslavsky G, Serre AC, Snyder EY, Yoo JJ, Furth M, Soker S, Atala A. Isolation of clonal amniotic stem cell lines with potential for therapy. *Nat Biotechnol* 2007;25(1):100–106.

33. Takeda J, Seino S, Bell GI. Human Oct3 gene family: cDNA sequences, alternative splicing, gene organization, chromosomal location, and expression at low levels in adult tissues. *Nucleic Acids Res* 1992;20:4613–4620.

34. Mosquera A, Fernandez JL, Campos A, *et al.* Simultaneous decrease of telomerase length and telomerase activity with ageing of human amniotic fluid cells. *J Med Genet* 1999;36:494–496.

35. Karsenty G. Role of Cbfa1 in osteoblast differentiation and function. *Semin Cell Dev Biol* 2000;11;343–346.

36. Olmsted-Davis EA, *et al.* Primitive adult hematopoietic stem cells can function as osteoblast precursors. *Proc Natl Acad Sci USA* 2003;100:15877–15882.

37. Kim JB, Wright HM, Wright M, Spiegelman BM. ADD1/SREBP1 activates PPARgamma through the production of endogenous ligand. *Proc Natl Acad Sci USA* 1998;95:4333–4337.

38. Ferrari G, *et al.* Muscle regeneration by bone marrow-derived myogenic progenitors. *Science* 1998;279:1528–1530.

39. Rosenblatt JD, Lunt AI, Parry DJ, Partridge TA. Culturing satellite cells from living single muscle fiber explants. *In Vitro Cell Dev Biol Anim* 1995;31:773–779.

40. Hamazaki T, *et al.* Hepatic maturation in differentiating embryonic stem cells *in vitro*. *FEBS Lett* 2001;497:15–19.

41. Dunn JC, Yarmush ML, Koebe HG, Tompkins RG. Hepatocyte function and extracellular matrix geometry: long-term culture in a sandwich configuration. *FASEB J* 1989;3:174–177.

42. Black IB, Woodbury D. Adult rat and human bone marrow stromal stem cells differentiate into neurons. *Blood Cells Mol Dis* 2001;27:632–636.

43. Barberi T, *et al.* Neural subtype specification of fertilization and nuclear transfer embryonic stem cells and application in Parkinsonian mice. *Nat Biotechnol* 2003;21:1200–1207.

6 Stem Cell Homing to Injury in Cellular Cardiomyoplasty

Adil Al Kindi, Dominique Shum-Tim
and Ray Chu-Jeng Chiu

Introduction

Until recently, it was thought that the body has no natural ability to regenerate dead cardiomyocytes following myocardial infarction (MI). MI involves a series of well-established pathophysiological processes characterized by death of cardiomyocytes and transient inflammatory response in the area involved in the tissue injury. Given the fact that cardiomyocytes are terminally differentiated, and *in situ* myocardial progenitor cells are very limited in quantity, following a MI the heart muscle primarily responds with cellular hypertrophy rather than proliferation. The area of necrotic tissue is replaced with non-contractile scar tissue, which disrupts normal contractile function and results in the dilatation of the heart, decreasing cardiac performance.

In response to this, scientists have tried to improve cardiac function by transplanting multipotent adult stem cells that are either capable of regenerating cardiomyocytes and/or contributing salutary paracrine effects. Different methods have been used to deliver such cells to the damaged heart, including direct intramyocardial injection, intracoronary administration, intravenous injection and recruiting the patient's own stem cells from his/her bone marrow (BM) by using various mobilizing signal agents. The later two approaches are based on the mobilization and homing of stem cells to specific injury sites. Knowledge for such an approach has accumulated over the years but the earliest and most extensive reports came from hematologists who studied the behavior of the hematopoietic stem cells in association with the BM transplant therapies.

The Prelude: Insights from Bone Marrow Transplantation

If stem cells are capable of homing to a specific site, then this ability should be most evident for the hematopoietic cells to migrate from the periphery to their natural home, i.e. the bone marrow. Devine *et al.* sought to determine whether baboon-derived mesenchymal stem cells (MSC), when systematically infused, were capable of homing to and engraft within the BM. Five baboons lethally irradiated received hematopoietic progenitor cells combined with either autologous or allogeneic mesenchymal stem cells. In all five animals studied, intravenously administered cells were found in the BM.[1] Primary BM-derived MSC upon their injection also home to other lymphohemopoietic organs. Twenty-four hours post-IV injection into lethally irradiated hosts, 65% of the MSCs could be recovered from the BM and spleen. No cells could be recovered from other lymphohemopoietic organs such as the thymus and lymph nodes.[2]

In another study using a murine model, labeled marrow stem cells were injected intravenously. Sections of the femur were examined under the microscope. They found that injected stem cells rapidly entered the BM and the distribution of the cells was not random but rather was dependent on their cellular maturity. The most immature cells concentrated to the endosteal region while the more mature ones were found in the central area of the marrow.[3]

However, in contrary to the above-mentioned distribution, Gao et al. reported that the majority of BM-derived MSCs transplanted into syngeneic recipients were detected in the liver and lungs while only a few were found in the bones and spleen.[4] In addition, MSCs seem to dramatically lose the ability to home and seed in the BM and spleen following repeated cultures.[2] This may have important clinical implications as cells cannot be cultured indefinitely in vitro if they are to be used for systemic transplantation.

The way the stem cells enter and leave the bone marrow has also been an area of extensive research. The details of which are beyond the scope of this section, but it will be discussed further in the "Adhesion to Cardiac Microvascular Endothelium" section. Several pieces of evidence also point towards the roles of cellular adhesion molecules and chemokine receptors. Integrins such as very late antigen-4 (VLA-4) and lymphocyte function associated antigen (LFA-1), as well as other adhesion receptors such as CD44, have also been shown to be important in stem cell homing to the bone marrow, presumably through the interaction with the ligands on bone marrow endothelium and stroma.[5,6]

Animal Studies

As described earlier, MSCs are capable of homing from the circulation to the BM. This fascinating property of stem cells has also been studied in the reverse direction, homing from the BM to areas of tissue inflammation and regeneration. This has been demonstrated in various settings and tissues, including ischemic cerebral tissue,[6] ischemic myocardium,[3] fractured bone, spleen and injured liver.[7]

In fact, our laboratory was one of the first to test the hypothesis that recruitment of MSCs from BM, which traffic to the myocardial infarct site to differentiate and participate in the post-infarct repair, is a part of natural pathophysiology in myocardial infarction. In one of our two studies summarized in the following sections, Bittira et al.[8] postulated that the inflammatory response associated with MI involves mobilization of not only scavenger cells, which remove tissue debris, but also marrow-derived adult stem cells and progenitor cells, which can participate in the post-infarct repair process.

Experimental study (I)

In this study, we isolated rat MSCs from the BM of Lewis rats and labeled them with Lac-Z using a retrovirus as a vector. The study included three groups.[8] In the first group (ten rats), six million MSCs were injected intravenously. In the second group (14 rats), rats underwent left coronary artery ligation to induce a myocardial infarction one week after intravenous (IV) injection of six million cells. In the third group (six rats), rats underwent a sham operation with no ligation one week after intravenous injection of MSCs.

The hearts were then harvested at different time points: one, two, four, six, and eight weeks. The hearts were stained with X-gal to identify Lac-Z positive injected cells. They were also immunohistochemically stained for cardiac cells specific markers; sarcomeric myosin heavy chain molecules, connexin-43, troponin I-c and α-smooth muscle actin.

Homing of stem cells to the BM was confirmed in group one where labeled MSCs were identified in the BM at various point times. MSCs were also identified in the bone marrow in both the coronary artery ligated and sham operated group.

In the coronary artery ligation group, X-gal staining of the heart showed blue discoloration localized in the area of scarred myocardium in all hearts (Fig. 1).

Immunohistochemical staining also identified the fate of these cells. Up to three weeks, MSCs in the heart showed an immature monocytes like appearance with a large nucleus to cytoplasm ratio. They did not reveal any evidence of differentiation. After three weeks, cells showed evidence of cardiomyogenic differentiation. The staining for troponin I-c was more pronounced in the cytoplasm of these cells at four to six weeks. Connexin-43, marker of gap junction formation, was evident at four weeks. Staining for α-smooth muscles identified the differentiation of labeled MSCs into smooth muscle phenotype, some of which appeared to be incorporated into the walls of arterioles.

Our study proved that MSCs are recruited to the infarcted myocardium. However, this natural process appears to be limited and short-lived explaining the very limited ability of the heart to regenerate cardiomyocytes following myocardial infarction. However, this study did not address the mechanism by which it occurs but since that time we strongly suspected that cytokines from the infarcted myocardium was behind the recruitment of the MSCs.

The main limitation of this study was lack of quantitative data. The study did not rule out the contribution of primitive cells in the BM in this process as there was no BM ablation performed before the IV transplantation of the cells.

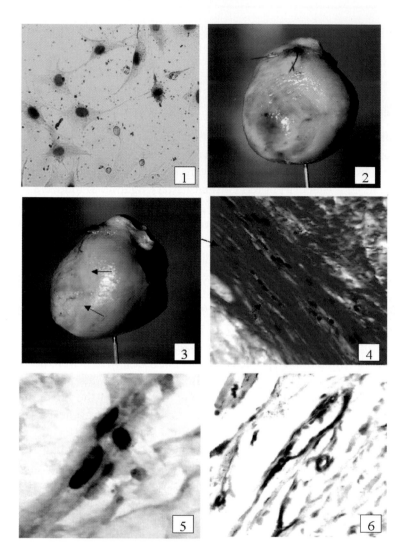

Fig. 1.   Mobilization and homing of bone marrow stromal cells in myocardial infarction. (1) Labeled MSCs prior to implantation in culture dish. Note 100% transfection efficiency. (2) Gross heart specimen after coronary ligation at four weeks stained with X-gal. Note blue color in scar distribution from MSC migration. (3) Oblique view of gross heart specimen in (2). Arrows denote border zone between scar and normal myocardium, showing absence of blue color in the non-infarcted posterior myocardium. (4) Cross-section of Picrosirius Red stained heart in (2) through scar region. Arrows denote labeled MSCs. (5) Cross-section of myocardial scar six weeks after coronary ligation, stained with cardiomyocyte-specific troponin I-c antibody. Brown color in the cytoplasm indicates positive stain. (6) Cross-section of an arteriole in a region of myocardial scar six weeks after coronary ligation. A labeled MSC can be found in the wall of the arteriole, positive for immunohistochemical stain for α-smooth muscle actin (brown). Counterstained with H & E. Reprinted with the permission of Elsevier.[8]

Experimental study (II)

In another study from our lab, Saito *et al.*[9] in addition to studying the immune tolerance of MSCs, he also studied the homing ability of these cells.

Bone marrow stromal cells were isolated from C57B1/6 mice and also retrovirally transduced with Lac-Z reporter gene for cell labeling. Three million cells were then injected intravenously for two consecutive days in adult Lewis rats. Fifteen rats had left coronary ligation after one week and another five only had a sham operation. Rats were then sacrificed at one day and two, four, eight, and 12 weeks after operation.

Mice MSCs were identified in the BM of all rats at all points of the study. Indicating that MCS injected intravenously will home naturally to the BM.

Saito *et al.* proved that, even though it was xenogeneic transplantation, mice MSCs are recruited from the BM to the heart in the event of a myocardial infarction. This recruitment occurred as early as the first day after the MI when he found mice MSCs in the perivascular zone of non-infarcted myocardium. However, after two weeks, most labeled MSCs were seen in the infarcted myocardium suggesting that they migrated to the injured site. Collected blood at various intervals showed mice MSCs in the bloodstream only after myocardial infarction proving that the blood stream is the route for mobilization after signaling release.

In a recent study that looked into the time dependent homing of intravenously injected stem cells, the number of MSCs increased significantly in the heart up to one week. However, in four and eight weeks, there was no significant difference in the number of homing cells to the heart. The MSCs did not show any differentiation in the first week and remained primitive. However, after four weeks, the MSCs stained positive for cardiac specific markers such as troponin, demin, smooth muscle, α-actin and CD31.[10]

The studies addressed in this section prove that MSCs home to the heart in the event of myocardial ischemia. However, they did not address the mechanism by which these cells mobilize from the BM, home through the circulation and reach and engraft in the heart.

The Signaling Molecules

The signaling between the heart and MSCs has been studied extensively. In order to understand these, it is important to realize that the heart after MI induces intense inflammatory reaction with different inflammatory cells infiltration including neutrophils, macrophages and mast cells. These together with myofibroblasts and cardiac cells produce inflammatory mediators including a number of cytokines. In the following sections, we will review the cytokines that have been shown to be associated with stem cell homing to the heart, and the mechanisms with which these signaling molecules mobilize the cells from the BM. Their use in the early clinical trials is also briefly discussed.

Stromal-derived factor-1 (SDF-1) and its receptor CXCR4

SDF-1 is a small molecule belonging to the chemokine family. It belongs to the Chemokine (C-X-C motif) ligand-12 (CXCL12) subfamily originally isolated from murine bone marrow stromal cells.[11] It is produced in three different forms: SDF-1α, SDF-1β and the recently discovered SDF-1γ[12] which are produced by alternate splicing of the same gene. SDF-1 genome is localized to chromosome 10q11.1.[12] SDF-1 is expressed by many organ tissues in the neutral inactive state including the liver, spleen, kidney, heart, brain, lung, and skeletal muscles.[12]

CXCR4, a seven-transmembrane spanning G protein-coupled receptor, is the only known receptor for SDF-1.[11] Its expression is dynamic and is affected by factors such as cytokines, chemokines, stromal cells, adhesion molecules and proteolytic enzymes.[13] In vitro, it can be up-regulated in about 40 hours.[13–15]

SDF-1 is continuously expressed by the heart in the resting non-ischemic state. In rats, it has been found that the ~5 kb mRNA transcript is abundantly expressed in the heart.[12] However, the role of this steady-state expression is not clear.

SDF-1 is produced primarily by BM stromal cells in the BM.[16] SDF-1 and its receptor CXCR4 play an important role in the retention of HSCs and other stem cells by the BM.[16–19] MSCs injected intravenously are capable of homing to the BM in the absence of peripheral injury.[8] Low levels of circulating SDF-1 produced by tissues other than the BM are not sufficient to mobilize the stem cells. Constitutive production of high levels of SDF-1 is essential to generate the reverse gradient and mobilize stem cells.[16]

One of the major signaling mechanisms for stem cells homing to the heart that have been studied is SDF-1.[17,20] It is one of the crucial chemoattractants for recruiting BM-derived cardiac progenitors in ischemic cardiomyopathy.[20] SDF-1 is also involved in stress-induced recruitment of stem cells to the liver and neointima.[21–23]

A number of studies have looked into the expression of SDF-1 after MI both in animals and humans.[12,17,21,24,25] Giordano et al.[21] studied the expression of SDF-1 mRNA and protein in the infarct and border zone of hearts in mice. They found that SDF-1 mRNA expression increased by $56.7 \pm 12.7\%$ in 48 hours. It peaked at 72 hours (95.7% increase) but then returned to baseline by seven days. This was confirmed in another study on rats,[21] where the SDF-1 levels also returned to baseline after seven days. However, in another study that was performed on rats,[12] SDF-1α peaked at one week and stayed until six weeks. SDF-1γ did not alter following infarction but it was abundantly expressed compared to isoforms SDF-1α, SDF-1β. SDF-1 is most highly expressed in the heart in the peri-infarct zone.[21] Such limited period of expression of this important chemokine following MI may explain why the heart has very limited ability to repair after MI using stem cells from the BM, given the very short duration of the signaling mechanism.

The surge in SDF-1 expression by the heart following acute myocardial ischemia is important in recruiting stem cells to the infarct area. There was an increase of 80%

in the homing of intravenously injected stem cells to the left ventricle following induced MI. This effect was reduced by 60% in the presence of CXCR4 antibodies.[21] In response to this over-expression of SDF-1, the homing stem cells increased by two folds. However, this over-expression had no effect on homing in the absence of MI, thus indicating SDF-1 is a necessary but not sufficient signal for stem cell recruitment to the heart.

Response of stem cells to SDF-1 is dose dependent.[11] Endothelial progenitor cells (EPCs) have shown a potent dose-dependent activity toward SDF-1 with the cell migration doubling when the concentration of SDF-1 is increased by ten folds. However, this concentration gradient of SDF-1 is not the only responsible factor for response to SDF-1.[12,24] The migratory response of bone marrow-mononuclear cells to SDF-1 was significantly reduced in patients with chronic ischemic heart disease. This was despite having a normal number of stem cells in the bone marrow to start with and with normal expression of CXCR4 receptor. The exact mechanism underlying impaired migratory response toward SDF-1 is unclear but perhaps due to alteration in intracellular pathway.[24] The response in animals was time gated also. BM-MNCs exhibited age related decrease in response to SDF-1 in mice. The response to SDF-1 was markedly diminished after one month of age, but not to human growth factor (HGF).[20]

Being the only known receptor to SDF-1, CXCR4 has been studied in relation to its interaction with SDF-1; 66% of EPCs cultured for seven days expressed CXCR4 receptor.[11] CD34+ cells that have been transduced with a gene to over express CXCR4 receptor have shown migration towards SDF-1 signal at a lower SDF-1 concentration. These CXCR4 over-expressed cells have also demonstrated increased proliferation and survival but did not lose their differentiation potential.[14] Cytokines involved in stem cell mobilization like SDF-1, SCF, and IL-6 can up-regulate CXCR4 expression on CD34+ cells.[16]

How SDF-1 releases BM stem cells have been studied on hematopoietic stem cells (HSC).[26] HSCs and other stem cells express c-kit, the receptor for KitL, suggesting that a common signaling cascade govern their proliferation and recruitment. When the bone marrow is exposed to SDF-1 and also VEGF, this induces the up-regulation of matrix metalloproteinase-9 (MMP-9). MMP-9 causes the release of soluble Kit-ligand (skitl). sKit is expressed on BM-derived progenitor that can give rise to cardiac muscles. sKitl permits the transform HSCs and other stem cells from the quiescent to the proliferative state. In the MMP-9$^{-/-}$ mice, release of sKitl and stem cell motility are impaired.

It is hypothesized that stem cell recruitment is a two-step process; stem cells adhere to the vascular endothelium in the injury zone followed by local chemotaxis to the site of engraftment. SDF-1 expression has been demonstrated to increase in the endothelium of the vasculature at the injury zone. Thus CXCR4-expressing cells bind to the endothelium in this zone.[21]

Other than functioning as a signaling cytokine, SDF-1 has been shown to have other functions that may prove important in myocardial regenerative therapy.[11,24]

When studied on EPC, SDF-1 reduced apoptosis from 26% to 7.1%. It also up-regulates endogenous VEGF expression which also plays an important role in chemotaxis and myocardial regeneration (see VEGF) In addition, evidence suggests that SDF-1 may have direct effects on vasculogenesis with increase in hind limb capillary density in SDF-1 treated rats by attracting EPCs.[11] Tachibana et al. reported that mice lacking SDF-1 have defective formation of large blood vessels.[27]

Interleukin-8 (IL-8)

IL-8 is a chemotactic cytokine belonging to the CXC (α) family.[28] Previously known as neutrophil activating protein (NAP-1), the gene encoding for IL-8 is located on the human chromosome 4q13–q21.[29,30] It is produced by a number of cells including monocytes, neutrophils, fibroblasts, endothelial cells, lung epithelial cells, mast cells, and keratinocytes.[29] A number of factors have been identified as stimulators for the release of IL-8 including lipopolysaccharide (LPS), tumor necrosis factor-α (TNF-α), granulocyte-macrophage colony stimulating factor (GM-CSF), IL-1, IL-2, and IL-3.[29,30]

In vivo, treatment of animals with IL-8 induces instant neutropenia followed by granulocytosis, neutrophil migration and infiltration, plasma exudation, and angiogenesis.[29] The initial neutropenia is due to sequestration of neutrophils into the lungs.[29,30] The $t_{1/2}$ of free plasma is about 9.9 ± 2.2 min.[30] IL-8 is a critical regulator of neutrophil influx and activation of inflammatory process but also exerts a pro-angiogenic effect.[31]

IL-8 mRNA is not expressed by normally perfused myocardium. Highest levels of induction are present in segments demonstrating the lowest levels of blood flow. However, further decrease in the blood flow result in less intense expression of IL-8 mRNA.[31] Itescu et al. demonstrated the type of cells that express IL-8.[32] IL-8 protein was selectively expressed by capillaries and arterioles within the infarct zone but not the vascular structures, cardiomyocytes or infiltrating rat mononuclear cells in the non-ischemic myocardium.

The response of bone marrow cellular elements to IL-8 has been studied with different types of cells. Myeloid progenitor cells have demonstrated a rapid and dramatic mobilization into the peripheral blood following a single injection of IL-8. This response was specific to IL-8 as it was blocked by treatment with polyclonal anti-IL8 antibody.[29]

In patients treated with G-CSF for allogeneic peripheral blood stem cell (PBSC) transplantation, the number of circulating CD34+ and CFU-GM correlated with the level of endogenously produced IL-8. Based on this study, it was suggested that IL-8 levels trigger the migration of progenitor cells from the BM to the circulation.[28] The role of IL-8 does seem to stop at stem cell mobilization, but also function as a chemotactic agent to ischemic myocardium. In vivo injection of IL-8 into rat's myocardium results in 2.3-fold increase in myocardial infiltration by CD34+ cells that were mobilized by G-CSF. However, this did not result in myocardial neovascu-larization, suggesting that differentiation of angioblasts to mature endothelial cells and induction of neovascularization in the heart requires other additional factors that are also produced in ischemia. Researchers then induced ischemia in rats and measured the amount of CD34+ trafficking in the heart. Then antibodies against CXCR1and CXCR2 (receptors specific for IL-8) were administered. Trafficking of CD34+ cells to the heart was reduced by 40%–60%. There was also a 25%–50% reduction in the amount of new capillaries formed.[32]

As with other chemokines, IL-8 also exhibits a dose-dependent response.[28,30,33] The response of neutrophils to increasing doses of IL-8 *in vitro* was almost linear with the percentage of responding neutrophils directly proportional to IL-8 concentration. There was also an increased adhesion of neutrophils to cardiac myocytes.[33] The number of circulating progenitor cells in the circulation is directly proportional to the levels of endogenous IL-8 levels. Patients with low levels of IL-8 after G-CSF induction mobilized cells poorly.[29] Also, there is a considerable individual variation in response to IL-8, ranging from ten to 100-fold increment to the same dose of IL-8.[29]

In a mouse model, infarcted myocardium demonstrated an increase in IL-8 mRNA and protein expression. The rise was seven folds and reached maximal levels at six to 12 hours after production of ischemia. This corresponded with a rise in serum levels of five folds that lasted for 48 hours.[32] The rise in ischemic myocardial expression of IL-8 had also been proven in another study. However, this study failed to demonstrate a significant rise in non-reperfused hearts. The maximum elevation of IL-8 mRNA in ischemic myocardial segments was demonstrated in reperfused hearts at three hours after reperfusion and persisted for 24 hours.[33] This study corre-lates well with a clinical study on patients with acute myocardial infarction (AMI). Patients with AMI had higher baseline serum IL-8 levels compared to patients with stable angina. However, after stenting, a more pronounced rise in the level of IL-8 was found in patients with AMI than in those with stable angina.[31]

Other than ischemia, endogenous production of IL-8 following stem cell mobi-lization with G-CSF has been looked into. Levels of IL-8 secretion increased signif-icantly on day 6 following G-CSF treatment and correlated with levels of circulating CD34+ cells.[31]

CD34+ cells that migrate to the heart following IL-8 chemotaxis not only con-tributed to new vessel formation but also exhibited the ability to suppress apoptosis of cardiomyocytes in the peri-infarct region.[32]

IL-8 enhances the adhesive potential of progenitor cells. It not only enhances the amount of mobilized cells but also their homing capacity within the myocardium and their potential regenerative capacity.[31]

The precise mechanisms responsible for stem cell mobilization from the BM remain unclear. In IL-8-induced mobilization, the enzyme matrix metalloproteinase-9 (MMP-9) has proven to be important.[34,35] MMP-9 degrades collagen type IV.[35] In rhesus monkeys injected with IL-8, there was a dramatic rise in the plasma levels of MMP-9 concomitant with the increase in the circulating progenitor cells.[32,34] Pretreatment with anti-MMP-9 antibody completely blocked the IL-8-induced mobilization.[34] There is also evidence for the importance of neutrophils in the mobilization of progenitor cells. IL-8-induced mobilization of progenitor cells was significantly reduced in neutropenic animals and recovered simultaneously with the recurrence of circulating polymorphonuclear cells (PMNs).[35] These results show that circulating PMNs are essential mediators of IL-8-induced stem cell mobilization.

Overall, IL-8 exhibits a strong potential for early and rapid mobilization of progenitor cells from the bone marrow. It also has a dose-dependent chemotaxis effect to progenitor cells for preferentially homing to the ischemic myocardium.

Tumor necrosis factor-alpha (TNF-α)

TNF-α is a multifunctional cytokine. When produced in low concentrations, TNF-α acts in an autocrine/paracrine manner and is thought to play an important role in regional tissue hemostasis as well as the regulation of local host defense responses. When produced in high concentrates, it produces devastating effects including cachexia, microvascular coagulation, and lethal hemodynamic collapse.[36,37] The activities of TNF-α are mediated through two functionally distinct receptors — TNF-RI and TNF-RII.[38] The extracellular domains of these two receptors are similar but the difference lies in the intracellular domains.[39]

TNF-α is produced by a gene located on chromosome 6p21.3. TNF-α has been shown to have an influence on the migration of a number of cells including neutrophils, vascular smooth muscle cells, Langerhan cells, dendritic cells, fibroblasts and stem cells.[38]

Cardiomyocytes are one of the cells in the body that have shown the ability to produce TNF-α under certain forms of stress. Stress factors that have been identified as stimulators of TNF-α expression include endotoxins, hemodynamic pressure overload, thermal injury and ischemia.[37,40–42] When freshly isolated hearts were stimulated with endotoxin *in vitro*, *de novo* TNF-α mRNA expression occurred within 30 minutes, and TNF-α protein production was detected within 60–75 minutes.[41] This was also produced *in vivo* using a guinea pig model where TNF-α was detected to be produced by cardiomyocytes following stimulation with endotoxin.[37] The onset and offset of TNF-α expression is rapid, with levels rising within 60 minutes and being able to return to baseline within 90 minutes of the effecting stress.[36] Patients with severe congestive heart failure, myocardial infarction, myocarditis, cardiomyopathy, cardiac transplant rejection and after cardiopulmonary bypass have elevated levels of TNF-α in plasma.[36]

In infarcted regions of rat myocardium, TNF-α gene expression peaks one week after infarction in rats. TNF-α mRNA is expressed by myocytes from one day to five weeks after MI.[38,42] The serum concentrations of TNF-α correlate with the cardiac expression, with serum levels of TNF-α peaking at seven days. TNF-α expression in the heart was maximal in the border zone.[38]

Mice over-expressing TNF-α develop a clinical syndrome consisting of decreased activity, tachypnea, hunched posture, poor grooming and death within days of onset of illness. The hearts of these mice exhibited severe impairment of cardiac function with marked ventricular dilatation and depressed ejection fraction.[37,41,42] The time to maximum cardiac dysfunction in rats is seven days, which correlates with peak blood levels of TNF-α.[42] The frequency of arrhythmia, poor hemodynamic parameters and infarct size were significantly reduced in TNF-α knockout mice compared with wild type mice.[43] These depressive effects of TNF-α on the myocardium have been blocked in animals by antibodies to TNF-RII.[42] However, despite being proven safe in phase I clinical trials, phase II trials of anti-TNF-α antibodies in humans have failed to produce any clinical improvement of CHF or decrease in mortality.[44]

The evidence for TNF-α as a mobilizing and chemotactic agent is not overwhelming as with others described earlier. Xiao *et al.* transfected TNF-α cDNA into rat cardiomyocytes and studied *in vitro* the effect of over-expression of TNF-α on the migration of embryonic stem cells (ESCs). They found that over-expression of TNF-α significantly enhanced migration of ESCs. The number of ESCs migrated were four folds higher than the control. This effect of TNF-α over-expression was attenuated with antibodies against Type II TNF-α receptors but not Type I.[38] In a study performed on patients with congestive heart failure (CHF), CD34+ cells were significantly decreased in patients with class IV and III CHF compared to those with class II and I.[45] The number of CD34+ cells correlated inversely with the level of serum TNF-α. Patients with advanced heart failure have higher serum levels of TNF-α. This may be due to the suppressive effect of TNF-α on the bone marrow which counteracts and overwhelms the triggers able to increase the numbers of circulating CD34+ cells.[45]

Granulocyte colony-stimulating factor (G-CSF) and stem cell factor (SCF)

G-CSF is a glycoprotein cytokine that is produced by a variety of cells in the body including endothelium, macrophages, fibroblasts, smooth muscle cells, mononuclear cells and other immune cells.[46] The gene for G-CSF is located on chromosome 17, locus q11.2–q12. In hematological diseases, G-CSF is a well-established stem-cell mobilizer for peripheral blood stem cell (PBSC) transplantation.

Stem cell factor (SCF), also known as c-kit ligand or steel factor, is a potent chemoattractant that stimulates directional motility of mucosal- and connective tissue-type mast cells. Several studies have looked into the expression of CSF and SCF constitutively and under stimulation.[46–49]

G-CSF and granulocyte-macrophage colony-stimulating factor (GM-CSF) were not detected in the non-infarcted myocardium. The expression of M-CSF was down-regulated by 38% in non-perfused MI after three days and persisted up to seven days.[46]

Constitutive SCF mRNA expression in the heart was noted.[47,49] The fate of SCF levels post-ischemic induction is controversial. Woldbaek *et al.* showed a 35% down-regulation of SCF after three days that still persisted at seven days following irreversible MI in mice.[47] However, Frangogiannis *et al.*, using a canine model of myocardial ischemia and reperfusion, demonstrated significant up-regulation of SCF mRNA after reperfusion intervals of 72–120 hours. No significant induction of SCF mRNA was noted with shorter reperfusion intervals (three to 48 hours). Further studies to identify the source of SCF using immunohistochemistry staining showed that it was localized predominantly in a small subset of macrophages infiltrating the healing myocardium. No SCF was found in endothelial cells or myofibroblasts. This may explain why SCF expression is down-regulated in ischemic but not reperfused heart, as the macrophages need to infiltrate the heart before there is expression of SCF. However, the expression of SCF can be suppressed by TNF-α, a cytokine that is known to be up-regulated in ischemia.[49]

CSF and SCF are used extensively by hematologists to mobilize bone marrow stem cells. In a study on non-human primate, administration of SCF and G-CSF resulted in an increase of circulating $CD34^+$ cells from day 3 and peaked between day 8 and day 12.[50] Administration of G-CSF following MI, resulted in an influx of inflammatory cells to the myocardium including WBCs, granulocytes, monocytes and mononuclear cells. This resulted in improvement in cardiac function and decrease in scar size in G-CSF treated group than in the control. In a mice model of MI, marked BM-derived cells were identified in the infarct border area of G-CSF- and SCF-treated groups either individually or in combination.[51] In another study, G-CSF treatment increased the number of bone marrow derived cells within the infarct zone by 12 folds compared to the infarcted control that did not receive G-CSF/SCF. Despite the increase in BM cells in the infarct zone, this study did not show any significant regeneration of new blood vessels or cardiomyocytes within the infarct zone.[48]

SCF over-expression following reperfusion has resulted in influx of mast cells into the myocardium starting at 72 hours after reperfusion and showing maximum accumulation in areas of collagen deposition.[49]

In clinical application of these cytokines, bone marrow stem cells were mobilized with G-CSF in patients with acute and chronic ischemia. The majority of mobilized cells were $CD34^+$ but mesenchymal stem cells (CD34-CD45-) were also mobilized.[52]

G-CSF and SCF mobilization of bone marrow stem cells produced positive effect in some studies but not in others. In a non-human primate model, mobilized stem cells produced an increase in the vascular density in the infarcted hearts treated with G-CSF and SCF but this did not translate into improvement in the myocardial function or attenuation of left ventricular dilatation. This maybe due to lack of evidence for the differentiation of mobilized cells to cardiomyocytes as there was no expression of connexin-43 detected. The drawback of this study is that they did not rule out the possibility that the regeneration of vascular structures could come from resident cardiac stem cells or the influence locally produced growth factors.[50] These findings were reproduced in another study on rats that also received G-CSF. However, in this study, treatment of the infarcted rat with skeletal myoblasts produced significant improvement in function, indicating that an additional factor is probably necessary for mobilized stem cells to engraft in the ischemic heart.[17] G-CSF alone maybe not sufficient to produce any improvement and may require additional cytokines such as VEGF to produce improvement in vasculogenisis.[53]

However, other studies proved the benefit of G-CSF/SCF treatment after MI. Animals treated with G-CSF/SCF individually or in combination demonstrated more capillary density, less apoptosis and better functional improvement compared with control.[51] Others demonstrated an increase in capillary density but no associated functional improvement.[49]

Clinically, patients that received an intracoronary infusion of G-CSF mobilized stem cells following acute ischemia were noted to have improvements in LV ejection fraction and a better myocardium perfusion on dobutamine infusion scans at six months follow-up.[54] However, there was a high rate of in-stent stenosis at the culprit lesion in G-CSF treated patients.

Adhesion molecules

We believe that the signaling factors described earlier are mainly involved in the mobilization of MSCs from the BM into the circulation and initiating their homing. We do not think that these signals released by the ischemic heart are the only mechanism to attract the MSCs to the site of cardiac injury. For if the MSCs are to be attracted precisely to the target site and engraft, there needs to be a concentration gradient of the signals in the blood stream from the heart to the bone marrow in order to guide the migration of these cells. In a rapidly circulating systemic blood flow, such a concentration gradient cannot be maintained. Thus, we support the notion that some of these signaling factors act locally to express adhesion molecules on the endothelial surface of micro-vessels at the site of myocardial ischemia, so that the MSCs mobilized from the marrow into the circulating blood can be trapped. In a process akin to that for leucocytes targeting infected sites in the body, the captured MSCs then encounter chemokines, which eventually activate integrins, resulting in firm arrest and then transendothelial migration into the extracellular space. This adhesion of MSCs to the site of interest is the first step in homing of those cells to the acutely damaged heart. In this section, we will review the evidence for the adhesion of MSCs to cardiac microvascular endothelium in the homing into the ischemic myocardium.

The interaction of hematopoietic stem cells (HSCs) with BM endothelium has been extensively studied. However, very few studies have looked into the interaction of MSCs outside the bone marrow endothelium. We will review a few studies that looked into the transmigration across the bone marrow endothelial cells (BMECs).

The aim is to establish a better understanding of this process before moving on to discuss transmigration across endothelial cells in the heart.

Studies have suggested the roles of vascular adhesion molecule-1 (VCAM-1).[55] and P-selectin[56] in the process of adhesion of HSCs to the bone marrow endothelial cells (BMECs). Anti-VCAM-1 antibodies almost completely inhibit the transendothelial migration of HSCs. Intracellular adhesion molecule-1 (ICAM-1) seems not to have the same importance on the adhesion and migration process.[57]

In the bone marrow, stem cells have been shown to first roll on endothelial E- and P-selectin. This is followed with activation of the stem cells by SDF-1 that is secreted from endothelial cells and triggers lymphocyte function associated antigen-1 (LFA-1)/ICAM-1 and very late antigen-4 (VLA-4)/VCAM-1 interactions to support firm adhesion to BMECs. HSCs exhibit a dose-dependent response to SDF-1 in their transendothelial BMECs migration. BMECs express SCF-1 mRNA and the migration is significantly inhibited by antibodies to SDF-1.[57,58] Cells that do not express sufficient CXCR4 will be detached from the endothelial walls by the shear stress. The stem cells adhering to the endothelium will, in response to SDF-1, extravasate and migrate through the underlying basal lamina in the extracellular membrane using VLA-4 and VLA-5 integrin receptors.[55]

SDF-1 has a major role in the mobilization and homing of MSCs. It also has been identified as a major player in the adhesion of stem cells in vascular endothelium. Its role has been studied in CD34$^+$ cells, and since MSCs express SDF-1 receptor CXCR4, we are extrapolating here the findings to CD 34^{--}MSCs. In the absence of selectins, SDF-1 promoted VLA-4 mediated tethering and firm adhesion to VCAM-1 under shear flow.[59] It also exhibited shear resistance adhesion to LFA-1 and very late antigen-5 (VLA-5) which was absent in the absence of SDF-1.[55] In hematopoietic progenitors, in the presence of surface-bound SDF-1, rolling on endothelial selectin rapidly developed firm adhesion to the endothelial surface mediated by an interaction between ICAM-1 and its integrin ligand.[55,59] The role of ICAM-1 in MSCs adhesion is very limited, nevertheless the role of SDF-1 may extend to this adhesion molecule as well.

In a study that looked into the effect of different cytokines in the adhesion of MSCs to CMVE in vitro, TNF-α and IL-1β increased adhesion. However, IL-3, IL-6, SCF, and SDF-1 did not affect the adhesion efficiency. This was confirmed in vivo where hearts injected with TNF-α treated cells contained three times more cells.[7,60] TNF-α and IL1β robustly induce VCAM-1 mRNA expression level. The expression of VLA-4 did not increase with these cytokines.[60] Interestingly, the expression of VCAM-1 is higher in the heart compared to the lung.[61]

The adhesion of MSCs to CMVE may be through VCAM-1 and ICAM-1 adhesion molecules. Pretreatment of CMVE with antibodies to VACM-1 or pre-treatment of MSCs to anti-VLA-4 partially abolishes the adhesion of MSCs to CMVE. This indicates that other integrins are used by MSCs to adhere to endothelial cells. However, blocking antibodies to ICAM-1 decreases but does not completely abolish the adhesion.[7,62]

The adhesion of MSCs to cardiac microvascular endothelium (CMVE) is dependent on the period that the cells are allowed to interact and the flow velocity.[7,60] The percentage adhesion increases with time and decreases exponentially with increasing velocity. The site for adhesion of the cells was in the capillaries and venules. No cells were observed in large vessels.[60] We could explain this by the fact that adhesion of MSCs to CMVE depends on low shear forces and in the large vessels the forces are high.

MSCs bind to endothelial cells in a P-selectin-dependent manner despite the fact that MSCs do not express P-selectin glycoprotein ligand-1 (PSGL-1). Pre-treatment with anti-P-selectin function-blocking antibody decreases MSCs binding to endothelial cells by 50%. MSCs in P-selectin$^{-/-}$ mice show significant reduction in the numbers binding to endothelial cells compared to wild type mice. It could be assumed that MSCs possess one or more functional P-selectin ligands which are different from PSGL-1 or CD 24.[7]

Summary and Conclusions

As is clear from the evidence put forward in this chapter, homing is an innate nature of the multipotent MSCs. Upon intravenous (IV) delivery, MSCs home mainly to the BM. Other hematopoietic organs, such as the liver and spleen, are also homing destinations for IV delivered MSCs.

In addition, MSCs are also recruited to organs of injury where they are needed for repair of the damaged tissue given their multipotent differentiation potential. There is sufficient evidence for MSCs homing to ischemic cerebral tissue, fractured bones and injured liver. In this chapter, we have presented evidence of the homing capability of MSCs to the acutely ischemic heart.

Different signaling cytokines reviewed in this chapter appear to be involved in the mobilization and homing process of MSCs from the BM to the ischemic myocardium. Probably the most important is SDF-1. SDF-1 has been found to surge following ischemia that correlated closely with homing of MSCs to the heart. The response of MSCs to SDF-1 has proven to be dose dependent. SDF-1 is also important locally as it has been shown to be a major factor in inducing adhesion molecules and also in local chemotaxis in the ischemic heart.

IL-8, which is only expressed by the heart following ischemia and reperfusion, is also an important signaling factor in the homing process. IL-8 dramatically mobilizes MSCs from the BM after its surge following reperfusion. This occurs early and rapid in the homing process. MSCs also show a dose-dependent chemotaxis to IL-8. The levels of secreted IL-8 correlate proportionally with the degree of ischemia but more importantly to the reperfusion process.

TNF-α is a factor that is also produced after myocardial ischemia. However, it has been proven that if over-expressed, it leads to a heart failure phenotype. It produced a state of chronic heart failure and maybe responsible for the BM suppression that is associated with this condition. The role of TNF-α is mostly local as it plays an important part in the regulation of adhesion molecules in CMVE.

SCF is another factor produced by the heart following reperfusion. It has a main role as a chemo-attractant agent for MSCs and other inflammatory cells into the heart. On the other hand, G-CSF is a very potent mobilizing agent from the BM. It is one of the very few agents that have reached clinical application in stem cell therapy of the ischemic and failing hearts. Some of the initial results were promising but still remain controversial at this time.

The VLA-4/VCAM-1 axis is probably the most important local adhesion and migration factor. This is strongly regulated by SDF-1 and has almost consistently been proven in all studies that it is the main adhesion molecule in MSCs homing to the heart. Other adhesion molecules that were reported to be important but not consistently so in all studies are ICAM-1 and P-selectin.

Overall, homing of MSCs from the BM to areas of ischemic myocardium is a process that occurs naturally. However, the signaling factors that regulate this process are probably short-lived and too weak to produce clinically significant effect. Thus, further studies are required to understand and augment this process clinically in order to produce benefits to the patients.

References

1. Devine SM, Bartholomew AM, Mahmud N, Nelson M, Patil S, Hardy W, et al. Mesenchymal stem cells are capable of homing to the bone marrow of non-human primates following systemic infusion. *Exp Hematol* 2001;29(2):244–255.
2. Rombouts WJ, Ploemacher RE. Primary murine MSC show highly efficient homing to the bone marrow but lose homing ability following culture. *Leukemia* 2003;17(1):160–170.
3. Nilsson SK, Johnston HM, Coverdale JA. Spatial localization of transplanted hemopoietic stem cells: inferences for the localization of stem cell niches. *Blood* 2001;97(8):2293–2299.
4. Gao J, Dennis JE, Muzic RF, Lundberg M, Caplan AI. The dynamic *in vivo* distribution of bone marrow-derived mesenchymal stem cells after infusion. *Cells Tissues Organs* 2001;169(1):12–20.
5. Papayannopoulou T, Craddock C, Nakamoto B, Priestley GV, Wolf NS. The VLA4/VCAM-1 adhesion pathway defines contrasting mechanisms of lodgement of transplanted murine hemopoietic progenitors between bone marrow and spleen. *Proc Natl Acad Sci USA* 1995; 92(21):9647–9651.
6. Kollet O, Spiegel A, Peled A, Petit I, Byk T, Hershkoviz R, et al. Rapid and efficient homing of human CD34(+)CD38(-/low)CXCR4(+) stem and progenitor cells to the bone marrow and spleen of NOD/SCID and NOD/SCID/B2m(null) mice. *Blood* 2001;97(10):3283–3291.
7. Ruster B, Gottig S, Ludwig RJ, Bistrian R, Muller S, Seifried E, et al. Mesenchymal stem cells display coordinated rolling and adhesion behavior on endothelial cells. *Blood* 2006; 108(12):3938–3944.
8. Bittira B, Shum-Tim D, Al-Khaldi A, Chiu RC. Mobilization and homing of bone marrow stromal cells in myocardial infarction. *Eur J Cardiothorac Surg* 2003;24(3):393–398.
9. Saito T, Kuang JQ, Bittira B, Al-Khaldi A, Chiu RC. Xenotransplant cardiac chimera: immune tolerance of adult stem cells. *Ann Thorac Surg* 2002;74(1):19–24.

10. Jiang W, Ma A, Wang T, Han K, Liu Y, Zhang Y, *et al.* Homing and differentiation of mesenchymal stem cells delivered intravenously to ischemic myocardium *in vivo*: a time-series study. *Pflugers Arch* 2006;453(1):43–52.

11. Yamaguchi J, Kusano KF, Masuo O, Kawamoto A, Silver M, Murasawa S, *et al.* Stromal cell-derived factor-1 effects on *ex vivo* expanded endothelial progenitor cell recruitment for ischemic neovascularization. *Circulation* 2003;107(9):1322–1328.

12. Pillarisetti K, Gupta SK. Cloning and relative expression analysis of rat stromal cell derived factor-1 (SDF-1)1: SDF-1 alpha mRNA is selectively induced in rat model of myocardial infarction. *Inflammation* 2001;25(5):293–300.

13. Lapidot T, Kollet O. The essential roles of the chemokine SDF-1 and its receptor CXCR4 in human stem cell homing and repopulation of transplanted immune-deficient NOD/SCID and NOD/SCID/B2m(null) mice. *Leukemia* 2002;16(10):1992–2003.

14. Kahn J, Byk T, Jansson-Sjostrand L, Petit I, Shivtiel S, Nagler A, *et al.* Overexpression of CXCR4 on human CD34+ progenitors increases their proliferation, migration, and NOD/SCID repopulation. *Blood* 2004;103(8):2942–2949.

15. Forster R, Kremmer E, Schubel A, Breitfeld D, Kleinschmidt A, Nerl C, *et al.* Intracellular and surface expression of the HIV-1 coreceptor CXCR4/fusin on various leukocyte subsets: rapid internalization and recycling upon activation. *J Immunol* 1998;160(3):1522–1531.

16. Hattori K, Heissig B, Tashiro K, Honjo T, Tateno M, Shieh JH, *et al.* Plasma elevation of stromal cell-derived factor-1 induces mobilization of mature and immature hematopoietic progenitor and stem cells. *Blood* 2001;97(11):3354–3360.

17. Askari AT, Unzek S, Popovic ZB, Goldman CK, Forudi F, Kiedrowski M, *et al.* Effect of stromal-cell-derived factor 1 on stem-cell homing and tissue regeneration in ischaemic cardiomyopathy. *Lancet* 2003;362(9385):697–703.

18. Levesque JP, Hendy J, Takamatsu Y, Simmons PJ, Bendall LJ. Disruption of the CXCR4/CXCL12 chemotactic interaction during hematopoietic stem cell mobilization induced by GCSF or cyclophosphamide. *J Clin Invest* 2003;111(2):187–196.

19. Petit I, Szyper-Kravitz M, Nagler A, Lahav M, Peled A, Habler L, *et al.* G-CSF induces stem cell mobilization by decreasing bone marrow SDF-1 and up-regulating CXCR4. *Nat Immunol* 2002;3(7):687–694.

20. Kucia M, Dawn B, Hunt G, Guo Y, Wysoczynski M, Majka M, *et al.* Cells expressing early cardiac markers reside in the bone marrow and are mobilized into the peripheral blood after myocardial infarction. *Cir Res* 2004;95(12):1191–1199.

21. Abbott JD, Huang Y, Liu D, Hickey R, Krause DS, Giordano FJ. Stromal cell-derived factor-1alpha plays a critical role in stem cell recruitment to the heart after myocardial infarction but is not sufficient to induce homing in the absence of injury. *Circulation* 2004;110(21):3300–3305.

22. Kollet O, Shivtiel S, Chen YQ, Suriawinata J, Thung SN, Dabeva MD, *et al.* HGF, SDF-1, and MMP-9 are involved in stress-induced human CD34+ stem cell recruitment to the liver. *J Clin Invest* 2003;112(2):160–169.

23. Schober A, Knarren S, Lietz M, Lin EA, Weber C. Crucial role of stromal cell-derived factor-1alpha in neointima formation after vascular injury in apolipoprotein E-deficient mice. *Circulation* 2003;108(20):2491–2497.

24. Heeschen C, Lehmann R, Honold J, Assmus B, Aicher A, Walter DH, *et al.* Profoundly reduced neovascularization capacity of bone marrow mononuclear cells derived from patients with chronic ischemic heart disease. *Circulation* 2004;109(13):1615–1622.

25. Wang Y, Haider H, Ahmad N, Zhang D, Ashraf M. Evidence for ischemia induced host-derived bone marrow cell mobilization into cardiac allografts. *J Mol Cell Cardiol* 2006;41(3):478–487.

26. Heissig B, Hattori K, Dias S, Friedrich M, Ferris B, Hackett NR, *et al.* Recruitment of stem and progenitor cells from the bone marrow niche requires MMP-9 mediated release of kit-ligand. *Cell* 2002;109(5):625–637.

27. Tachibana K, Hirota S, Iizasa H, Yoshida H, Kawabata K, Kataoka Y, *et al.* The chemokine receptor CXCR4 is essential for vascularization of the gastrointestinal tract. *Nature* 1998;393(6685): 591–594.

28. Watanabe T, Kawano Y, Kanamaru S, Onishi T, Kaneko S, Wakata Y, *et al.* Endogenous interleukin-8 (IL-8) surge in granulocyte colony-stimulating factor-induced peripheral blood stem cell mobilization. *Blood* 1999;93(4):1157–1163.

29. Laterveer L, Lindley IJ, Hamilton MS, Willemze R, Fibbe WE. Interleukin-8 induces rapid mobilization of hematopoietic stem cells with radioprotective capacity and long-term myelolymphoid repopulating ability. *Blood* 1995;85(8):2269–2275.

30. Laterveer L, Lindley IJ, Heemskerk DP, Camps JA, Pauwels EK, Willemze R, *et al.* Rapid mobilization of hematopoietic progenitor cells in rhesus monkeys by a single intravenous injection of interleukin-8. *Blood* 1996;87(2):781–788.

31. Schomig K, Busch G, Steppich B, Sepp D, Kaufmann J, Stein A, *et al.* Interleukin-8 is associated with circulating CD133+ progenitor cells in acute myocardial infarction. *Eur Heart J* 2006;27(9):1032–1037.

32. Kocher AA, Schuster MD, Bonaros N, Lietz K, Xiang G, Martens TP, *et al.* Myocardial homing and neovascularization by human bone marrow angioblasts is regulated by IL-8/Gro CXC chemokines. *J Mol Cell Cardiol* 2006;(4):455–464.

33. Kukielka GL, Smith CW, LaRosa GJ, Manning AM, Mendoza LH, Daly TJ, *et al.* Interleukin-8 gene induction in the myocardium after ischemia and reperfusion *in vivo*. *J Clin Invest* 1995; 95(1):89–103.

34. Pruijt JF, Fibbe WE, Laterveer L, Pieters RA, Lindley IJ, Paemen L, *et al.* Prevention of interleukin-8-induced mobilization of hematopoietic progenitor cells in rhesus monkeys by inhibitory antibodies against the metalloproteinase gelatinase B (MMP-9). *Proc Natl Acad Sci USA* 1999;96(19):10863–10868.

35. Kohtani T, Abe Y, Sato M, Miyauchi K, Kawachi K. Protective effects of anti-neutrophil antibody against myocardial ischemia/reperfusion injury in rats. *Eur Surg Res* 2002;34(4):313–320.

36. Nakano M, Knowlton AA, Dibbs Z, Mann DL. Tumor necrosis factor-alpha confers resistance to hypoxic injury in the adult mammalian cardiac myocyte. *Circulation* 1998;97(14):1392–1400.

37. Bryant D, Becker L, Richardson J, Shelton J, Franco F, Peshock R, *et al.* Cardiac failure in transgenic mice with myocardial expression of tumor necrosis factor-alpha. *Circulation* 1998;97(14):1375–1381.

38. Chen Y, Ke Q, Yang Y, Rana JS, Tang J, Morgan JP, *et al.* Cardiomyocytes overexpressing TNF-alpha attract migration of embryonic stem cells via activation of p38 and c-Jun amino-terminal kinase. *FASEB J* 2003;17(15):2231–2239.

39. Zhang Y, Harada A, Bluethmann H, Wang JB, Nakao S, Mukaida N, *et al.* Tumor necrosis factor (TNF) is a physiologic regulator of hematopoietic progenitor cells: increase of early hematopoietic progenitor cells in TNF receptor p55-deficient mice *in vivo* and potent inhibition of progenitor cell proliferation by TNF alpha *in vitro*. *Blood* 1995;86(8):2930–2937.

40. Kapadia S, Lee J, Torre-Amione G, Birdsall HH, Ma TS, Mann DL. Tumor necrosis factor-alpha gene and protein expression in adult feline myocardium after endotoxin administration. *J Clin Invest* 1995;96(2):1042–1052.

41. Sivasubramanian N, Coker ML, Kurrelmeyer KM, MacLellan WR, DeMayo FJ, Spinale FG, *et al.* Left ventricular remodeling in transgenic mice with cardiac restricted overexpression of tumor necrosis factor. *Circulation* 2001;104(7):826–831.

42. Berthonneche C, Sulpice T, Boucher F, Gouraud L, de Leiris J, O'Connor SE, *et al.* New insights into the pathological role of TNF-alpha in early cardiac dysfunction and subsequent heart failure after infarction. *Am J Physiol* 2004;287(1):H340–H350.

43. Maekawa N, Wada H, Kanda T, Niwa T, Yamada Y, Saito K, *et al.* Improved myocardial ischemia/reperfusion injury in mice lacking tumor necrosis factor-alpha. *J Am Coll Cardiol* 2002;39(7):1229–1235.

44. Mann DL, McMurray JJ, Packer M, Swedberg K, Borer JS, Colucci WS, *et al.* Targeted anti-cytokine therapy in patients with chronic heart failure: results of the Randomized Etanercept Worldwide Evaluation (RENEWAL). *Circulation* 2004;109(13):1594–1602.

45. Valgimigli M, Rigolin GM, Fucili A, Porta MD, Soukhomovskaia O, Malagutti P, *et al.* CD34+ and endothelial progenitor cells in patients with various degrees of congestive heart failure. *Circulation* 2004;110(10):1209–1212.

46. Woldbaek PR, Hoen IB, Christensen G, Tonnessen T. Gene expression of colony-stimulating factors and stem cell factor after myocardial infarction in the mouse. *Acta Physiol Scand* 2002;175(3):173–181.

47. Orlic D, Kajstura J, Chimenti S, Jakoniuk I, Anderson SM, Li B, *et al.* Bone marrow cells regenerate infarcted myocardium. *Nature* 2001;410(6829):701–705.

48. Kanellakis P, Slater NJ, Du XJ, Bobik A, Curtis DJ. Granulocyte colony-stimulating factor and stem cell factor improve endogenous repair after myocardial infarction. *Cardiovasc Res* 2006;70(1):117–125.

49. Frangogiannis NG, Perrard JL, Mendoza LH, Burns AR, Lindsey ML, Ballantyne CM, *et al.* Stem cell factor induction is associated with mast cell accumulation after canine myocardial ischemia and reperfusion. *Circulation* 1998;98(7):687–698.

50. Norol F, Merlet P, Isnard R, Sebillon P, Bonnet N, Cailliot C, *et al.* Influence of mobilized stem cells on myocardial infarct repair in a nonhuman primate model. *Blood* 2003;102(13):4361–4368.

51. Ohtsuka M, Takano H, Zou Y, Toko H, Akazawa H, Qin Y, *et al.* Cytokine therapy prevents left ventricular remodeling and dysfunction after myocardial infarction through neovascularization. *FASEB J* 2004;18(7):851–853.

52. Kastrup J, Ripa RS, Wang Y, Jorgensen E. Myocardial regeneration induced by granulocyte-colony-stimulating factor mobilization of stem cells in patients with acute or chronic ischaemic heart disease: a non-invasive alternative for clinical stem cell therapy? *Eur Heart J* 2006;27(23):2748–2754.

53. Ripa RS, Wang Y, Jorgensen E, Johnsen HE, Hesse B, Kastrup J. Intramyocardial injection of vascular endothelial growth factor-A165 plasmid followed by granulocyte-colony stimulating factor to induce angiogenesis in patients with severe chronic ischaemic heart disease. *Eur Heart J* 2006;27(15):1785–1792.

54. Kang HJ, Kim HS, Zhang SY, Park KW, Cho HJ, Koo BK, *et al.* Effects of intracoronary infusion of peripheral blood stem-cells mobilised with granulocyte-colony stimulating factor on left ventricular systolic function and restenosis after coronary stenting in myocardial infarction: the MAGIC cell randomised clinical trial. *Lancet* 2004;363(9411):751–756.

55. Peled A, Kollet O, Ponomaryov T, Petit I, Franitza S, Grabovsky V, *et al.* The chemokine SDF-1 activates the integrins LFA-1, VLA-4, and VLA-5 on immature human CD34(+) cells: role in transendothelial/stromal migration and engraftment of NOD/SCID mice. *Blood* 2000;95(11):3289–3296.

56. Frenette PS, Subbarao S, Mazo IB, von Andrian UH, Wagner DD. Endothelial selectins and vascular cell adhesion molecule-1 promote hematopoietic progenitor homing to bone marrow. *Proc Natl Acad Sci USA* 1998;95(24):14423–14428.

57. Imai K, Kobayashi M, Wang J, Ohiro Y, Hamada J, Cho Y, *et al.* Selective transendothelial migration of hematopoietic progenitor cells: a role in homing of progenitor cells. *Blood* 1999;93(1):149–156.

58. Netelenbos T, van den Born J, Kessler FL, Zweegman S, Merle PA, van Oostveen JW, *et al.* Proteoglycans on bone marrow endothelial cells bind and present SDF-1 towards hematopoietic progenitor cells. *Leukemia* 2003;17(1):175–184.

59. Peled A, Grabovsky V, Habler L, Sandbank J, Arenzana-Seisdedos F, Petit I, *et al*. The chemokine SDF-1 stimulates integrin-mediated arrest of CD34(+) cells on vascular endothelium under shear flow. *J Clin Invest* 1999;104(9):1199–1211.
60. Segers VF, Van Riet I, Andries LJ, Lemmens K, Demolder MJ, De Becker AJ, *et al*. Mesenchymal stem cell adhesion to cardiac microvascular endothelium: activators and mechanisms. *Am J Physiol* 2006;290(4):H1370–H1377.
61. Lim YC, Garcia-Cardena G, Allport JR, Zervoglos M, Connolly AJ, Gimbrone MA, Jr., *et al*. Heterogeneity of endothelial cells from different organ sites in T-cell subset recruitment. *Am J Pathol* 2003;162(5):1591–1601.
62. Boyce JA, Mellor EA, Perkins B, Lim YC, Luscinskas FW. Human mast cell progenitors use alpha4-integrin, VCAM-1, and PSGL-1 E-selectin for adhesive interactions with human vascular endothelium under flow conditions. *Blood* 2002;99(8):2890–2896.

7 A Molecular Imaging Perspective on Cardiac Repair with Stem Cells

Kishore Bhakoo

Introduction

Stem cell research is undergoing a significant evolution from being a discipline of the basic sciences to being recognised as a potential component of clinical therapy. Cell transplants to replace cells lost due to injury or degenerative diseases, for which there are currently no cures, are being pursued in a wide range of experimental models.

It is now clear that the migration dynamics of stem cells will determine the extent of tissue regeneration at the site of implantation and surrounding tissue. Methods for monitoring implanted stem cells non-invasively *in vivo* will greatly facilitate the clinical realisation and optimisation of the opportunities for cell-based therapies. Due to the seamless integration into the host parenchyma, and migration over long distances, cell grafts cannot be detected based on their mass morphology. To monitor cell migration and positional fate after transplantation, current methods use either reporter genes or chimeric animals. These methods are cumbersome, involve sacrifice of the animal and removal of tissue for histological procedures, and cannot be translated to human studies.[1,2] Moreover, this approach lacks the temporal analysis of the donor cells, so, in practices its uses are limited. The monitoring of cellular grafts, non-invasively, is an important aspect of the ongoing safety assessment of cell-based therapies. Molecular imaging is potentially well suited for such an application. For transplanted stem cells to be visualised and tracked by imaging technologies, they need to be tagged so that they are 'visible'. Moreover, imaging and biosensor technologies are moving from the diagnostic arena towards therapeutic and interventional roles.

The ideal imaging modality should provide integrated information relating to the entire process of cell engraftment, survival, and functional outcome. The imaging technologies of MRI, radionuclide and optical imaging are emerging as key modalities for *in vivo* molecular imaging because of their ability to detect molecular events. However, in order to exploit sensitivity, specificity, temporal and spatial resolutions offered by these modalities they require an interaction with respective contrast agents. Effective use of the tools of molecular imaging requires knowledge at several levels including the basis of detection of the imaging modality, mechanism of contrast agent interactions and the biological environment.[3]

Radionuclide Imaging

Nuclear medicine is the main form of molecular and cellular imaging in current clinical use. Because of their exquisite (10^{-11} to 10^{-12} mol/l) sensitivity, PET and SPECT imaging modalities are able to detect tracer quantity of radioisotopes for studying biological processes in living subjects. Technological developments of both PET and SPECT have led to the implementation of specialised systems for small animal imaging with much greater spatial resolution (1–2 mm),[4–6] which has dramatically advanced the field of *in vivo* cell tracking in animal models.

Radioisotopes with a relatively long half-life have been used to track cells during a period of several hours or even days; for instance the use of In-111 ($T_{1/2} = 2.8$ days) for SPECT and Cu-64 ($T_{1/2} = 12.7$ hours) for PET. The biodistribution of In-111-labelled stem cells after intravenous, intramyocardial, intracoronary, and interstitial retrograde coronary venous delivery has been reported.[7] It was found that most delivered cells were not retained in the heart for each delivery route, which raises a concern that pro-angiogenic or non-intentional effects of these cells may have a negative impact on tissues or organs that are not the target of these cells.

One concern of using radioactive labelling agents is the potential radiation damage to the cell, as demonstrated in a study, in which haematopoietic progenitor cells were labelled with In-111-labelled-oxine and injected into the cavity of the left ventricle heart in a rat model of myocardial infarct; although gamma imaging revealed homing of the progenitor cells to infarcted myocardium, significant impairment of proliferation and function of labelled cells was observed. Moreover, rapid efflux of labels out of cells over the course of time leads to label loss from viable cells.

Optical Imaging

This imaging modality measures the intensity of the emitted light at the surface of the animal with a detector system (generally CCD cameras in a black box). Two distinct contrast mechanisms can be used for molecular and genomic imaging studies, one involving fluorescence and the other bioluminescence. Generally both can be implemented in optical imaging systems. In fluorescence imaging an external light source is used to excite fluorescent molecules inside the subject.[8] Green fluorescent protein (GFP) and its variants are the most commonly used optical reporters. Conversely in bioluminescence there is no need for external light stimulation. In this case, *in vivo* imaging involves introducing reporter genes that encode for enzymes (known as luciferases) that can catalyse a light-producing chemical reaction.[9] However, optical imaging is generally limited to use in small rodents, due to its limited penetration depth.

In spite of the opaque nature of tissue, optical imaging is an increasingly important technique for molecular imaging in small animals. It is also being developed for clinical use, in particular using near infra red (NIR) wavelengths (~650–900 nm), which can theoretically travel 5–15 cm through tissue.[10]

A notable theoretical advantage of optical techniques is the fact that multiple probes with different spectral characteristics can be used for multi-channel imaging, similar to *in vivo* karyotyping.[11]

Optical imaging is increasingly used in animal models of disease. In humans the poor penetration will limit its applications, though the development of NIR intravascular imaging catheters and endoscopes may allow this approach to delve deeper into the body. The other drawback of optical imaging is its resolution, which is good near the surface, but poorer for deeper structures. New approaches such as fluorescence molecular tomography, which allows volumetric reconstruction of the source of fluorescent light, can improve spatial resolution to 1 mm at the surface to 3 mm in the centre of small animals, as well as increasing sensitivity (as low as 200 femtomoles).[10,12]

Hence, attachment of a fluorophore to a stem cell would allow detection of stem cells by histology using conventional microscopes. Both luminescence and fluorescence, have been used for stem cell tracking *in vivo*.[13,14]

Ultrasound

The ultrasound allows real-time imaging with a high resolution ($<50\,\mu$m) in the absence of ionising radiation. Ultrasound imaging relies on the acoustic properties of the tissue causing reflection of a high frequency ultrasound (40 MHz) beam. Contrast agents are being developed that promise to give US a major role in molecular imaging and therapeutics. The main contrast agent used is microbubbles formed from albumin, polymers or lipids and containing perfluorocarbon or air. These bubbles adsorb and scatter the sound waves, and in turn the bubbles can resonate and emit secondary sound waves at differing multiples of the original input frequency. This 'non-linear' property of microbubbles makes them a powerful contrast agent as only the microbubbles are detected at these bubble specific frequencies, making the contrast such that single bubbles can be detected. Targeting is possible by decorating the surface of bubble with targeting molecules such as antibodies or other ligands.

Ultrasound has a number of advantages, including safety and portability. The short half-life and the benign composition of the bubbles allow rapid repeated scans. However, their large size (10–$300\,\mu$m) limits its application to intravascular investigations.

MRI Imaging

MRI provides good anatomical information (resolution $<100\,\mu$m). Conventionally MRI relies on differences in proton density and the local magnetic environment of the hydrogen atoms to detect differences between and within tissues. Images can be enhanced using contrast agents. MR only indirectly detects most contrast agents; they are 'seen' as a result of their affect on water molecules. These characteristics have led to this imaging modality to become the method of choice for following anatomical changes in soft tissue. Furthermore, MRI has the ability to measure physiological parameters such as water diffusion and blood oxygenation. Current developments of contrast agents are based on the use of gadolimium chelates or ultra-small superparamagnetic iron oxide particles (USPIOs) as contrastophores. Despite the inherent reduced sensitivity of MR compared to techniques using radionuclides (SPECT, PET), MRI is more sensitive to cell detection due to the higher concentration of contrastophores.

There are numerous ways of increasing the signal-to-noise ratio in micro-MRI when imaging small animals, and thus achieving near microscopic resolution. These include working at relatively high-magnetic fields (4.7–14 T), using hardware and software customised to the small size of animals of interest, and the relative flexibility of much longer acquisition times during imaging. MRI can provide detailed morphologic and functional information and, therefore, seems ideally suited to integrate efficacy assessments with the capability for cell tracking. Yet studies show that the lowest detectable number of cells is 10^5 with the use of conventional MRI scanners without any sequence modification. This threshold of detection can be lowered using high-field magnets (11.7 T) such that single cells containing a single iron particle can be detected and tracked as well as other such approaches have appeared in recent literature.[15,16]

MRI is now a rapidly evolving molecular and cellular imaging strategy. The era of modern molecular imaging began with the first successful demonstration of a functionalised Gd-complex to visualise gene expression.[17] Since then successful linking of receptor molecules to contrast agents for a broad variety of applications has been published.

For stem cells to be visualised and tracked by MRI, they need to be tagged so that they are 'MR visible'. At present there are two types of MR contrast agent used clinically. These are gadolinium-analogues (e.g. Gd^{3+}-DTPA) or iron oxide nanoparticles. However, these reagents were designed as blood-pool contrast agents and are impermeable to cells. Thus, several approaches have been deployed to enhance cell labelling to allow *in vivo* cell tracking by conjugating MRI contrast agents to a range of ancillary molecules to enhance their uptake. With the growing array of cell labelling techniques, cells tagged with various monocrystalline MR probes have been evaluated both *in vitro* and *in vivo*.[18–20]

The development of methods for monitoring stem cell grafts non-invasively, with sufficiently high sensitivity and specificity to identify and map the fate of transplanted cells, is an important aspect of application and safety assessment of stem cell therapy. MRI methods are potentially well suited for such applications as this produces non-invasive 'images' of opaque tissues or structures inside the body.

Intracellular MRI Contrast Agents

Recent work in the design of MRI contrast agents has opened the possibility of combining the spatial resolution available in MRI for anatomic imaging with the ability to 'tag' cells, and thus enable non-invasive detection and study of cell migration from the site of implantation. *In vivo* monitoring of stem cells after grafting is essential for understanding their migrational dynamics, which is an important aspect in determining the overall therapeutic index in cell therapies. Despite recent advances in both the synthesis of paramagnetic molecules and the basic cell biology of stem cells, methods for achieving effective cell labeling using molecular MR-tags are still in their infancy.

MRI contrast agents are used extensively in the clinic to improve sensitivity/selectivity in order to facilitate diagnosis of an underlying pathology. MRI contrast agents either alter the T1 and/or T2 relaxation time, making the local tissue hyper- or hypo-intense, respectively. The contrast agents used most extensively can be classed as either paramagnetic or superparamagnetic.

Examples of paramagnetic agents are Fe^{3+}, Mn^{2+}, and Gd^{3+}. The effect of the magnetic moment in solution results in a dipolar magnetic interaction between the paramagnetic ion and neighbouring water molecules. Fluctuations in this magnetic interaction causes the decrease in T1/T2 relaxation time.[21] Paramagnetic compounds produce, predominantly, T1 effect, giving a hyperintense region.

The other class of contrast agents is superparamagnetic reagents. These consist of an iron oxide core, typically 4–10 nm in diameter, where several thousand iron atoms are present. A biocompatible polymer surrounds the core to provide steric and/or electrostatic stabilisation. This is required due to the large surface area to volume of the nanoparticles. If no stabilisation is present, the particles spontaneously precipitate out of solution due to colloidal instability. The polymers used to stabilise the iron oxide core are typically polysaccharide based (dextran, starch) but others have also been used, e.g. polyethylene glycol (PEG).

There are two types of superparamagnetic contrast agents, superparamagnetic iron oxide (SPIO) and ultrasmall superparamagnetic iron oxide (USPIO). The difference between the two is that the SPIOs consist of several magnetic cores surrounded by a polymer matrix, whereas USPIOs are individual cores surrounded by a polymer. Superparamagentic contrast agents provide predominantly a T2 effect, but smaller particles have shown to act as a T1 agent.[21] A new class of USPIO has been produced known as cross-linked iron oxide (CLIO), whereby the dextran coat of the USPIO is cross-linked, and then subjected to amination in the presence of ammonia to produce amine terminated nanoparticles suitable for conjugation.[22]

However, none of the contrast agents used in the clinic were designed to go into cells. To cross the cell membrane, contrast agents must either be used in conjunction with a transfection agent or conjugated to a biological entity, such as a peptide translocating domain. One molecule that is emerging as a useful reagent to enhance cellular uptake is the HIV-1 TAT peptide.[23] TAT is capable of translocating exogenous moieties into cells.[24] TAT peptide has been used to modify the surface of paramagnetic nanoparticles.

However, these peptides label cells non-specifically. In order to facilitate the labelling of specific cell types, contrast agents conjugated with antibodies to cell surface antigens have also been deployed with some success.[25] Moreover, it is important that the presence intracellular contrast agent does not affect proliferation, differentiation, integration, or cellular activity.

MRI Tracking of Stem Cells in the Heart

Myocardial infarction is by nature an irreversible injury. The extent of the infarction depends on the duration and severity of the perfusion defect.[26] Beyond contraction and fibrosis of myocardial scar, progressive ventricular remodelling of non-ischaemic myocardium can further reduce cardiac function in the weeks to months after initial event.[27]

Many of the therapies available to clinicians today can significantly improve the prognosis of patients following an acute myocardial infarction.[27,28] However, no pharmacological or interventional procedure used clinically has shown efficacy in replacing myocardial scar with functioning contractile tissue. Cellular agents such as foetal cardiomyocytes,[29] mesenchymal stem cells (MSCs),[30] endothelial progenitor cells,[31] and skeletal myoblasts,[32] or embryonic stem cells[33] have already shown some efficacy to engraft in the infarct, differentiate toward a cardiomyocyte phenotype by expressing cardiac specific proteins, and preserve left ventricular function and inhibit myocardial fibrosis. Moreover, *in vitro* exposure of stem cells, specifically mesenchymal stem cells, to specific signal molecules prior to transplantation into infarcted myocardium allows the pre-differentiation into cardiomyocytes,[34] and may facilitate a successful engraftment.[35]

However, verifying the status of the transplanted stem cells in animal models has been performed with histological analysis. Clinical data on stem cell transplantation is in its infancy and is very limited. But preliminary clinical data have showed that stem cell transplantation for the treatment of ischaemic heart failure is feasible and promising.[36] MR was used in some clinical studies to assess the improvement of contractile function after cell transplantation,[37] but without any possibility to visualise the transplanted stem cells. Therefore, the ability to pre-label stem cells with MRI-visible contrast agents[38] should enable serial tracking of transplanted stem cells, non-invasively by MRI, with high-spatial resolution and allow limited quantification. The programmable nature of the imaging planes allows reproducible and volumetric coverage of the heart. Moreover, this technology scales well with subject size ranging from mouse to human.

The detectability of labelled cells depends on a number of factors, including, e.g. how much iron accumulates inside each cell (labelling efficiency) or the number of cells implanted. Feridex-labelled stem cells were readily detected by MRI[39] when injected directly into myocardium of a swine. For stem cells labelled with MPIO particles, detection of 10^5 cells implanted into a pig heart was reported at a 1.5-T scanner.[40] The detection of a small number of cells or single cells by MRI has been achieved in the rat and mouse brain. In contrast, detection of cells in the heart by MRI is an even greater challenge because of cardiac motion and limited intracellular concentration of imaging probes.

Visualisation of magnetically labelled endothelial progenitor cells transplanted intra-myocardially for therapeutic neovascularisation in infarcted rats has been demonstrated with *ex vivo* MRI at 8.5T on T2-weighted images.[41]

Garot *et al.*[42] have demonstrated the feasibility of *in vivo* MRI tracking of skeletal muscle-derived myogenic precursor cells (MPC) pre-loaded with iron oxide nanoparticles (Endorem®) injected into healthy and infarcted porcine myocardium. Iron-loaded cells in the infarcted region were all detected by T2-weighted spin-echo MRI at 1.5T. In addition, MRI guided catheter has been deployed for the targeted injection of the labelled cells, into the ischaemic myocardium, using Gd-DTPA delayed-enhancement of the site. Furthermore, post-mortem analysis demonstrated the presence of iron-loaded MPC at the centre and periphery of the infarcted tissue as predicted by MRI.

Mesenchymal stem cells (MSCs) derived from bone marrow have been detected and tracked by MRI for up to three weeks.[40,43,44] More recently, Stuckey *et al.*, were able to track iron oxide labelled rat MSCs for 16 weeks after administration into the infarcted rat heart.[45] Allogeneic ferumoxides[43] or iron fluorescent particle[40,44] were given by intramyocardial injection in a pig model of myocardial infarction. A minimum quantity of MSCs per injection was required to be MR-detectable on T2*-[40] or T2-weighted images[43] as hypointense lesions. Indeed, Kraitchman *et al.* have shown that using a limited number of MSC injection per animal, only a part (~70%) of the injections performed in each animal can be visualised.

These promising experiments demonstrate the need for future studies to delineate the fate of injected stem cells by incorporating non-invasive tagging methods to monitor myocardial function following cells engraftment in the myocardial infarction. Consequently, MRI may lead to a better understanding of the myocardial pathophysiology as well as the assessing of proper implantation and the effects of stem cell therapy by allowing a multimodal approach to evaluating anatomy, function, perfusion, and regional contractile parameters in a single non-invasive examination.

Multimodality

The challenge for understanding the very complex phenomena behind the stem cell therapy is multimodality. In fact, no unique technique is capable of addressing all the critical issues and their complementary role has to be fully exploited. So a multidisciplinary approach is needed to validate the therapy.

Most available molecular imaging techniques are applicable not only in animals but also in humans, but some problem arise. Optical techniques are not applicable to humans (and large animals); so an intelligent combination of it with MRI and radionuclide (SPECT or PET) is needed to validate the human technologies and protocols to close existing gaps between basic science and clinical trials of cardiac cell therapy. The goal of imaging is to monitor the stem cells trafficking, homing and fate, viability and differentiation providing also quantifiable information. Imaging-based cell-tracking methods can potentially evaluate the short-term distribution of infused cells or their long-term survival and cardiac differentiation status. To this end we have recently demonstrated the feasibility of using multiple imaging modalities to report on the efficacy of stem cell therapy in an animal model of myocardial infarction.[46] Therefore, these methods would play an indispensable role in detailed preclinical studies to optimise the cell type, delivery methods, and strategies for enhancing cell survival.

Conclusions

Molecular and cellular imaging has enormous potential for transplantation. In the first place, this will largely be as a research tool in both animal and clinical studies — providing early response markers for trials and allowing the development of new therapies. Eventually these applications may translate into more routine clinical applications, allowing monitoring of individual patients. While this review can provide some ideas on the potential, it cannot however, show the difficulties of developing this technology for a particular application. As well as the input of clinicians and biologists there is a need for chemists to help develop the contrast agents, physicists for refining the imaging process itself and mathematicians to improve the post-acquisition data analysis and handling. None of these elements is trivial. However, in the next decade some of the approaches outlined in this review may become essential tools for research and management of transplants.

Safety and ethical issues imposed on a human study would limit the clinical translation of those methods that require extensive manipulation of cells; however, most direct labelling methods have already been applied in patients or would be suitable for the clinic.

These promising experiments demonstrate the need for future studies to delineate the fate of injected stem cells by incorporating non-invasive tagging methods to monitor myocardial function following cells engraftment in the myocardial infarction. Consequently, MRI may lead to a better understanding of the myocardial pathophysiology as well as the assessing of a proper implantation and the effects of stem cell therapy by allowing a multimodal imaging approach to evaluating anatomy, function, perfusion, and regional contractile parameters in a single non-invasive examination.

In conclusion, the use of stem cells therapy to treat cardiac diseases is a realistic possibility in the near future. However, the need for non-invasive imaging techniques is a prerequisite in order to monitor these transplants and to determine clinical efficacy. Examination by MRI ensures that the stem cells are not only injected to the lesion site, but it also allows the monitoring of inappropriate cellular migration, and furthermore identify damage to surrounding tissues.

Methods for monitoring implanted stem cells non-invasively *in vivo* will greatly facilitate the clinical realisation and optimisation of the opportunities for stem cell-based therapies. In writing this short chapter, we have concentrated on the application of MRI to tracking of stem cells in only the repair of lesioned cardiac tissues. There are, however, numerous examples where similar methodologies of cell tracking can aid in clinical diagnosis or can be used to trace other cells types, such as those in the nervous system and from a tumour or following an inflammatory response.[47,48]

References

1. Lagasse E, Connors II, Al Dhalimy M, Reitsma M, Dohse M, Osborne L, *et al.* Purified hematopoietic stem cells can differentiate into hepatocytes *in vivo. Nat Med* 2000;6:1229 1234.
2. Orlic D, Kajstura J, Chimenti S, Jakoniuk I, Anderson SM, Li B, *et al.* Bone marrow cells regenerate infarcted myocardium. *Nature* 2001;410(6829):701–705.
3. Bengel FM, Schachinger V, Dimmeler S. Cell-based therapies and imaging in cardiology. *Eur J Nucl Med Mol Imaging* 2005;32 (Suppl 2):S404–S416.
4. Massoud TF, Gambhir SS. Molecular imaging in living subjects: seeing fundamental biological processes in a new light. *Genes Dev* 2003;17:545–580.
5. Weber DA, Ivanovic M. Ultra-high-resolution imaging of small animals: implications for pre-clinical and research studies. *J Nucl Cardiol* 1999;6:332–344.
6. Acton PD, Kung HF. Small animal imaging with high resolution single photon emission tomography. *Nucl Med Biol* 2003;30:889–895.
7. Hou D, Youssef EA, Brinton TJ, Zhang P, Rogers P, Price ET, *et al.* Radiolabeled cell distribution after intramyocardial, intracoronary, and interstitial retrograde coronary venous delivery: implications for current clinical trials. *Circulation* 2005;112(Suppl 9):I150–I156.
8. Becker A, Hessenius C, Licha K, Ebert B, Sukowski U, Semmler W, *et al.* Receptor-targeted optical imaging of tumors with near-infrared fluorescent ligands. *Nat Biotechnol* 2001; 19:327–331.
9. Sato A, Klaunberg B, Tolwani R. *In vivo* bioluminescence imaging. *Comp Med* 2004;54:631–634.
10. Weissleder R, Ntziachristos V. Shedding light onto live molecular targets. *Nat Med* 2003; 9:123–128.
11. Zacharakis G, Kambara H, Shih H, Ripoll J, Grimm J, Saeki Y, *et al.* Volumetric tomography of fluorescent proteins through small animals *in vivo. Proc Natl Acad Sci USA* 2005; 102:18252–18257.
12. Ntziachristos V, Tung CH, Bremer C, Weissleder R. Fluorescence molecular tomography resolves protease activity *in vivo. Nat Med* 2002;8:757–760.
13. Wu JC, Inubushi M, Sundaresan G, Schelbert HR, Gambhir SS. Optical imaging of cardiac reporter gene expression in living rats. *Circulation* 2002;105:1631–1634.
14. Niyibizi C, Wang S, Mi Z, Robbins PD. The fate of mesenchymal stem cells transplanted into immunocompetent neonatal mice: implications for skeletal gene therapy via stem cells. *Mol Ther* 2004;9:955–963.
15. Heyn C, Bowen CV, Rutt BK, Foster PJ. Detection threshold of single SPIO-labeled cells with FIESTA. *Magn Reson Med* 2005;53:312–320.

16. Shapiro MG, Atanasijevic T, Faas H, Westmeyer GG, Jasanoff A. Dynamic imaging with MRI contrast agents: quantitative considerations. *Magn Reson Imaging* 2006;24:449–462.
17. Louie AY, Huber MM, Ahrens ET, Rothbacher U, Moats R, Jacobs RE, *et al*. *In vivo* visualization of gene expression using magnetic resonance imaging. *Nat Biotechnol* 2000;18:321–325.
18. Josephson L, Tung CH, Moore A, Weissleder R. High-efficiency intracellular magnetic labeling with novel superparamagnetic-Tat peptide conjugates. *Bioconjug Chem* 1999;10:186–191.
19. Bhorade R, Weissleder R, Nakakoshi T, Moore A, Tung CH. Macrocyclic chelators with paramagnetic cations are internalized into mammalian cells via a HIV-tat derived membrane translocation peptide. *Bioconjug Chem* 2000;11:301–305.
20. Lewin M, Carlesso N, Tung CH, Tang XW, Cory D, Scadden DT, *et al*. Tat peptide-derivatized magnetic nanoparticles allow *in vivo* tracking and recovery of progenitor cells. *Nat Biotechnol* 2000;18:410–414.
21. Merbach AE, Toth E. *The Chemistry of Contrast Agents in Medical Magentic Resonance Imaging*. John Wiley and Sons, 2001.
22. Wunderbaldinger P, Josephson L, Weissleder R. Crosslinked iron oxides (CLIO): a new platform for the development of targeted MR contrast agents. *Acad Radiol* 2002;9(Suppl 2):S304–S306.
23. Vives E, Brodin P, Lebleu B. A truncated HIV-1 Tat protein basic domain rapidly translocates through the plasma membrane and accumulates in the cell nucleus. *J Biol Chem* 1997;272:16010–16017.
24. Caron NJ, Torrente Y, Camirand G, Bujold M, Chapdelaine P, Leriche K, *et al*. Intracellular delivery of a Tat-eGFP fusion protein into muscle cells. *Mol Ther* 2001;3:310–318.
25. Kang HW, Josephson L, Petrovsky A, Weissleder R, Bogdanov A, Jr. Magnetic resonance imaging of inducible E-selectin expression in human endothelial cell culture. *Bioconjug Chem* 2002;13:122–127.
26. Reimer KA, Lowe JE, Rasmussen MM, Jennings RB. The wavefront phenomenon of ischemic cell death. 1. Myocardial infarct size vs duration of coronary occlusion in dogs. *Circulation* 1977;56:786–794.
27. Pfeffer MA. Left ventricular remodeling after acute myocardial infarction. *Annu Rev Med* 1995;46:455–466.
28. Ryan TJ, Antman EM, Brooks NH, Califf RM, Hillis LD, Hiratzka LF, *et al*. Update: ACC/AHA guidelines for the management of patients with acute myocardial infarction. A report of the American College of Cardiology/American Heart Association Task Force on Practice Guidelines (Committee on Management of Acute Myocardial Infarction). *J Am Coll Cardiol* 1999;34:890–911.
29. Etzion S, Battler A, Barbash IM, Cagnano E, Zarin P, Granot Y, *et al*. Influence of embryonic cardiomyocyte transplantation on the progression of heart failure in a rat model of extensive myocardial infarction. *J Mol Cell Cardiol* 2001;33:1321–1330.
30. Shake JG, Gruber PJ, Baumgartner WA, Senechal G, Meyers J, Redmond JM, *et al*. Mesenchymal stem cell implantation in a swine myocardial infarct model: engraftment and functional effects. *Ann Thorac Surg* 2002;73:1919–1925.
31. Kawamoto A, Gwon HC, Iwaguro H, Yamaguchi JI, Uchida S, Masuda H, *et al*. Therapeutic potential of *ex vivo* expanded endothelial progenitor cells for myocardial ischemia. *Circulation* 2001;103:634–637.
32. Jain M, DerSimonian H, Brenner DA, Ngoy S, Teller P, Edge AS, *et al*. Cell therapy attenuates deleterious ventricular remodeling and improves cardiac performance after myocardial infarction. *Circulation* 2001;103:1920–1927.
33. Min JY, Yang Y, Converso KL, Liu L, Huang Q, Morgan JP, *et al*. Transplantation of embryonic stem cells improves cardiac function in postinfarcted rats. *J Appl Physiol* 2002;92:288–296.
34. Makino S, Fukuda K, Miyoshi S, Konishi F, Kodama H, Pan J, *et al*. Cardiomyocytes can be generated from marrow stromal cells *in vitro*. *J Clin Invest* 1999;103:697–705.

35. Tomita S, Li RK, Weisel RD, Mickle DA, Kim EJ, Sakai T, *et al.* Autologous transplantation of bone marrow cells improves damaged heart function. *Circulation* 1999;100(Suppl 19): II247–II256.

36. Hamano K, Nishida M, Hirata K, Mikamo A, Li TS, Harada M, *et al.* Local implantation of autologous bone marrow cells for therapeutic angiogenesis in patients with ischemic heart disease: clinical trial and preliminary results. *Jpn Circ J* 2001;65:845–847.

37. Britten MB, Abolmaali ND, Assmus B, Lehmann R, Honold J, Schmitt J, *et al.* Infarct remodeling after intracoronary progenitor cell treatment in patients with acute myocardial infarction (TOPCARE-AMI): mechanistic insights from serial contrast-enhanced magnetic resonance imaging. *Circulation* 2003;108:2212–2218.

38. Bulte JW, Douglas T, Witwer B, Zhang SC, Strable E, Lewis BK, *et al.* Magnetodendrimers allow endosomal magnetic labeling and *in vivo* tracking of stem cells. *Nat Biotechnol* 2001; 19:1141–1147.

39. Moelker AD, Baks T, van den Bos EJ, van Geuns RJ, de Feyter PJ, Duncker DJ, *et al.* Reduction in infarct size, but no functional improvement after bone marrow cell administration in a porcine model of reperfused myocardial infarction. *Eur Heart J* 2006;27:3057–3064.

40. Hill JM, Dick AJ, Raman VK, Thompson RB, Yu ZX, Hinds KA, *et al.* Serial cardiac magnetic resonance imaging of injected mesenchymal stem cells. *Circulation* 2003;108:1009–1014.

41. Weber A, Pedrosa I, Kawamoto A, Himes N, Munasinghe J, Asahara T, *et al.* Magnetic resonance mapping of transplanted endothelial progenitor cells for therapeutic neovascularization in ischemic heart disease. *Eur J Cardiothorac Surg* 2004;26:137–143.

42. Garot J, Unterseeh T, Teiger E, Champagne S, Chazaud B, Gherardi R, *et al.* Magnetic resonance imaging of targeted catheter-based implantation of myogenic precursor cells into infarcted left ventricular myocardium. *J Am Coll Cardiol* 2003;41:1841–1846.

43. Kraitchman DL, Heldman AW, Atalar E, Amado LC, Martin BJ, Pittenger MF, *et al. In vivo* magnetic resonance imaging of mesenchymal stem cells in myocardial infarction. *Circulation* 2003;107:2290–2293.

44. Dick AJ, Guttman MA, Raman VK, Peters DC, Pessanha BS, Hill JM, *et al.* Magnetic resonance fluoroscopy allows targeted delivery of mesenchymal stem cells to infarct borders in Swine. *Circulation* 2003;108:2899–2904.

45. Stuckey DJ, Carr CA, Martin-Rendon E, Tyler DJ, Willmott C, Cassidy PJ, *et al.* Iron particles for noninvasive monitoring of bone marrow stromal cell engraftment into, and isolation of viable engrafted donor cells from, the heart. *Stem Cells* 2006;24:1968–1975.

46. Chapon C, Jackson J, Aboagye E, Herlihy A, Jones W, Bhakoo K. An *in vivo* multimodal imaging study using MRI and PET of stem cell transplantation after myocardial infarction in rats. *Mol Imaging Biol* (in press).

47. Taupitz M, Schmitz S, Hamm B. [Superparamagnetic iron oxide particles: current state and future development]. *Rofo Fortschr Geb Rontgenstr Neuen Bildgeb Verfahr* 2003;175:752–765.

48. Yeh TC, Zhang W, Ildstad ST, Ho C. *In vivo* dynamic MRI tracking of rat T-cells labeled with superparamagnetic iron-oxide particles. *Magn Reson Med* 1995;33:200–208.

8 Marrow Stromal Cells as Universal Donor Cells for Cardiac Regenerative Therapy: Fact or Fancy?

Jun Luo, Dominique Shum-Tim
and Ray Chu-Jeng Chiu

Introduction

Recently there has been an explosive advance in our knowledge on stem cells for their use as donor cells for myocardial regenerative therapies. Associated with such advances are some unexpected and controversial findings which defy current scientific dogmas. One of such dilemma is a series of observations indicating that certain populations of multipotent stem cells are immune privileged, able to survive and differentiate in immuno-compatibility unmatched allogeneic or even xenogeneic transplant recipients. Such findings challenge the traditional paradigm of immunological self- and non-self recognition concept initially promulgated by Medawar and Billingham decades ago.

Although the idea that certain unmatched stem cell populations may possess unique immune tolerance upon transplantation remains highly controversial, it seems there is a *nature's experiment,* so to speak, which might shed light on this issue. For example, it has been known for some time that mothers have cells originating from their fetuses circulating in their blood for years. Evidence that such allogeneic cells may differentiate inside the mother's bodies without being rejected in spite of the fact that 50% of genes in such fetal cells are derived from the father, is also surprising. Such phenomena are not readily explainable by the current immunological theories, and often dismissed as small number of such cells simply escaping the mother's immune surveillance.

In this review, we will briefly describe some of these observations on maternal/fetal microchimerism, since they may share the underlying mechanisms of immune tolerance with "universal donor cells" for cellular cardiomyoplasty to be discussed in this chapter. Then in order to illustrate the scientific data underlying this still controversial issue, we will describe in detail one of our recent experimental studies in which porcine marrow stromal cells (MSC) were implanted into rat hearts following coronary artery ligations without immunosuppression, and briefly review the literature on such studies. Finally we will summarize some relevant immuno-logical studies, and posit a mechanistic synthesis in an attempt to explain these observations. Such a hypothesis may eventually be refined, modified or even rejected with additional data and scrutiny. Nevertheless we hope that systematic consideration with open mind of such phenomena observed will be beneficial in fostering further advance in this important field, since the availability of immune tolerant "universal donors" will have vast clinical implications.

Maternal-Fetal Cell Microchimerism: A Whisper of the Nature?

It has been known for several decades that fetal cells are able to enter maternal blood circulation during pregnancy[1] and persist for many years post-partum.[2] The fetal cells begin to enter maternal circulation at four to six weeks of gestation, and by 34 weeks all pregnant mothers have detectable fetal cells in their blood.[3] Most such studies used the detection of Y chromosome in male fetus cells from the mother's blood to confirm the presence of microchimeric fetal cells. In recent years, such cells have been isolated from the pregnant mothers to be used for prenatal genetic diagnoses.[4] These microchimeric cells have been found, both experimentally[5,6] and clinically,[7-9] to be able to engraft various maternal tissues such as the bone marrow, spleen and liver. Furthermore, these cells are also found to be able to undergo *in situ* differentiation to become mature thyrocytes in humans[10] and even into various neural cells in a mouse model.[11]

Immunologically, it is strange why these MHC mismatched cells can survive in the blood and tissue of the mother, since it is well known that child-to-mother organ transplants, such as transplanting a child's skin or kidney, to his/her mother cannot be successful without immunosuppression. In fact, there has been speculation that these microchimeric cells are insidiously rejected by the immune system of the mother, such that they play a pathogenic role for the autoimmune diseases of the mother. Thus, a number of chimeric fetus-derived cells found in the tissues affected by various autoimmune diseases had been reported, such as in scleroderma,[12] primary biliary cirrhosis[9] Sjoegren syndrome,[13] systemic lupus erythematosus[14] thyroid autoimmune diseases,[15] etc. Reviewing the results of these studies, however, Khosrotehrani and Bianchi[16] concluded that strong evidence now exists that microchimerism is not always found in association with autoimmune diseases. In contrast, there are epidemiological and biological evidences suggesting that these fetus-derived cells may provide protective effects in the affected tissue. They cited references[10,17,18] on the clinical protective effects of pregnancy itself, and the observed increase in chimeric cell quantity in non-autoimmune tissue injuries. Based on such findings they proposed the hypothesis that the fetus-derived cells in the maternal circulation can be mobilized to the site of tissue injury where they take part in tissue repair. It is fascinating to see that this notion, originating from a totally unrelated avenue of research, reached a conclusion virtually identical to the current thoughts that marrow derived adult stem cells may be mobilized to reach the injured tissue, and participate in its repair process.[19]

However, to date, the fundamental question of why these chimeric cells are not immunologically rejected, even after they have differentiated into mature cell phenotypes, remain unanswered.

Stem Cell Therapy and Immune Rejection

Although there has been an explosive advance in the knowledge of stem cell biology in recent years, and the potential of using such pluri-/multipotent cells for the regenerative therapy of various diseases is widely appreciated, the immunological behavior of such cells have not been fully explored. It has been generally assumed that the stem cells will induce host immunological response in parallel to that follows any tissue transplantation. In order to avoid the need for immunosuppression of the cell recipient, a couple of strategies have been adopted, depending on the origin of the donor stem cells. Since autologous embryonic stem cells (ESC) will not be available for regenerative therapy of any patient, allogeneic transplant becomes mandatory. Although the truly pluripotent ESCs derived from blastosycst can survive in an allogeneic recipient because they poorly express MHC antigens, they will form a teratoma, a distinctive tumor containing cell types from all three embryonic layers.[20] To avoid this, the ESCs are induced in culture prior to transplantation to differentiate into the desired phenotypes. Such mature cells however become recognizable by the immune system of the recipients, and face rejection as in allogeneic tissue (e.g. skin) transplantation. As is well known, this is why "therapeutic cloning," employing "nuclear transfer" technique, is being thought as the holy grail for regenerative therapy since such "autologous equivalent cells" will be immunologically accepted by the intended recipient.

Another source of pluri- or multipotent cells for regenerative therapy comes from "adult stem cells," found in the bone marrow, fat, nerve, heart and other tissues. Many of these cells have been reported to be capable of participating in the repair process of damaged tissues, through various mechanisms such as transdifferentiation,[21–23] fusion[24] and paracrine[25] effects. For such cells, immunological rejection issue can be circumvented by using autologous donor cells available from the patient himself or herself. The stem cells originating from the umbilical cord blood, amniotic fluid or placenta may be used as the autologous cells for the future use by an infant when these cells are collected and preserved at birth. However, although the use of autologous donor cells has obvious advantages, there are limitations as well. Since the processes of isolating and preparing the donor cells for transplantation take time, autologous transplants may limit their use in urgent situations, such as for patients with acute myocardial infarction. If the tissue damage is due to genetic defects, autologous cells may not be desirable because they will share the same genetic abnormalities.

Furthermore, it has been reported that adult stem cells obtained from senile and/or debilitated patients may be functionally compromised.[26,27] These deficiencies may be addressed with the surprising recent findings of unique immune tolerance of marrow-derived stromal stem cells (MSCs), as described in the following section.

Marrow Stromal Cells for Cellular Cardiomyoplasty: An Experimental Study in Porcine/Rat Xenotransplant Model Without Immunosuppression

There has been increasing evidence in recent literature, including those from our group,[28–31] that histocompatibility mismatched MSCs can be tolerated without immunosuppression in fully immunocompetent adult rodents. Yet, there are still controversies as to whether such universal compatibility may be extrapolated beyond rodent species,[32] as our earlier studies involved a mouse cells to rat heart xenotransplant model. There had been reports of successful allogeneic porcine cell transplant experiments for myocardial infarction,[30,31] but the degree of their MHC mismatch was not well defined. Thus, in our more recent study described in detail in the following sections, we pushed the envelope, so to speak, and carried out an experiment to examine the responses of xenogeneic porcine MSCs transplanted into the heart of infarcted rat hearts without immunosuppression.

Materials and methods

Animal model

Male Yorkshire pigs and female Lewis rats were used respectively as donors and recipients in this study. All animals received humane care in compliance with the "Guide to the Care and Use of Experimental Animals" of the Canadian Council on Animal Care.

Porcine cell isolation and labeling

Bone marrow cells were harvested from the male porcine femurs and cultured with Dulbecco's Modified Eagle's Medium (DMEM) containing selected lots of 10% fetal bovine serum and antibiotics as described previously.[33] Briefly, whole bone marrow cells were plated on culture dishes and the non-adherent cells were discarded by changing culture medium using Caplan's method.[34] Porcine fibroblasts were harvested by outgrowth from a piece of chest skin taken from the same male porcine donor.

After obtaining 90% of confluence *in vitro*, Lac-Z reporter gene was transfected into the porcine MSCs and fibroblasts with Lac-Z-GP AM12 amphotropic retrovirus as described previously.[35] At the end of transfection, a sample of these cells were stained with solution containing 2% 5-bromo-4-chloro-3-indoyl-Z-D-galactoside

(X-gal), dimethyl formamide, 20 mM $K_3Fe(CN)_6$, 20 mM $K_4Fe(CN)_6$. $3H_2O$ and 2 mM magnesium chloride to confirm proper labeling. The sample cells were then incubated at 37°C, pH 7.8 to 8 and protected from light for 16 hours. The presence of blue-labeled cells was confirmed under an inverted microscope (BX-FLA Olympus, Tokyo, Japan).

Lac-Z positive cells were then collected and suspended in DMEM medium at a concentration of 2×10^7 cells/ml. Cell suspension (0.15 ml) was then prepared in a 28-gauge insulin syringe and placed on ice until cell implantation.

Experimental design

Fifty-eight female Lewis rats (250–300 g) were used as recipient animals, and randomized into two groups. In Group I ($n = 42$), male porcine MSCs (3×10^6), suspended in DMEM, were injected directly into rat myocardium at the anterior region of the left ventricle via a left anterolateral thoracotomy. After implantation, recipient rats were randomly sacrificed at ten minutes ($n = 9$), one week ($n = 9$), four weeks ($n = 9$), six weeks ($n = 9$), and 24 weeks ($n = 6$). Four heart samples from recipients at each time point were used to purify genomic DNA for polymerase chain reaction (PCR) analysis. The same amount of labeled male porcine fibroblasts were similarly implanted in Group II ($n = 16$), and heart samples were harvested at ten minutes ($n = 4$), four days ($n = 4$), one week ($n = 4$) and three weeks ($n = 4$). Two heart samples at each time point were used to purify genomic DNA for PCR analysis.

Transplantation of porcine cells

After general anesthesia was induced, female Lewis rats were intubated and mechanically ventilated. A left anterolateral thoracotomy was performed under sterile conditions and the heart was exposed by pericardiotomy. Labeled male porcine MSCs or fibroblasts (3×10^6) were injected directly into the anterior region of the rat left ventricle and randomized into one of the two groups mentioned above. After confirming no active bleeding, the chest incision was closed with 4–0 absorbable Vicryl® suture and animals survived until elective termination at pre-determined time points.

Histology and immunohistochemistry

The recipient rats were sacrificed under general anesthesia at various time points as mentioned in the experimental design. Through a sternotomy incision, the hearts were perfused *in situ* with PBS and 2% paraformaldehyde. The hearts were then harvested and kept in 2% paraformaldehyde PBS at 4°C. The staining for β-galactosidase activity was performed as described earlier, but with the addition of 0.02% Nonidet P-40 and 0.01% deoxycholate to the staining solution for 16 hours (pH 7.8) at 37°C. After X-gal staining, they were embedded in paraffin and serial coronal sections of 5-μm thick were mounted on glass slides for different stainings. One serial section from each heart specimen was stained with hematoxylin and eosin and another serial section was used for immunohistochemical staining for connexin-43 (Zymed Laboratories Inc, San Francisco, CA), and troponin I-C (Santa-Cruz Biotechnology Inc, Santa-Cruz, CA). Briefly, after de-parafinization, sections were placed in boiled citrate buffer (pH 6.0). After blocking in normal serum, sections were treated with the respective monoclonal antibodies overnight and with secondary antibodies on the following day. Diaminobenzidine (DAB), which produces a brown color, was then used as a chromogen for light microscopy. Counter-staining of sections by hematoxylin was also performed. Cells derived from implanted labeled MSCs or fibroblasts were identified by their blue nuclei.

DNA purification and cDNA synthesis

Genomic DNA was purified from recipients' left ventricular myocardium and porcine MSCs using DNeasy® (QIAGEN, Valencia, CA) according to the manufacturer's instructions. DNA was eluted in 100 ul TE, pH 8.0. DNA yield was determined a minimum of four times by absorbance reading at 260/280 nm (GeneQuant spectrophotometer, Pharmacia Biotech, Piscataway, NJ). DNA was diluted into 25 ng/μl and stored at $-80°$C until the assay was carried out.

In order to synthesize first strand cDNA template for primer quality control test, total RNA was extracted from female rat heart and male porcine MSCs using Trizol (Invitrogen, Carlsbad, CA) according to manufacturer's instructions. First strand cDNA was synthesized with SuperScript II Rnase H(-) reverse transcriptase, and Oligo (dT) 12–18 primer (Invitrogen Carlsbad) with the following condition: total reaction volume was 11 μl containing 2 μl RNA (15 μg), 1 μl Oligo (dT) 12–18 (500 μg/ml), 4 μl 5X first strand buffer, 2 μl 0.1M DTT and 1 μl 10 mM dNTPs. RT products (cDNA 150ng/μl) were stored at $-80°$C for PCR primers optimization procedure.

Primers design and probes selection

Based on the GeneBank sequence (NCBI, Washington), the specific porcine SRY primers were designed according to porcine cDNA sequence and elongating to 20 nucleotides each (Forward: 5′-GGCTTTCTGTTCCTGAG CAC-3′, Reverse: 5′-CTGGGATGCAAGTGGAAAAT-3′, product size: 247 bp). A pair of primers for rat aconitase gene was designed according to rat genomic DNA sequence for use at higher annealing temperatures (Forward: 5′-TTTCAAACCCTGTCAACAAATG-3′, Reverse: 5′-CTTCCAAGTGAGCGAAGACC-3′, product size: 119 bp). Master Mix Sybr Green Kit (Applied Biosystem Inc., Foster City, CA) was used in the real-time PCR procedure serving as a real-time fluorescence reporter by labeling the double-stranded DNA in the amplification system. PCR threshold cycle (C_T) was recorded when the florescence was first detected and was dependent on or correlated to the DNA concentration.

To test the specificity of primers, conventional PCR was performed with combined DNA or cDNA templates purified from pig and rat, under the following conditions: 10 ng DNA template in the reaction mixture (containing 200 μM each of the respective nucleotides, 300 nM of each primer, 3.5 mM MgCl$_2$) was denatured for 15 minutes at 95°C followed by 35 cycles amplification, each containing 30-second annealing at 58°C, 30-second polymerization at 72°C (holding temperature 4°C). For PCR, the Taq DNA polymerase and related reagents from Invitrogen (Carlsbad, CA) were used. Amplified DNA fragments were analyzed by electrophoresis through 1.5% agarose gels containing ethidium-bromide, and subsequently visualized through ultraviolet light and the bands captured by a digital camera.

Conventional PCR analysis

In order to confirm that the histological findings of Lac-Z positive cells were not artifacts, we have elected to use PCR analysis to determine the gender-mismatched DNA markers between the donors-recipients. The presence of living porcine cells in female recipient rats' myocardium was confirmed by PCR amplifying a 247 bp fragment of the sex determined region on the porcine Y chromosome using designed primers. As a control, a pair of primers for rat aconitase gene was used in a serial of parallel PCR reactions. Genomic DNA template from both male porcine marrow stromal cells and normal female rat heart tissues were processed in parallel as a positive and negative control, respectively.

The PCR reaction mixture contains 150 nM of each primer, and 100 ng of sample DNA template with the thermal protocol described in primer specificity test.

Construction of standard curves

DNA used for the purpose of construction of standard curves by real-time PCR were purified from male porcine MSCs and female Lewis rat heart tissue without porcine cell transplantation. After measuring DNA concentration by optical density, standard mixtures of porcine-rat DNA were made from 1:2500 to 1:10. Three quality controls were used: PCR reagents without template DNA, female porcine DNA, and female rat DNA.

Porcine SRY gene quantitative analysis with real-time PCR

Evaluation of porcine SRY gene levels in rat heart DNA extract was achieved by real-time quantitative PCR kinetics using the Sybr Green I chemistry in the ABI PRISM® 7900HT Sequence Detection System (Applied Biosystem Inc., Foster City, CA).

The PCR reaction mixture contained 200 μM each of the respective nucleotides, 300 nM of each primer, 3.5 mM MgCl$_2$, 0.04 μM Rox, 1:40000000 Sybr Green, and 100 ng of sample DNA template. The thermal protocol was as follows: 15 minutes at 95°C, followed by 40 cycles (30 seconds at 95°C, 30 seconds at 58°C, 30 seconds at

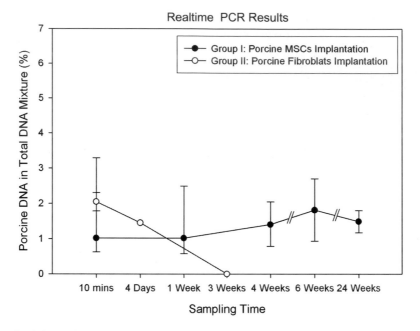

Fig. 1. Real-time PCR results. In contrary to the stable porcine SRY gene level in group I, a rapid loss of porcine SRY gene within days was observed in Group II.

72°C where the fluorescence signal was acquired). A melting curve of PCR products (59°C–90°C) was also performed to ensure the absence of artifacts. A DNA mixing curve prepared form known male porcine and female rat DNA was included in every assay as a standard to quantify the unknown samples. To decrease the effect of intra-test variability at very low concentration of porcine DNA, five standard samples were used at each dilution point in the standard curve and five replicates were tested for all of the unknown samples.

Statistical analysis of real-time PCR data

To calculate the porcine SRY gene levels (fold induction), data were analyzed using the SDS 2.1 software (Applied Biosystem Inc., Foster City, CA), in which the mathematical model used was based on mean threshold cycle differences between the sample and the calibration curve.[36] To compare between each analyzed target, median PCR efficiency value (C_T) obtained from different experiments was used to plot the curve (Fig. 1).

Results

All rats survived the cell implantation procedure until the time scheduled for elective sacrifice. Gross examination of the recipients' hearts did not reveal any structural abnormalities except for adhesions at the implantation sites.

Cell culture and labeling

In MSCs culture flasks, the hematopoietic cells that did not adhere to the culture flasks were successfully removed by changing the medium. Both porcine MSCs and skin fibroblasts in the culture medium were attached to the plastic culture flasks and developed spindle-shaped morphology. Following three to four passages of subculture, all the cells in culture flasks were successfully transfected by replication defective retrovirus carrying the Lac-Z reporter gene. *In vitro* immunohistochemical staining with X-gal showed the transfection efficiency was nearly 100% (Fig. 2A).

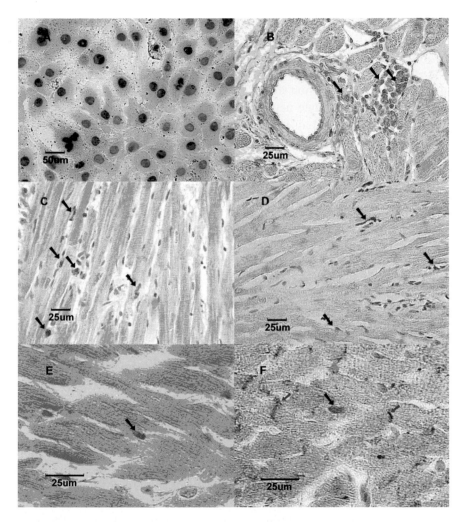

Fig. 2. Histologic and immunohistochemical analysis (Group I). 100% Lac-Z labeled MSCs (blue) was shown before implantation in culture dish (**A**). Myocardium samples were harvested at ten minutes (**B**), four weeks (**C, D**) and 24 weeks (**E, F**) after MSCs implantation. Black arrows showed porcine MSCs in myocardium. Brown color in (D) indicates Troponin I-C positive cells. Red arrows in (F) showed connexin-43 positive staining.

Histologic and immunohistochemical analysis of cell engraftment

Gross heart specimens from Group I showed a blue discoloration at the sites of MSCs implantation in the anterior left ventricle, whereas in the immunologic control Group II, no obvious discoloration was detected.

Serial section from the recipients' left ventricle confirmed that the blue color visible in the gross sample was indicative of β-gal positive stained cells which were distributed in all the layers of the heart muscles. Most of the cells that clustered around the needle track from the implantation sites maintained the original MSCs morphology (Fig. 2B), whereas the cells scattered in the myocardium adopted the morphology similar to the host cardiomyocyte (Fig. 2C).

The surviving MSCs in Group I stained positively for the cardiomyocyte-specific protein troponin I-C starting from four weeks after cell implantation (Fig. 2D). The β-gal positive cells could be detected up to 24 weeks after MSC implantation despite the lack of immunosuppression (Fig. 2E). Some of them appeared to be integrated into the native cardiomyocyte network and expressed the cardiac specific protein connexin-43 (Fig. 2F), a major component of cardiomyocyte intercalated discs. Throughout the duration of the study, no obvious inflammatory infiltrates could be observed in the heart specimens taken from ten minutes to 24 weeks post-transplant in Group I (Figs. 2B–F).

In the immunologic control Group II, histological examination of the left ventricle revealed extensive mononuclear cellular infiltration within one day (Fig. 3A), and more aggressive inflammatory infiltration was observed on the fourth day (Fig. 3B) and the seventh day (Fig. 3C) with a rapid lose of labeled fibroblasts within a week in all the recipients (Fig. 3D).

PCR analysis

DNA purified from the recipient's hearts was examined by conventional PCR and real-time PCR for porcine SRY to confirm the histological findings that some of the cells were indeed originated from donor MSCs that were labeled with Lac-Z initially. Pig SRY gene primers were highly specific for porcine genomic DNA or cDNA only (Fig. 4A). Porcine SRY gene was detected in the sample from the first day to the 24th week in Group I (Fig. 4B). Yet, no porcine SRY gene could be detected in any of the samples tested after porcine fibroblast implantation in Group II beyond the fourth day (Fig. 4B). This finding correlated precisely with histological findings.

The DNA standard curve used as a standard or reference for unknown DNA samples (Fig. 5C) was constructed with real-time PCR. Results indicated that amplification occurred in the whole porcine rat DNA dilutions serial even in the presence of a large amount of rat DNA (Fig. 5A). The one-peak dissociation curve showed single SRY product in the reaction system (Fig. 5B), which indicated high specificity and efficiency of porcine SRY gene primers. Quantification data showed a stable amount of porcine gene in Group I that indicated the consistent retention of living

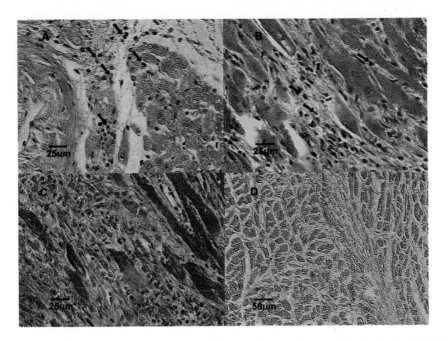

Fig. 3.   Histologic analysis (Group II). Myocardium samples were harvested at the first day (**A**), fourth day (**B**), seventh day (**C**), and the third week (**D**) after porcine fibroblast implantation. Black arrows showed porcine fibroblasts in myocardium. Note the presence of extensive inflammatory reaction.

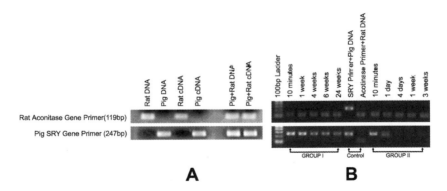

Fig. 4.   Conventional PCR results. Pig SRY gene primers were highly specific for pig genomic DNA or cDNA only (**A**). Pig SRY gene was detected up to 24 weeks in Group I, but no more than four days in Group II (**B**).

porcine MSCs in the rat heart tissues. Yet, in Group II, a rapid loss of porcine SRY gene was observed in all specimens tested (Fig. 1).

Discussion

Liechty *et al.*[37] first reported that human MSCs after intraperitoneal implantation in a late gestational stage fetal sheep which has gained immune competence, could

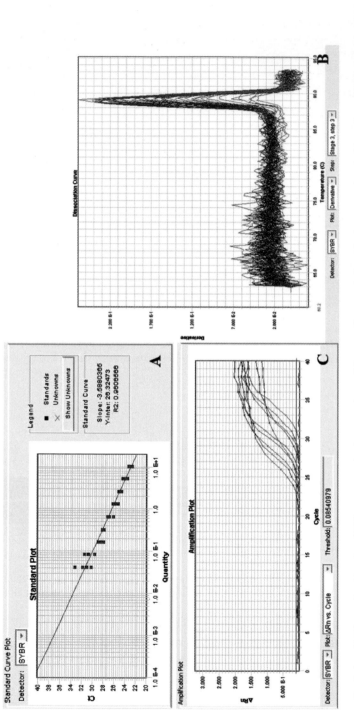

Fig. 5. Real-time PCR results. Porcine SRY gene standard curve showed inverse linear relationship between threshold cycle and the porcine DNA concentration (R2 = 0.96, $P < 0.001$, **A**). The single dissociation peak was shown in (**B**). In (**C**), the blue amplification plot was standard serial, and red plot showed porcine SRY product from the sample.

engraft into various tissue organs including myocardium. In a swine and a rat model, Amado *et al.*[30] and Dai *et al.*[31] respectively reported that allogeneic MSC implantation in myocardium could survive without immunosuppressive therapy. Moreover, these cells were shown to differentiate and contributed to functional improvement of the host myocardium. Recently, our group also demonstrated that mouse MSCs transplanted intravenously into adult rat recipients were recruited and engrafted into infarcted rat myocardium, and formed a stable xenogeneic chimera.[28] We also found direct intramyocardial implantation of mouse MSCs into ischemic rat heart contributed to improved cardiac function.[29]

In the present study, the mature differentiated porcine skin fibroblasts (Group II) were rapidly rejected in the immune-competent rat myocardial environment with massive inflammatory mononuclear cellular infiltration at the sites of implantation (Figs. 3B and 3C). No fibroblast could be detected beyond four days after implantation. This suggested that the xenogeneic terminally differentiated fibroblasts were acutely rejected, as can be expected. In contrast, porcine MSCs from the same donors could survive in a xenogeneic myocardium for as long as six months without obvious inflammatory reaction or rejection (Group I), which was consistent with our earlier findings in the mice-to-rat xenotransplantation model.[28,29] Thus our findings reported here support the hypothesis that MSCs were uniquely immune privileged cells,[38] even across a vast phylogenic distance between pig and rat.

We further employed real-time PCR assay for the porcine Y chromosome specific SRY gene[39,40] to quantify the survival of male porcine cells after implantation into the female recipient rats. This technique provides a more reliable quantitative estimate of the surviving donor cells than counting cells on histological sections.[39]

Consistent with our histological findings in the MSCs transplantation group, porcine SRY gene could be detected from the recipients' hearts for up to six months (Fig. 3B), while in the control Group II which received porcine fibroblasts, no porcine SRY gene could be detected beyond four days (Figs. 3C and 3D). Moreover, real-time PCR showed that the amount of porcine DNA in MSCs implantation group was stable over time for at least six months (Fig. 1A), while in Group II, a rapid decrease followed by complete absence of porcine DNA within days was noted.

Interestingly, our immunohistochemical sections revealed that after porcine MSCs were implanted into rat heart myocardium, they started to express cardiac cell markers troponin I-C in the fourth week (Fig. 2D) and samples from the sixth week showed labeled cells were integrated into the rat cardiomyocytic network structure through intercalated discs (Fig. 2F). This demonstrated the feasibility of "milieu dependent differentiation" in a xenogeneic microenvironment. Fukuhara *et al.*[41] provided such evidence *in vitro* by co-culturing labeled mouse MSCs with neonatal rat cardiomyocytes. In this study, when the direct physical contact was allowed between these two types of cells, mouse MSCs formed synthicium with the neonatal rat cardiomyocytes and successively expressed cardiac-specific cell markers troponin-I, myosin heavy chain and connexin-43, and started to contract synchronously with the cardiomyocytes. However, if the co-cultured MSCs were separated from neonatal cardiomyocytes using a filter which still permitted the passage of macromolecules in the culture media, MSCs failed to show myogenic differentiation.

Mechanisms of MSC Immune Tolerance

In vitro and *in vivo* studies

Recently, efforts are being made to understand the mechanisms underlying this unique MSC immune tolerance.[42] It is known that MSCs weakly express major histocompatibility class I (MHC) antigens, while class II antigens are found intracellularly. In addition, they can directly modulate immune responses to induce tolerance. Le Blanc and co-workers[43] showed that MSCs do not elicit alloreactive lymphocyte proliferative response and modulate immune responses. Aggarwal *et al.*[44] have shown in the *in vitro* mixed lymphocyte culture (MLC) studies of human MSCs and various purified immune cells, that human MSCs could alter cytokine secretion profiles of dendritic cells, naïve and effector T-cells as well as natural killer cells to induce more anti-inflammatory or tolerant phenotypes (see Table 1). We suspect such anti-inflammatory capability of MSCs may play an important role in the so called "paracrine effects" associated with MSC cell therapy for acute myocardial infarction postulated by many investigators. In *in vivo* studies, investigators had reported that MSCs could delay the rejection of skin allografts experimentally[45] and MSC allograft have been used clinically to improve engraftment of bone marrow transplants, and reduce graft versus host reactions.

The "stealth immune tolerance" hypothesis

While the immunomodulary effects of multipotent MSCs are now well-documented, it provides no explanation as to why tolerance persists even after implanted xeno-/allogeneic stem cells differentiated into their targeted tissue phenotypes. Recently, we suggested that this dilemma may be addressed by the "danger model" theory of immune response, originally proposed by Metzenger[46] in 1992. Since the differentiation of multipotent cells takes place over many days and weeks, the "danger signals" associated with the surgical trauma of transplantation should have subsided. Such danger signals consist of molecules released by the damaged or stressed cells. They could be picked up by the antigen presenting cells (APCs) such as dendric cells to express co-stimulant molecules, which are required to activate the T-cells recognized by the specific MHC antigens. The activated T-cells then proliferate and mount immune rejection. Thus if the surgically induced danger signal has subsided by the time MSCs have fully differentiated, T-cells will not be activated. Consequently the newly differentiated cells derived from the MSCs are tolerated and survive, which is somewhat similar to how a developing neoplasm escapes immune surveillance. When the fetal cells enter the maternal blood, there is no associated surgical or other tissue injuries which release danger signals, thus according to this

Table 1. *In vitro* functional studies of MSCs, using MLC, summarizing the immunosuppressive mediators of the MSCs.

Marrow stromal cells	Cytokines	APCs	Precursor cells	Mature cells	Cytokines
	TNF-α ↓; IL-12 ↓ →	Mature DC-1 (CD-40) →	Naïve T-cell (CD 40L; CCL2) ↑	T_H^1-cell ↑	IFN-γ ↓
	- - - - - - - -	- - - - - - - -	↑		
MHC1 PGE-2					
CD-102 VEGF					
CD-90 IL-6	IL-10 ↑ →	DC Mature/immature (TGF-β) →	Naïve T-cell (CD 28; IL-10) ↑	↑ T_{reg}-cell ↑	IL-10 ↑
				→	
CD-73 IL-8					
CD-54	IL-10 ↑ →	Mature DC-2 (MHC II) →	Naïve T-cell (TCR) ↑	T_H^2-cell ↑	IL-4 ↑
	- - - - - - - -	- - - - - - - -	↑		
	- - - - - - - -	- - - - - - - -	↑	NK-cell ↑	IFN-γ ↓

Source: Adapted from Ref. 44.

scenario, allows the establishment of microchimerism. The dilemma why a fetus can be tolerated *in utero* by the mother's immune system but the skin from this baby after birth grafted to the mother will be rejected can be understood within this paradigm.

The original "danger model" theory of Metzenger did not elucidate how APCs can recognize certain molecules to represent "dangers." With the better understanding of evolutionarily developed "innate immune response" mechanisms in recent years, it appears that such molecules are recognized by the TLR (Tolus-Like Receptors, which are part of the "innate immune system") present on the surface of APCs and other cells, which then may signal the genes to express co-stimulant molecules.[47] There were studies which specifically identified the receptor TLR-4 to play such a role,[48] although other TLR receptors may be also found to be involved by future studies.

The "stealth immune tolerance" hypothesis[38] thus posits that for the MHC mismatched donor cells to survive successfully as universal donors, they have to fulfill two unique characteristics: first, the undifferentiated donor cell is either unrecognizable, i.e. lack of MHC expression; and/or *capable of directly inducing T-cell tolerance* by sending cytokine signals. Second, subsequent *insidious differentiation* of the donor cells allow them to *escape the effects of "danger signal"* associated with surgical trauma as co-stimulant, which is required to activate the T-cell population to mount immune rejection. There are some evidence that MSCs undergoing early phase of differentiation, starting to express some phenotype-specific markers, can still maintain their capacity to induce immune tolerance,[49] and thus able to allow more time for the danger signals (molecules released from damaged or stressed cells) to subside. In this scenario, by the time the MSCs have fully differentiated and eventually lose the capacity to induce immune tolerance, the danger signals will be gone, and they will no longer be susceptible to immune rejection.

One of the mechanisms for MSC immune tolerance which has not being discussed in detail here is the possible role of the regulatory T-cells (T_{reg}-cells), although the ability of the MSCs to augment T_{reg}-cells was noted in Table 1. Increased or stimulated T_{reg}-cells will enhance immunosuppression, thus prevent immune rejection. Future studies will likely shed more light on this important mechanism[50] (Table 2).

Is "stealth immune tolerance" phenomenon unique to MSCs?

As discussed earlier on maternal-fetal cell microchimerism, the long term survival of mismatched fetal cells within the mother and their *in situ* differentiations suggest many similarities between the behavior of such cells and the MSCs. We postulate that the fetal cells entered the maternal circulation without inducing danger signal expression, thus achieved "stealth immune tolerance" and survive. Whether such cells of fetal origin can induce T-cell tolerance directly by secreting various cytokines, as had been confirmed in MSCs, is not clear at this time.

Table 2.   Summary of the "Stealth Immune Tolerance" Hypothesis.[38]

"Stealth Immune Tolerance" hypothesis

Part I: Pluripotent stem cell stage:
            Immuno-modulatory capabilities (confirmed by the *in vitro* MLC data).

Part II: Upon differentiation of these cells
            Two *simultaneous* signals at immunological synapsis required to initiate
            immune response:
            1) Recognition signal (MHC molecules) and
            2) Activation signal (co-stimulant).*

* Surgical injury associated with cell transplantation procedure releases "Danger
            signal"......via TLR receptors...expression of co-stimulant molecules.

"Stealth hypothesis": Gradual differentiation of MSCs enables the *dissociation of
            recognition signal from activation signal*, inducing immuno-suppression.

Other inhibitory mechanisms, such as by augmented regulatory T-cells, may also
            play a role.

On the other hand, several multipotent cell populations derived from a number of different sources have been shown to share some of the immunological characteristics of MSCs described above. These include such cells isolated from the fat tissue (i.e. ASC, adipose-derived stem cells),[51] the umbilical cord blood, and the placenta.[52] In the *in vitro* MLC studies, these cells demonstrate capability to suppress MLC response, although few *in vivo* studies have been reported to date.

Embryonic stem cells (ESC) have been characterized by their response to their implantation *in vivo*. It is known that ESCs injected into a fetus will differentiate into many phenotypes *in situ* to participate in the growth of various organs,[37] but when implanted into an adult animal, they will grow into a teratoma, which is a tumor consists of a disorganized collection of different tissues. With appropriate culture conditions, the ESCs can be directed to differentiate into a desired phenotype, thus for myocardial repair, the usual approach is to have them differentiate *in vitro* to become cardiomyocytes before implanting them into the damaged myocardium. Under this scenario, the fully differentiated immunologically mismatched cardiomyocytes should be rejected, unless the original ESCs were from cloned eggs whose cell nuclei came from the designated recipient. Interestingly, in spite of such consensus, Min *et al.*[53] in a series of xenogeneic transplant studies reported that cardiomyocytes derived *in vitro* from mouse ESCs differentiated and survived for up to 32 weeks inside rat hearts without immunosuppression. The explanation for such observations is not clear, but it is known some rodent models of allogeneic cardiomyocyte transplants are "low-responding,"[54] perhaps in part related to the weak expression of histocompatibility antigens in such cells. This was one of the reasons why we undertook our pig to rat xenotransplant study, which has been described in detail earlier.

In a more recent study, Swijnenberg *et al.*[20] experimented with an unmatched allogeneic transplant model and injected cells from an ESC cell line without prior *in vitro* induction of these cells to differentiate. They found the formation of intramyocardial teratomas which were not immunologically tolerated, and rejected with vigorous infiltration within four weeks. We posit that such an observation does not refute the "stealth immune tolerance" hypothesis, since ESCs do express low level of MHC Class I molecules, which should become fully expressed as they differentiate into teratoma cells. Although human ESCs are reported to possess certain immune-privileged properties,[55] and their differentiated derivatives are less susceptible for immune rejection than adult cells,[56] they may not share the active immunosuppressive capability of various adult and fetal mesenchymal stem cell[52] described earlier in this review. Perhaps such active suppression is required to prevent the expression of co-stimulant molecules of APCs which are required to activate full immune response. Further studies are clearly needed to elucidate the molecular basis of such observations.

Conclusions

In recent years, MSCs had been used clinically to improve engraftment in bone marrow transplants, and to reduce the graft-versus-host responses. Based on these clinical and some experimental data discussed above, the Food and Drug Administration in the United States approved an Osiris Inc. sponsored Phase I multi-center clinical trials of allogeneic MSC therapy for patients with acute myocardial infarctions. This may lead to additional efficacy trials in the near future. The establishment of MSCs as effective "universal donor cells" will dramatically expand the therapeutic potential for cellular cardiomyoplasty. By being able to escape immune rejection, we may be able to achieve some of the ambitious goals of "therapeutic cloning," the holy grail of regenerative cell therapy.

References

1. Herzenberg LA, Bianchi DW, Schroder J, *et al.* Fetal cells in the blood of pregnant women: detection and inrichment by fluorescence-activated cell sorting. *Proc Natl Acad Sci USA* 1979;76:1453–1455.
2. Bianchi DW, Zickwolf GK, Weil GJ, *et al.* Male fetal progenitor ells persist in maternal blood for as long as 27 years postpartum. *Proc Natl Acad Sci USA* 1996;93:705–708.
3. Ariga H, Ohto H, Busch MP, *et al.* Kinetics of fetal cellular and cell-free DNA in the circulation during and after pregnancy: implications for noninvasive prenatal diagnosis. *Transfuion* 2001;41:1524–1530.
4. Bianchi DW. Current knowledge about fetal blood cells in the maternal circulation. *J Pediat Med* 1998;26:175–185.
5. Liegeois A, Gaillard MC, Ouvre E, *et al.* Microchimerism in pregnant mice. *Transplant Proc* 1981;13:1250–1252.

6. Philip PJ, Ayraud N, Masseyeff R. Transfer, tissue localization and proliferation of fetal cells in pregnant mice. *Immunol Lett* 1982;4:175–178.

7. Johnson KL, Nelson JL, Furst DE, *et al*. Fetal cell microchimerism in tissue from multiple sites in women with systemic sclerosis. *Arthritis Rheum* 2001;44:1848–1854.

8. Invernizzi P, De Andreis C, Sirchia SM, *et al*. Blood fetal microchimerism in primary biliary cirrhosis. *Clin Exp Immunol* 2000;122:418–422.

9. Tanaka A, Lindor K, Gish R, *et al*. Fetal microchimerism alone does not contribute to the induction of primary biliary cirrhosis. *Hepatology* 1999;30:833–838.

10. Srivatsa B, Srivasta S, Johnson KL, *et al*. Microchimerism of presumed fetal origin in thyroid specimens from women: a case-control study. *Lancet* 2001;358:2034–2038.

11. Tan XY, Liao H, Sun L, *et al*. Fetal microchimerism in the maternal mouse brain: a novel population of fetal progenitor or stem cells able to cross the blood–brain barrier? *Stem Cells* 2005;23:1443–1452.

12. Nelson JL, Furst DE, Maloney S, *et al*. Microchimerism and HLA-compatible relationships of pregnancy in scleroderma. *Lancet* 1998;351:559–562.

13. Endo Y, Negishi I, Ishikawa O. Possible contribution of microchimerism to the pathogenesis of Sjogren's syndrome. *Rheumatology* 2002;41:490–495.

14. Johnson KL, McAlindon TE, Mulcahy E, Bianchi DW. Microchimerism in a female patient with systemic lupus erythematosus. *Arthritis Rheum* 2001;43:4107–4111.

15. Klintschar M, Schwaiger P, Mannweiler S, *et al*. Evidence of fetal microchimerism in Hashimoto's thyroiditis. *J Clin Endocrinol Metab* 2001;86:2494–2498.

16. Khosrotehrani K, Bianchi DW. Fetal cell microchimerism: helpful or harmful to the parous woman? *Curr Opin Obstet Gynecol* 2003;15:195 199.

17. Artlett CM, Rasheed M, Russo-Stieglitz KE, *et al*. Influence of prior pregnancies on disease course and cause of death in systemic sclerosis. *Ann Rheum Dis* 2002;61: 346–350.

18. Johnson KL, Samura O, Nelson JL, *et al*. Significant fetal cell microchimerism in a non-transfused woman with hepatitis C: evidence of long-term survival and expansion. *Hepatology* 2002;36:1295–1297.

19. Chiu RCJ. Adult stem cell therapy for heart failure. *Expert Opin Biol Ther* 2003;3: 216–225.

20. Swijnenburg RJ, Tanaka M, Vogel H, Baker J, Kofidis T, Gunawan F, Lebl DR, Caffarelli AD, de Bruin JL, Fedoseyeva EV, Robbins RC. Embryonic stem cell immunogenicity increases upon differentiation after transplantation into ischemic myocardium. *Circulation* 2005;112:I166–I172.

21. Pittenger MF, Martin BJ. Mesenchymal stem cells and their potential as cardiac therapeutics. *Circ Res* 2004;95:9–20.

22. Deb A, Wang S, Skelding KA, Miller D, Simper D, Caplice NM. Bone marrow-derived cardiomyocytes are present in adult human heart: a study of gender-mismatched bone marrow transplant patients. *Circulation* 2003;107:1247–1249.

23. Kajstura J, Rota M, Whang B, *et al*. Bone marrow cells differentiate in cardiac cell lineages after infarction independently of cell fusion. *Circ Res* 2005;96:127–137.

24. Terada N, Hamazaki T, Oka M, Hoki M, Mastalerz DM, Nakano Y, Meyer EM, Morel L, Petersen BE, Scott EW. Bone marrow cells adopt the phenotype of other cells by spontaneous cell fusion. *Nature* 2002;416(6880):542–545.

25. Togel F, Fu Z, Weiss K, *et al*. Administered mesenchymal stem cells protect against ischemic acute renal failure through differentiation-independent mechanisms. *Am J Physiol Renal Physiol* 2005;289:F31–F42.

26. Sethe S, Scutt A, Stolzing A. Aging of mesenchyymal stem cells. *Ageing Res Rev* 2006;5:91–116.

27. Heeschen C, Lehmann R, Honold J, Assmus B, Aicher A, Walter DH, Martin H, Zeiher AM, Dimmeler S. Profoundly reduced neovascularization capacity of bone marrow mononuclear cells derived from patients with chronic ischemic heart disease. *Circulation* 2004;109:1615–1622.

28. Saito T, Kuang JQ, Bittira B, Al-Khaldi A, Chiu RCJ. Xenotransplant cardiac chimera: immune tolerance of adult stem cells. *Ann Thorac Surg* 2002;74(1):19–24.
29. MacDonald DJ, Saito T, Shum-Tim D, Duong M, Bernier PL, Chiu RCJ. Persistence of marrow stromal cells implanted into acutely infarcted myocardium: observations in a xenotransplant model. *J Thorac Cardiovasc Surg* 2005;130(4):1114–1121.
30. Amado LC, Saliaris AP, Schuleri KH, St John M, Xie J-S, Cattaneo S, Durand DJ, Fitton T, Kuang JQ, Stewart G, Lehrke S, Baumgartner WW, Martin BJ, Heldman AW, Hare JM. Cardiac repair with intramyocardial injection of allogeneic mesenchymal stem cells after myocardial infarction. *Proc Natl Acad Sci USA* 2005;102:11474–11479.
31. Dai W, Hale SL, Martin BJ, Kuang J-Q, Dow JS, Wold LE, Kloner RA. Allogeneic mesenchymal stem cell transplantation in postinfarcted rat myocardium: short- and long-term effects. *Circulation* 2005;112:214–223.
32. Grinnemo KH, Mansson A, Dellgren G, Klingberg D, Wardell E, Drvota V, Tammik C, Holgersson J, Ringden D, Sylven C, Le Blanc K. Xenoreactivity and engraftment of human mesenchymal stem cells transplanted into infarcted rat myocardium. *J Thorac Cardiovasc Surg* 2004;127(5): 1293–1300.
33. Bittira B, Shum-Tim D, Al-Khaldi A, Chiu RC. Mobilization and homing of bone marrow stromal cells in myocardial infarction. *Eur J Cardiothorac Surg* 2003;24(3):393–398.
34. Caplan AI. Mesenchymal stem cells. *J Orthop Res* 1991;9(5):641–650.
35. Jaalouk DE, Eliopoulos N, Couture C, *et al.* Glucocorticoid-inducible retrovector for regulated transgene expression in genetically engineered bone marrow stromal cells. *Hum Gene Ther* 2000;11(13):1837–1849.
36. Pfaffl MW. A new mathematical model for relative quantification in real-time RT-PCR. *Nucleic Acids Res* 2001;29(9):e45.
37. Liechty KW, MacKenzie TC, Shaaban AF, *et al.* Human mesenchymal stem cells engraft and demonstrate site-specific differentiation after *in utero* transplantation *in sheep. Nat Med* 2000;6(11):1282–1286.
38. Chiu RC. "Stealth immune tolerance" in stem cell transplantation: potential for "universal donors" in myocardial regenerative therapy. *J Heart Lung Transplant* 2005;24(5):511–516.
39. Becker M, Nitsche A, Neumann C, *et al.* Sensitive PCR method for the detection and real-time quantification of human cells in xenotransplantation systems. *Br J Cancer* 2002;87(11): 1328–1335.
40. Nitsche A, Becker M, Junghahn I, *et al.* Quantification of human cells in NOD/SCID mice by duplex real-time polymerase-chain reaction. *Haematologica* 2001;86(7): 693–699.
41. Fukuhara S, Tomita S, Yamashiro S, *et al.* Direct cell-cell interaction of cardiomyocytes is key for bone marrow stromal cells to go into cardiac lineage *in vitro. J Thorac Cardiovasc Surg* 2003;125(6):1470–1480.
42. Ryan JM, Barry FP, Murphy JM, Mahon BP. Mesenchymal stem cells avoid allogeneic rejection. *J. Inflammation* 2005;2:8.
43. Le Blanc K, Tammik C, Rosendahl K, *et al.* HLA expression and immunologic properties of differentiated and undifferentiated mesenchymal stem cells. *Exp Hematol* 2003;31(10):890–896.
44. Aggarwal S, Pittenger MF. Human mesenchymal stem cells modulate allogeneic immune cell responses. *Blood* 2005;105(4):1815–1822.
45. Bartholomew A, Sturgeon C, Siatskas M, *et al.* Mesenchymal stem cells suppress lymphocyte proliferation *in vitro* and prolong skin graft survival *in vivo. Exp Hematol* 2002;30(1):42–48.
46. Matzinger P. The danger model: a renewed sense of self. *Science* 2002;296(5566): 301–305.
47. Iwasaki A, Medzhitov R. Toll-like receptor control of the adaptive immune responses. *Nat Immunol* 2004;5:987–995.

48. Prince JM, Levy RM, Yang R, *et al.* Toll-like receptor-4 signaling mediates hepatic injury and systemic inflammation in hemorrhagic shock. *J Am Coll Surg* 2006;202:407–417.

49. Liu CT, Yang YJ, Yin F, *et al.* The immunobiological development of human bone marrow mesenchymal stem cells in the course of neuronal differentiation. *Cell Immunol* 2006;244: 19–32.

50. Jiang H, Chess L. Regulation of immune responses by T cells. *N Engl J Med* 2006;354:1166–1176.

51. Puissant B, Barreau C, Bourin P, *et al.* Immunomodulatory effect of human adipose tissue-derived adult stem cells: comparison with bone marrow mesenchymal stem cells. *Br J Haematol* 2005;129:118–129.

52. Chang CJ, Yen ML, Chen YC, Chien CC, Huang HI, Bai CH, Yen BL. Placenta-derived multipotent cells exhibit immunosuppressive properties that are enhanced in the presence of interferon-γ. *Stem cell* 2006;24:2466–2477.

53. Min JY. Uamg Y, Sullivan MF, *et al.* Long-term improvement of cardiac function in rats after infarction by transplantation of embryonic stem cells. *J Thorac Cardiovasc Surg* 2003;125: 361–369.

54. Tsai MK, Ho HN, Chien HF, *et al.* The role of B7 ligands (CD80 and CD86) in CD152-mediated allograft tolerance: a crosscheck hypothesis. *Transplantation* 2004;77:48–54.

55. Li L, Baroja ML, Majumdar A, *et al.* Human embryonic stem cells possess immune-privileged properties. *Stem Cells* 2004;22:448–456.

56. Drukker M, Katchman H, Katz G, *et al.* Human embryonic stem cells and their differentiated derivatives are less susceptible for immune rejection than adult cells. *Stem Cells* 2006;24: 221–229.

9   The Preclinical Basis to Find the Right Choice of Stem Cells for Cardiac Transplantation

Matthias Siepe

Introduction

Stem cells are the building blocks through which organs are developed and maintained. A lot of researchers predict that stem cells will prove terrifically useful in clinical medicine. Possible uses include applications in diseases associated with cell deficiency. As one of the least regenerative organs in the body, the damaged heart may benefit greatly from addition of new functioning cells. Cardiovascular researchers have risen to this challenge and, as a result, cardiac repair is arguably the most advanced program in the emerging field of regenerative medicine. Progress has been rapid, from the beginnings with skeletal myoblasts or cardiomyoblasts, moving to multipotent adult stem cells and, finally, to embryonic stem cells. Figure 1A highlights the growing interest in cardiac stem cell research by simple counting relevant PubMed hits. The interest in this field seems to explode during the last ten years. From the articles found throughout the years, around two-thirds of the papers included animal models. Small animal models attained highest interest while large animals were used sparely (Fig. 1B). However, in 2004, a great proportion of studies conducted included human cells or were clinical studies while main characteristics of the mode of action in the most frequently used stem cell types is unclear.

This chapter reviews the relevant animal models that contributed to the great achievements in this research field. For the autologous stem cell types, this chapter explains (1) the proof of concept in small animals or *in vitro*, (2) the search for the underlying mechanism, and (3) possible translational steps before initiating clinical studies.

Skeletal Myoblasts (SM)

Proof of concept

The idea of using skeletal muscle to support the injured heart evolved well before the introduction of cell therapy as a treatment option. In 1987, the latissimus dorsi

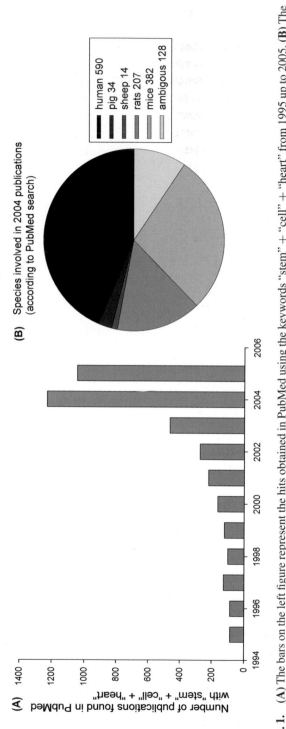

Fig. 1. (**A**) The bars on the left figure represent the hits obtained in PubMed using the keywords "stem" + "cell" + "heart" from 1995 up to 2005. (**B**) The fills of the pie figure on the right represent the frequency of different species studies found among the 2004 hits.

muscle, after being preconditioned by chronic pacing, was surgically wrapped around the failing heart to provide contractile support to the left ventricle — a procedure described by Alain Carpentier's group as dynamic cardiomyoplasty.[1] Within a few years, cardiomyoplasty was introduced on a cellular level when cells derived from the C_2C_{12} skeletal myoblast cell line were successfully transplanted into normal, uninjured mouse hearts.[2] Myoblasts are derived from satellite cells in the peripheral muscle, first described in 1961 as cells that can fully regenerate damaged skeletal muscle.[3] Skeletal myoblasts (SM) expand and form neofibers after muscle injury. Thus, it is not surprising that they were the first candidates for cardiac repair. In 1994, Zibaitis *et al.* reported the first successful experimental transplantation of skeletal myoblasts into injured heart.[4] The promising finding that transplanted skeletal myoblasts survived and formed striated muscle grafts within the hostile damaged cardiac tissue was followed by several independent experimental studies investigating the engraftment potential of these cells. When Taylor *et al.* first showed in 1998 that the successful engraftment of SM into injured myocardium improved left ventricular contractile performance and attenuated left ventricular remodeling, it was deemed novel.[5] Other groups and especially Menasché and colleagues confirmed the observation and optimized the preclinical setting for myoblast transplantation in small animals.[6–8]

The search for the underlying mechanism (and possible side effects)

How these successfully engrafted skeletal myoblasts improved function was unclear. It appeared that the muscle cells could improve contractility of the previously scarred heart without strictly transdifferentiating into cardiomyocytes. Furthermore, SM appeared to yield two populations of cells in injured heart — myogenin-positive transplanted skeletal muscle-like cells in the center of the scar and a second population of myogenin-negative more primitive cardiac muscle-like cells "recruited" around the scar periphery.[9] The transplanted skeletal myoblasts appeared to adapt to the surrounding myocardium by forming slow twitch myofibers that were electrically isolated from host cardiomyocytes[10] and yet improved cardiac performance. These results, showing an improvement in cardiac function without full integration of transplanted cells into host-myocardium and with recruitment of endogenous cells, raised questions about potential mechanisms that underlie functional benefit. Numerous mechanisms have been suggested, ranging from passive changes in wall stress or cardiac geometry to active mechanically induced contraction of the injected cells, but the exact mechanism(s) underlying the beneficial effect is still under debate. It is likely that the benefits involve both a direct effect of the transplanted muscle cells on performance and an indirect "paracrine" and trophic effect on endogenous cell recruitment and on remodeling of the failing heart. As the mechanisms underlying the positive effects of myoblast transplantation are still not fully understood, there is some debate as to whether myoblasts improve contraction or just prevent further deterioration of the injured heart — despite much preclinical data suggesting

Table 1. Cell type selection.

Cell type	Proliferation potential	Differentiation to cardiomy-ocytes	Large number available	Accepted for clinical trials	Selected references
Skeletal myoblasts	+	−	+	+	2–18, 20, 21, 39, 41
Mesenchymal stem cells	+	(+)	+	+	23, 25–51, 63–66
Endothelial precursor cells	(−)	(−)	(−)	+	24, 52–59
Hematopoetic stem cells	(+)	(−)	−	+	22, 60–62
Cord blood cells	+	(+)	(−)	(+)	67–70
Cardiac precursor cells	(−)	++	−	−	71–77
Embryonic stem cell	++	+	(−)	−	78–94

a direct positive effect of the cells above that seen with other or sham treatments. Despite these remaining questions, approximately 15 years of preclinical data have shown that autologous skeletal myoblast transplantation can augment both diastolic and systolic myocardial performance in a number of animal species after both acute and chronic injury (Table 1 adjusted to Ref. 11). These preclinical data opened the field of cardiovascular disease to this new therapeutic approach, showing that the centuries-old idea of organ repair could become clinical reality.

Although the preclinical and initial clinical data appear encouraging, myoblast transplantation is not without potential limitations. The first limitation is associated with any autologous cell that has to be expanded in the laboratory (i.e. myoblasts, mesenchymal stem cells, endothelial progenitor cells) that the use of autologous cells necessitates sufficient time between injury and injection to allow cell expansion *in vitro*. In normal healthy donors and many heart failure patients, this ranges from several days to several weeks depending on the cell type, which does not seem problematic if treatment is not urgent. In the case of acute MI where early treatment may be beneficial, either alternative cells can be chosen, or if myoblasts are truly superior, allogeneic cells may ultimately offer a solution. Of note, the two- to three-week period after acute myocardial injury is one of heightened inflammation, which in preclinical studies has been associated with increased loss of cells after transplantation. Thus, delayed autologous delivery could be beneficial even in this setting.

A second potential limitation to any cardiac cell therapy is the (in)ability of the transplanted cells to electrically integrate with native myocardium. Myoblasts are the most well-studied cell type, yet it is not clear if, and how, they electrically integrate into surrounding myocardium. Nor is it completely clear what impact integration may have on either function or cardiac rhythm. Menasché's group reports no electrical integration of myoblasts preclinically.[12] Likewise, Suzuki *et al.* report that in the absence of connexin-43 overexpression myoblasts do not couple well with surrounding heart.[13] Yet, as mentioned earlier, cells appear to contract in synchrony with surrounding tissue and contribute to cardiac performance. Early clinical data suggest that myoblast transplantation may be associated with a transient period of electrical instability within the first weeks after cell transplantation. How this occurs in the absence of cell coupling is unclear. The running randomized clinical trial should make it possible by interpretation of Holter data, to really assess whether myoblast transplantation increases the risk of arrhythmias beyond that related to the underlying heart failure substrate. Finally, the choice of patient population may also play a role in electrical instability.

A possible mechanism by which the cells might act on the surrounding tissue without being relevant as contractile units might be the parakrine pathway. This mechanism suggests that cytokines continuously released by the transplanted cells play a critical role. One possibility is the direct interaction with the remodeling process by these cytokines. Yacoub's group postulate that IL-1 is involved in this pathway.[14,15] Finally, our recent investigation points to a parakrine pathway as well. We implanted myoblast seeded scaffolds on the surface of a myocardial infarction zone. The cells remained either undifferentiated in the polyurethane scaffold[16] or differentiated entrapped in Matrigel (Fig. 2). Using both approaches we were able to improve function to the same degree as direct injection of the cells could.[17,18] We interpret this data in a way that the seeded scaffold acts like a cytokine factory.

Translational steps before and while initiating clinical studies

Skeletal myoblasts were the first to enter clinical arena after completion of a decade of experimental testing resulting in at least 40 studies. These studies primarily involved small animal myocardial infarction models. Due to problems with both transferability and the validity of hemodynamical measurements in small animal models,[19] it is recommended to perform large animal studies before initiating clinical trials. For skeletal myoblasts, large animal studies confirmed the preclinical finding before Menasché initiated the clinical trial. His laboratory chose a sheep model with chronic ischemic heart failure and myoblast transplantation and observed the functional consequences in the long term.[20] They proclaimed that this study "further legitimates the implementation of clinical trials". Another safety step before using skeletal myoblast transplantation in a larger clinical scale was performed by Pagani *et al.*[21] They injected autologous myoblasts during LVAD implantation and analyzed the resected hearts after the patients had received heart transplant (Fig. 3). By using this approach it is possible to search for the long-term fate of the cells and

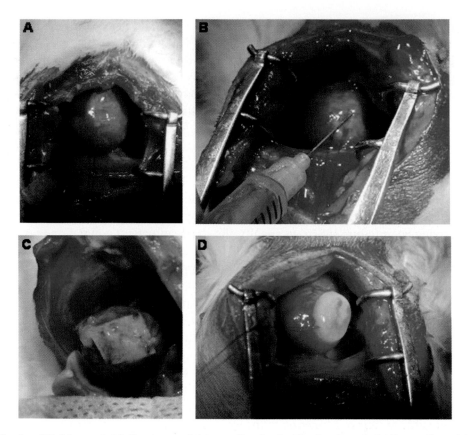

Fig. 2. **(A)** Myocardial infarction model in small animals. The Lewis rat's LAD was ligated 14 days before. Note the huge scar area. **(B)** Classical intramyocardial injection of the stem cells: 150 μl of cell enriched medium is injected using four to six microinjections with an insulin syringe. **(C)** A myoblast-seeded polyurethane patch was implanted four weeks before. Using this approach, numerous undifferentiated myoblasts were applied to the scar area. This application technique resulted in a functional benefit for the heart which is an indirect hint towards a paracrine action of the cells.[15] **(D)** The applied Matrigel contained differentiated myotubes. This approach resulted in an equivalent effect as direct injection.[18]

possible tumor formation without the risk of damaging a vitally important organ function.

Bone Marrow Stem Cells

Proof of concept

Fueled in part by hopes for cardiac transdifferentiation, as well as by the considerable body of data supporting angiogenic activity (see below) bone marrow studies moved remarkably quickly from small animals to clinical trials. The rapid progress

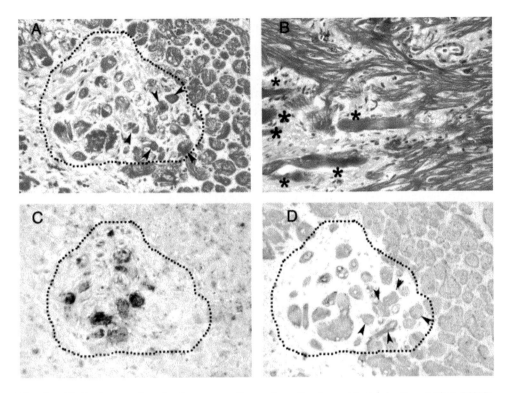

Fig. 3. Skeletal myotubes in the myocardium after direct myoblast injection during LVAD-implantation. The patient was transplanted eight months after injection and histological analysis performed on the excised heart. The dotted lines in A, C, and D mark the graft. Staining was performed using trichrome (**A and B**), Skeletal-specific fast myosin MY-32 (**C**) and myosin heavy chain beta slow (**D**). The alignment between grafted myofibers and host myofibers is striking.[17]

of bone marrow studies was facilitated by the extensive clinical experience already in place for marrow as a cell-based therapy. In addition, the large numbers of unfractionated cells that could be obtained with little processing also reduced both cost and regulatory burden. In this regard, it is important to remember that bone marrow is a heterogeneous tissue, containing rare hematopoietic,[22] mesenchymal[23] stem cells (around 0.01% of the total cell population), and endothelial progenitor cells[24] as well as large populations of committed progenitor cells and their highly differentiated progeny. Similar progenitor cell populations have also been isolated from umbilical cord blood. When considering studies involving unfractionated marrow, it is therefore important to keep in mind that this population contains > 99.9% committed cells, the composition of which may vary from study to study.

The initial report about the functional relevance of cardiac cell therapy using bone marrow cells was published in 2001 by Anversa's group.[25] Following this study several clinical studies were conducted in parallel to extensive preclinical work to identify the right cell mix and the mechanisms involved.

The search for the underlying mechanism (and possible side effects)

Mesenchymal stem cells (MSCs)

MSCs are rare multipotent progenitor cells also known as bone marrow stromal cells. In the past, MSCs were shown to differentiate into a number of mesenchymal cell types including adipocytes, fat, bone, cartilage, and skeletal muscle both *in vitro* and in infarcted rat myocardium.[26,27]

in vitro studies have focused on transdifferentiation of MSCs to cardiomyocytes. It was possible to identify cardiac specific surface markers after culture enrichment with 5-azacytidine.[28] The capacity seems to depend on the donor age.[29] An effective transdifferentiation in functional cardiomyocytes including contractile function and typical action potential was seldom reported and seems to be a rare finding.[30] Conversely, other authors were only able to obtain functional cardiomyocytes using a co-culture with cardiac cells.[31–33]

In vivo, there is also some evidence that, on injection into myocardium, these MSCs can differentiate into cardiomyocytes.[25,34] However, other studies suggest that this can only happen when MSCs are in contact with native cardiomyocytes[35] and does not happen in infarct interior. These data imply that the optimal time course of therapy using MSCs may be early in the disease state, when surviving cardiomyocytes are present in the infarct. Despite their inability to form cardiomyocytes to a significant degree, transplanted MSCs have been shown to engraft at high numbers in infarcted heart and lead to a number of functional benefits including increased neovascularization and improved regional contractility and global diastolic function.[36–38] In fact, the study of Thompson *et al.* suggests that MSCs and skeletal myoblasts improve function after ischemia-induced cardiac injury to a similar degree when compared in a side-by-side fashion,[39] while the results of other groups question the capability of truly regenerating infarcted myocardial tissue.[40,41] Moreover, the issue of the degree of transdifferentiation into cardiomyocytes still remains to be controversial.[42–44]

Furthermore, several studies suggest that MSCs can home to sites of injury following injection into the coronary or even peripheral vasculature.[35,45] However, it has also been reported that intracoronary administration of MSCs can actually cause micro-infarcts and promote damage of otherwise healthy myocardium which has led to some reticence about vascular delivery of these cells.[46]

More recently, MSCs have been touted as immune privileged cells capable of allogeneic administration *in vivo* with few or no negative consequences.[47,48] This is possibly the most tantalizing aspect of these cells for cardiac repair, the potential to develop an off-the-shelf product for use in many patients.

The mechanism by which these cells might improve function is still unclear because a transdifferentiation to a sufficient number of contractile units is more than questionable. One probable way to counteract remodeling is the neovascularization initiated by the MSCs transplantation.[49] Similar to the skeletal myoblasts, parakrine mechanisms may also play a critical role after MSC transplantation. Through an involvement of the Akt-pathway the cytokines VEGF, HGF and FGF-2 released by MSCs my help to protect cardiomyocytes from cell death.[44,50]

Another possible mechanism affects the recruitment of either bone marrow cells or of residual cardiac progenitors by excretion of G-CSF and bone morphogenetic protein-2 (BMP), respectively.[51]

Endothelial progenitor cells

EPCs are bone marrow-derived cells that are mobilized into peripheral blood and believed to participate in neoangiogenesis.[24] Recent research has shown that the number of EPCs present in vascular circulation is increased in patients following acute myocardial infarction.[52] EPCs are thought to respond to ischemic damage in the heart (and other tissues) and migrate to damaged areas to induce formation of neovasculature. In support of this, it was shown that when EPCs were injected into either the tail vein or the left ventricular cavity of rats with ischemic myocardial injury, a greater than twofold increase in the accumulation of infused EPCs was observed when compared with animals undergoing sham surgery.[53] Furthermore, rats undergoing intravenous injection of EPCs following ischemic myocardial injury showed improvements in ventricular dimensions, fractional shortening, and regional wall motion when compared with control animals injected with culture media.[54,55] Although the mechanism of EPC benefit has not been clearly elucidated, it is likely that improvements seen in this study were at least in part secondary to improved myocardial perfusion and subsequent rescue of hibernating myocardium. To date, EPCs have not been shown to participate in myogenesis within injured myocardium, but several paracrine roles have recently been attributed to the cells.[56]

Human EPCs are typically thought to originate from cells expressing CD133 (AC133), CD34, VEGF-R2, and occasionally CD14. The number EPCs present in human patients decreases with age and mirrors a rapid increase in cardiovascular deaths. It has been suggested that this correlation is due to EPCs' role in maintaining vascular health.[57] Recent data have shown that the number of circulating EPCs, and their ability to migrate, decreases in patients with increased risk of coronary artery disease, including myocardial infarction.[58] These data suggest that a reduction in the number and/or functional capacity of EPCs may be a critical factor in ischemic cardiovascular events. Rauscher *et al.* further showed in a genetic mouse model, that this finding is also relevant in aging.[59]

Hematopoetic stem cells

Historically, HSCs have been thought of as cells which differentiate only into various blood and immune cells (i.e. red and white blood cells). These cells are typically identified by labeling with CD34 and/or AC133 for human cells, and in mice are examined to be negative for a cocktail of mature hematopoetic cell lineage markers (lin-negative) and sometimes positive for stem cell antigen-1 (Sca-1) and c-kit (CD117). Over the past few years, it has been shown that HSCs have the ability under the right conditions to differentiate into various cell types, including cardiomyocytes.[60] Although HSCs can become cardiomyocytes under strict laboratory conditions, there is as yet no evidence that they can transdifferentiate into cardiomyocytes when transplanted into infarcted myocardium.[61,62] Perhaps because of this lack of differentiation in infarct, recent studies suggest that HSCs may not have the potential of some other cell types discussed here to improve LV function following transplantation into infarcted myocardium.[61]

Translational steps before and while initiating clinical studies

The step from the initial observation of the possible clinical relevance to a first wave of clinical trials was rapid. Unfortunately, useful translational studies are rare. To date, no investigation on the fate of transplanted bone marrow cells during LVAD implantation and later heart transplantation are published. An Investigator group at the Columbia University is currently starting a trial supported by the National Heart, Lung, and Blood Institute (NHLBI): "Bone Marrow Cell Transplantation to Improve Heart Function in Individuals With End-Stage Heart Failure." However, the angiogenetic effects of bone marrow cells have been investigated in few trials by stem cells injection in ischemic limbs and hands.[63–66] The therapeutic effect in these trials was attributed to the natural ability of marrow cells to supply endothelial progenitor cells and to secrete various angiogenic factors or cytokines. In these types of studies the contralateral limb may be a natural control. Tateishi-Yuyama *et al.* were able to perform a histological analysis of treated and non-treated limb after cardiac related death of one patient showing effective neovascularization.[66] All of the mentioned studies concluded that the therapy is safe meaning there was no notice of a tumor formation.

Cord Blood Cells

A relatively new source for stem or progenitor cells is umbilical cord blood, which contains most if not all of the bone marrow-derived cell types discussed above. Cord blood cells are easily obtained, albeit not in large volumes, have the potential to develop into multiple lineages, do not pose the ethical questions raised by embryonic stem cells, and are less immunogenic than their bone marrow counterparts — which means a larger percentage of the population could receive cells from matched donors. Further, if cord blood cells are isolated and stored at birth, these cells could provide an autologous source of stem cells to treat later myocardial damage. Current studies in animal models show that unfractionated cord blood cells injected directly into infarct show the potential to improve LV ejection fraction, anteroseptal wall thickening, and $dP/dt(max)$, while decreasing infarct size.[67] In addition, intravenous injection of cord blood cells in mice following ligation-induced ischemic cardiac injury showed approximately 20% higher capillary density in infarct border zones when compared with untreated animals.[68] However, contractility in this myocardial infarction model could only be positively affected by bone marrow cells, but not by the cord blood cells.[69] Furthermore, only weak evidence yet suggests that cord blood cells injected into infarcted portions of the myocardium have the potential to differentiate into mature cardiomyocytes.[70] Overall, however, it appears that cord blood cells may provide an interesting cell of choice for further study in the treatment of myocardial injury.

Cardiac Precursor Cells

Within the past several years, human cardiac stem cells have been identified that must now be considered as a potential cell for cardiac repair. Although the evidence for cardiac repair with these cells is limited, their potential to mature into cardiomyocytes makes them an attractive candidate. Again, these cells have primarily been isolated from neonatal hearts and to a very limited extent from adult myocardium.[71–73] Their preclinical use is intriguing and suggests that the future of cardiac repair may involve endogenous mobilization or recruitment of these stem cells – if they can be found to exist at reasonable numbers in adult myocardium.

Cardiac-derived stem cells can be isolated from neonatal rat hearts using LIM-homeodomain transcription factor islet-1 (isl1). It is possible to expand these cells *in vitro* when coupled with a cardiac mesenchymal feeder layer. Further, when these cells are co-cultured with neonatal cardiomyocytes, they are able to electrically integrate with myocardial cells *in vitro* by forming gap junctions.[73] Cardiac-derived stem cells isolated from adult hearts, including those from acutely infarcted, failing, and even uninjured transplant hearts, have been identified by their expression of c-kit, MRD1, and Sca-1 and by their lack of expression of hematopoetic lineage markers.[74] These cells show the ability to differentiate down myocyte, smooth muscle cell, and endothelial cell pathways, but their ability to form mature cells of these types (or cardiomyocytes) is unknown. Endogenous Sca-1-positive cardiac-derived stem cells may be able to differentiate into functional cardiomyocytes, but their differentiation potential within infarct scar is, as yet, unknown.[75] To date, methods for harvest, expansion, and *in vitro* growth of these precursors are still limited. This, combined with their unknown differentiation potential, makes their clinical use at this time difficult. Nonetheless, their biology is interesting and bears watching for future developments. Cardiac progenitor cells expanded from endomyocardial biopsies and predifferentiated *in vitro* could become very strong candidates for cardiac repair. Especially, because animals studies on cardiomyoblast transplantation using neonatal or fetal murine cells have shown a functional regeneration of the heart by an actual neomyogenesis.[76,77]

Embryonic Stem Cells

Embryonic stem cells (ESCs) appeared as promising candidates for cardiac repair in the mid-1990s. However, there were several difficulties that slowed their use for cardiac application although it had been described previously that these cells could give rise to cardiomyocytes.[78] Some other groups have shown that cardiomyocytes can be reliably obtained from the inner cell mass of preimplantation murine and human embryos (summarized in Ref. 79). Cardiomyocytes derived from ESCs express cardiac molecular markers such as sarcomeric myosin heavy chain, cardiac troponin I, GATA4, and Nkx2.5. Additional ultrastructural studies showing myofibrillar assembly and formation of intercalated disks indicate potential for a high degree of maturity.[80] Electromechanical integration by coupling and electrophysiological specialization have been observed as well.[81,82] The pluripotency of ESCs is their greatest advantage, but it also makes the isolation of the cells of interest more challenging. Current protocols yield cardiomyocytes in sufficient quantities for most basic research needs. However, a limitation of the use of ESCs for therapeutic purpose is the inefficiency with which cardiomyocytes are generated (around 1% of cells in differentiation culture). Strategies based on developmental paradigms have used directed differentiation to increase cardiomyocyte generation.[83] However, no method has been able to fully recapitulate the complex mixture of cardiogenic factors and environmental cues that efficiently induce cardiomyogenesis during embryonic development.

Alternatively, transplantation of undifferentiated ESC into the heart might provide the accurate signals to induce cardiogenesis and a functional improvement.[84,85] However, this approach leads to tumor formation.[86] The tumor formation is arguably the greatest risk associated with ESC-based therapy. Therefore, predifferentiation and purity are prerequisites to the application of cell-based therapies using ESCs. Klug *et al.*[87] reported a genetic technique for selecting highly purified cardiomyocytes, using the cardiac-specific α-myosin heavy chain promoter to drive expression of an antibiotic resistance gene. They showed that cardiomyocytes derived by this approach formed stable grafts in uninjured mouse hearts without tumor formation. Other groups have enriched for cardiomyocytes using restricted promoters in combination with fluorescence-based cell sorting. The promise of the genetic selection strategy has been further validated by Zandstra *et al.*[88]

Some studies used murine ESCs. However, this strategy has been difficult because mouse-derived ESC-cardiomyocytes have low-mitotic rates.[89] Conversely, human ESC-cardiomyocytes have a surprisingly high proliferative capacity.[90-92] The advantages of this capacity are multiple. For example, the biological mechanisms of proliferation control can be investigated using human cells. *In vivo* studies have raised hope that grafts of transplanted human ESC-cardiomyocytes can fully replace a damaged myocardial region.[93] However, ethical constrains forbid more extensive research on human ESC in several countries. Therefore, some groups search for other sources of cells with pluripotent potential.[94]

Closing Remarks

Unless researchers will establish ways to identify and expand human cardiac precursor cells in sufficient numbers or establish ways for a safe use of embryonic stem cells, the focus will remain on the autologous cell types mainly from the bone marrow and the peripheral muscle for clinical cell transplantation. However, their mode of action to regenerate myocardium is not understood in detail. Obviously, more preclinical research is needed to enhance the potential for improvement. Especially, the mechanisms by which the different cell types exert their effect need further clarification.

Other techniques to regenerate the damaged myocardium have been proposed. Gene therapy, growth factor application, and tissue engineering in combination with stem cell transplantation are thought to improve the effectiveness of stem cells. We are still at the beginning of cardiac stem cell therapy. If more randomized clinical studies confirm the safety and effectiveness of autologous cardiac stem cell transplantation, it pushes basic researchers even more to ascertain the accurate selection of stem cells

References

1. Chachques JC, Grandjean PA, Tommasi JJ, Perier P, Chauvaud S, Bourgeois I, *et al.* Dynamic cardiomyoplasty: a new approach to assist chronic myocardial failure. *Life Support Syst* 1987;5(4):323–327.
2. Koh GY, Klug MG, Soonpaa MH, Field LJ. Differentiation and long-term survival of C2C12 myoblast grafts in heart. *J Clin Invest* 1993;92(3):1548–1554.
3. Mauro A. Satellite cell of skeletal muscle fibers. *J Biophys Biochem Cytol* 1961;9:493–495.
4. Zibaitis A, Greentree D, Ma F, Marelli D, Duong M, Chiu RC. Myocardial regeneration with satellite cell implantation. *Transplant Proc* 1994;26(6):3294.
5. Taylor DA, Atkins BZ, Hungspreugs P, Jones TR, Reedy MC, Hutcheson KA, *et al.* Regenerating functional myocardium: improved performance after skeletal myoblast transplantation. *Nat Med* 1998;4(8):929–933.
6. Pouzet B, Vilquin JT, Hagege AA, Scorsin M, Messas E, Fiszman M, *et al.* Intramyocardial transplantation of autologous myoblasts: can tissue processing be optimized? *Circulation* 2000;102(19 Suppl 3):210–215.
7. Pouzet B, Ghostine S, Vilquin JT, Garcin I, Scorsin M, Hagege AA, *et al.* Is skeletal myoblast transplantation clinically relevant in the era of angiotensin-converting enzyme inhibitors? *Circulation* 2001;104(12 Suppl 1):223–228.
8. Pouzet B, Vilquin JT, Hagege AA, Scorsin M, Messas E, Fiszman M, *et al.* Factors affecting functional outcome after autologous skeletal myoblast transplantation. *Ann Thorac Surg* 2001;71(3):844–850.
9. Atkins BZ, Lewis CW, Kraus WE, Hutcheson KA, Glower DD, Taylor DA. Intracardiac transplantation of skeletal myoblasts yields two populations of striated cells *in situ*. *Ann Thorac Surg* 1999;67(1):124–129.

10. Murry CE, Wiseman RW, Schwartz SM, Hauschka SD. Skeletal myoblast transplantation for repair of myocardial necrosis. *J Clin Invest* 1996;98(11):2512–2523.

11. Siepe M, Heilmann C, von SP, Menasche P, Beyersdorf F. Stem cell research and cell transplantation for myocardial regeneration. *Eur J Cardiothorac Surg* 2005;28(2):318–324.

12. Leobon B, Garcin I, Menasche P, Vilquin JT, Audinat E, Charpak S. Myoblasts transplanted into rat infarcted myocardium are functionally isolated from their host. *Proc Natl Acad Sci USA* 2003;100(13):7808–7811.

13. Suzuki K, Brand NJ, Allen S, Khan MA, Farrell AO, Murtuza B, *et al.* Overexpression of connexin 43 in skeletal myoblasts: relevance to cell transplantation to the heart. *J Thorac Cardiovasc Surg* 2001;122(4):759–766.

14. Murtuza B, Suzuki K, Bou-Gharios G, Beauchamp JR, Smolenski RT, Partridge TA, *et al.* Transplantation of skeletal myoblasts secreting an IL-1 inhibitor modulates adverse remodeling in infarcted murine myocardium. *Proc Natl Acad Sci USA* 2004;101(12):4216–4221.

15. Suzuki K, Murtuza B, Beauchamp JR, Brand NJ, Barton PJ, Varela-Carver A, *et al.* Role of interleukin-1beta in acute inflammation and graft death after cell transplantation to the heart. *Circulation* 2004;110(11 Suppl 1):219–224.

16. Siepe M, Giraud MN, Tevaearai H, Liljensten E, Nydegger U, Menasche P, *et al.* Construction of skeletal myoblast-based polyurethane scaffolds for myocardial repair. *Artif Organs* 2007;31(6):425–433.

17. Siepe M, Giraud MN, Pavlovic M, Receputo C, Beyersdorf F, Menasche P, *et al.* Myoblast-seeded biodegradable scaffolds to prevent post-myocardial infarction evolution toward heart failure. *J Thorac Cardiovasc Surg* 2006;132(1):124–131.

18. Giraud MN, Ayuni E, Cook S, Siepe M, Carrel T, Tevaearai H. Hydrogel-based engineered skeletal muscle grafts (ESMGs) normalize heart function early after myocardial infarction. *Artif Organs*, in press.

19. Siepe M, Rüegg M, Giraud MN, Python J, Carrel T, Tevaearai HT. Effect of acute body positional changes on the hemodynamics of rats with and without myocardial infarction. *Exp Physiol* 2005;90(4):627–634.

20. Ghostine S, Carrion C, Souza LC, Richard P, Bruneval P, Vilquin JT, *et al.* Long-term efficacy of myoblast transplantation on regional structure and function after myocardial infarction. *Circulation* 2002;106(12 Suppl 1):131–136.

21. Pagani FD, DerSimonian H, Zawadzka A, Wetzel K, Edge AS, Jacoby DB, *et al.* Autologous skeletal myoblasts transplanted to ischemia-damaged myocardium in humans. Histological analysis of cell survival and differentiation. *J Am Coll Cardiol* 2003;41(5):879–888.

22. Abkowitz JL, Catlin SN, McCallie MT, Guttorp P. Evidence that the number of hematopoietic stem cells per animal is conserved in mammals. *Blood* 2002;100(7):2665–2667.

23. Pittenger MF, Martin BJ. Mesenchymal stem cells and their potential as cardiac therapeutics. *Circ Res* 2004;95(1):9–20.

24. Kalka C, Masuda H, Takahashi T, Kalka-Moll WM, Silver M, Kearney M, *et al.* Transplantation of *ex vivo* expanded endothelial progenitor cells for therapeutic neovascularization. *Proc Natl Acad Sci USA* 2000;97(7):3422–3427.

25. Orlic D, Kajstura J, Chimenti S, Jakoniuk I, Anderson SM, Li B, *et al.* Bone marrow cells regenerate infarcted myocardium. *Nature* 2001;410(6829):701–705.

26. Jiang Y, Jahagirdar BN, Reinhardt RL, Schwartz RE, Keene CD, Ortiz-Gonzalez XR, *et al.* Pluripotency of mesenchymal stem cells derived from adult marrow. *Nature* 2002;418(6893):41–49.

27. Verfaillie CM, Pera MF, Lansdorp PM. Stem cells: hype and reality. *Hematol Am Soc Hematol Educ Program* 2002;369–391.

28. Xu W, Zhang X, Qian H, Zhu W, Sun X, Hu J, *et al.* Mesenchymal stem cells from adult human bone marrow differentiate into a cardiomyocyte phenotype *in vitro*. *Exp Biol Med (Maywood)* 2004;229(7):623–631.

29. Zhang H, Fazel S, Tian H, Mickle DA, Weisel RD, Fujii T, *et al.* Increasing donor age adversely impacts beneficial effects of bone marrow but not smooth muscle myocardial cell therapy. *Am J Physiol Heart Circ Physiol* 2005;289(5):H2089–H2096.

30. Makino S, Fukuda K, Miyoshi S, Konishi F, Kodama H, Pan J, *et al.* Cardiomyocytes can be generated from marrow stromal cells *in vitro*. *J Clin Invest* 1999;103(5):697–705.

31. Fukuhara S, Tomita S, Yamashiro S, Morisaki T, Yutani C, Kitamura S, *et al.* Direct cell-cell interaction of cardiomyocytes is key for bone marrow stromal cells to go into cardiac lineage *in vitro*. *J Thorac Cardiovasc Surg* 2003;125(6):1470–1480.

32. Rangappa S, Entwistle JW, Wechsler AS, Kresh JY. Cardiomyocyte-mediated contact programs human mesenchymal stem cells to express cardiogenic phenotype. *J Thorac Cardiovasc Surg* 2003;126(1):124–132.

33. Wang T, Xu Z, Jiang W, Ma A. Cell-to-cell contact induces mesenchymal stem cell to differentiate into cardiomyocyte and smooth muscle cell. *Int J Cardiol* 2006;109(1):74–81.

34. Kawada H, Fujita J, Kinjo K, Matsuzaki Y, Tsuma M, Miyatake H, *et al.* Nonhematopoietic mesenchymal stem cells can be mobilized and differentiate into cardiomyocytes after myocardial infarction. *Blood* 2004;104(12):3581–3587.

35. Strauer BE, Brehm M, Zeus T, Kostering M, Hernandez A, Sorg RV, *et al.* Repair of infarcted myocardium by autologous intracoronary mononuclear bone marrow cell transplantation in humans. *Circulation* 2002;106(15):1913–1918.

36. Kocher AA, Schuster MD, Szabolcs MJ, Takuma S, Burkhoff D, Wang J, *et al.* Neovascularization of ischemic myocardium by human bone-marrow-derived angioblasts prevents cardiomyocyte apoptosis, reduces remodeling and improves cardiac function. *Nat Med* 2001;7(4):430–436.

37. Schuster MD, Kocher AA, Seki T, Martens TP, Xiang G, Homma S, *et al.* Myocardial neovascularization by bone marrow angioblasts results in cardiomyocyte regeneration. *Am J Physiol Heart Circ Physiol* 2004;287(2):H525–H532.

38. Thompson RB, van den Bos EJ, Davis BH, Morimoto Y, Craig D, Sutton BS, *et al.* Intracardiac transplantation of a mixed population of bone marrow cells improves both regional systolic contractility and diastolic relaxation. *J Heart Lung Transplant* 2005;24(2):205–214.

39. Thompson RB, Emani SM, Davis BH, van den Bos EJ, Morimoto Y, Craig D, *et al.* Comparison of intracardiac cell transplantation: autologous skeletal myoblasts versus bone marrow cells. *Circulation* 2003;108(Suppl 1):II264–II271.

40. Agbulut O, Mazo M, Bressolle C, Gutierrez M, Azarnoush K, Sabbah L, *et al.* Can bone marrow-derived multipotent adult progenitor cells regenerate infarcted myocardium? *Cardiovasc Res* 2006;72(1):175–183.

41. Ott HC, Bonaros N, Marksteiner R, Wolf D, Margreiter E, Schachner T, *et al.* Combined transplantation of skeletal myoblasts and bone marrow stem cells for myocardial repair in rats. *Eur J Cardiothorac Surg* 2004;25(4):627–634.

42. Amado LC, Schuleri KH, Saliaris AP, Boyle AJ, Helm R, Oskouei B, *et al.* Multimodality noninvasive imaging demonstrates *in vivo* cardiac regeneration after mesenchymal stem cell therapy. *J Am Coll Cardiol* 2006;48(10):2116–2124.

43. Heil M, Ziegelhoeffer T, Mees B, Schaper W. A different outlook on the role of bone marrow stem cells in vascular growth: bone marrow delivers software not hardware. *Circ Res* 2004;94(5):573–574.

44. Noiseux N, Gnecchi M, Lopez-Ilasaca M, Zhang L, Solomon SD, Deb A, *et al.* Mesenchymal stem cells overexpressing Akt dramatically repair infarcted myocardium and improve cardiac function despite infrequent cellular fusion or differentiation. *Mol Ther* 2006;14(6):840–850.

45. Bittira B, Shum-Tim D, Al-Khaldi A, Chiu RC. Mobilization and homing of bone marrow stromal cells in myocardial infarction. *Eur J Cardiothorac Surg* 2003;24(3):393–398.

46. Vulliet PR, Greeley M, Halloran SM, MacDonald KA, Kittleson MD. Intra-coronary arterial injection of mesenchymal stromal cells and microinfarction in dogs. *Lancet* 2004;363(9411): 783–784.

47. Inoue S, Popp FC, Koehl GE, Piso P, Schlitt HJ, Geissler EK, *et al.* Immunomodulatory effects of mesenchymal stem cells in a rat organ transplant model. *Transplant* 2006;81(11):1589–1595.

48. Jiang XX, Zhang Y, Liu B, Zhang SX, Wu Y, Yu XD, *et al.* Human mesenchymal stem cells inhibit differentiation and function of monocyte-derived dendritic cells. *Blood* 2005;105(10):4120–4126.

49. Kamihata H, Matsubara H, Nishiue T, Fujiyama S, Tsutsumi Y, Ozono R, *et al.* Implantation of bone marrow mononuclear cells into ischemic myocardium enhances collateral perfusion and regional function via side supply of angioblasts, angiogenic ligands, and cytokines. *Circulation* 2001;104(9):1046–1052.

50. Gnecchi M, He H, Noiseux N, Liang OD, Zhang L, Morello F, *et al.* Evidence supporting paracrine hypothesis for Akt-modified mesenchymal stem cell-mediated cardiac protection and functional improvement. *FASEB J* 2006;20(6):661–669.

51. Kruithof BP, van WB, Somi S, Kruithof-de JM, Perez Pomares JM, Weesie F, *et al.* BMP and FGF regulate the differentiation of multipotential pericardial mesoderm into the myocardial or epicardial lineage. *Dev Biol* 2006;295(2):507–522.

52. Shintani S, Murohara T, Ikeda H, Ueno T, Honma T, Katoh A, *et al.* Mobilization of endothelial progenitor cells in patients with acute myocardial infarction. *Circulation* 2001;103(23): 2776–2779.

53. Aicher A, Brenner W, Zuhayra M, Badorff C, Massoudi S, Assmus B, *et al.* Assessment of the tissue distribution of transplanted human endothelial progenitor cells by radioactive labeling. *Circulation* 2003;107(16):2134–2139.

54. Kawamoto A, Gwon HC, Iwaguro H, Yamaguchi JI, Uchida S, Masuda H, *et al.* Therapeutic potential of *ex vivo* expanded endothelial progenitor cells for myocardial ischemia. *Circulation* 2001;103(5):634–637.

55. Masuda H, Kalka C, Asahara T. Endothelial progenitor cells for regeneration. *Hum Cell* 2000;13(4):153–160.

56. Urbich C, Knau A, Fichtlscherer S, Walter DH, Bruhl T, Potente M, *et al.* FOXO-dependent expression of the proapoptotic protein Bim: pivotal role for apoptosis signaling in endothelial progenitor cells. *FASEB J* 2005;19(8):974–976.

57. Khakoo AY, Finkel T. Endothelial progenitor cells. *Annu Rev Med* 2005;56:79–101.

58. Schmidt-Lucke C, Rossig L, Fichtlscherer S, Vasa M, Britten M, Kamper U, *et al.* Reduced number of circulating endothelial progenitor cells predicts future cardiovascular events: proof of concept for the clinical importance of endogenous vascular repair. *Circulation* 2005;111(22):2981–2987.

59. Rauscher FM, Goldschmidt-Clermont PJ, Davis BH, Wang T, Gregg D, Ramaswami P, *et al.* Aging, progenitor cell exhaustion, and atherosclerosis. *Circulation* 2003;108(4):457–463.

60. Yeh ET, Zhang S, Wu HD, Korbling M, Willerson JT, Estrov Z. Transdifferentiation of human peripheral blood CD34+-enriched cell population into cardiomyocytes, endothelial cells, and smooth muscle cells *in vivo*. *Circulation* 2003;108(17):2070–2073.

61. Deten A, Volz HC, Clamors S, Leiblein S, Briest W, Marx G, *et al.* Hematopoietic stem cells do not repair the infarcted mouse heart. *Cardiovasc Res* 2005;65(1):52–63.

62. Murry CE, Soonpaa MH, Reinecke H, Nakajima H, Nakajima HO, Rubart M, *et al.* Haematopoietic stem cells do not transdifferentiate into cardiac myocytes in myocardial infarcts. *Nature* 2004;428(6983):664–668.

63. Esato K, Hamano K, Li TS, Furutani A, Seyama A, Takenaka H, *et al.* Neovascularization induced by autologous bone marrow cell implantation in peripheral arterial disease. *Cell Transplant* 2002;11(8):747–752.

64. Koshikawa M, Shimodaira S, Yoshioka T, Kasai H, Watanabe N, Wada Y, *et al.* Therapeutic angiogenesis by bone marrow implantation for critical hand ischemia in patients with peripheral arterial disease: a pilot study. *Curr Med Res Opin* 2006;22(4):793–798.

65. Miyamoto M, Yasutake M, Takano H, Takagi H, Takagi G, Mizuno H, *et al.* Therapeutic angiogenesis by autologous bone marrow cell implantation for refractory chronic peripheral arterial disease using assessment of neovascularization by 99mTc-tetrofosmin (TF) perfusion scintigraphy. *Cell Transplant* 2004;13(4):429–437.

66. Tateishi-Yuyama E, Matsubara H, Murohara T, Ikeda U, Shintani S, Masaki H, *et al.* Therapeutic angiogenesis for patients with limb ischaemia by autologous transplantation of bone-marrow cells: a pilot study and a randomised controlled trial. *Lancet* 2002;360(9331):427–435.

67. Henning RJ, bu-Ali H, Balis JU, Morgan MB, Willing AE, Sanberg PR. Human umbilical cord blood mononuclear cells for the treatment of acute myocardial infarction. *Cell Transplant* 2004;13(7–8):729–739.

68. Ma N, Stamm C, Kaminski A, Li W, Kleine HD, Muller-Hilke B, *et al.* Human cord blood cells induce angiogenesis following myocardial infarction in NOD/scid-mice. *Cardiovasc Res* 2005;66(1):45–54.

69. Ma N, Ladilov Y, Moebius JM, Ong L, Piechaczek C, David A, *et al.* Intramyocardial delivery of human CD133+ cells in a SCID mouse cryoinjury model: bone marrow vs. cord blood-derived cells. *Cardiovasc Res* 2006;71(1):158–169.

70. Kogler G, Sensken S, Airey JA, Trapp T, Muschen M, Feldhahn N, *et al.* A new human somatic stem cell from placental cord blood with intrinsic pluripotent differentiation potential. *J Exp Med* 2004;200(2):123–135.

71. Anversa P, Nadal-Ginard B. Myocyte renewal and ventricular remodelling. *Nature* 2002; 415(6868):240–243.

72. Beltrami AP, Barlucchi L, Torella D, Baker M, Limana F, Chimenti S, *et al.* Adult cardiac stem cells are multipotent and support myocardial regeneration. *Cell* 2003;114(6):763–776.

73. Laugwitz KL, Moretti A, Lam J, Gruber P, Chen Y, Woodard S, *et al.* Postnatal isl1+ cardioblasts enter fully differentiated cardiomyocyte lineages. *Nature* 2005;433(7026):647–653.

74. Urbanek K, Torella D, Sheikh F, De AA, Nurzynska D, Silvestri F, *et al.* Myocardial regeneration by activation of multipotent cardiac stem cells in ischemic heart failure. *Proc Natl Acad Sci USA* 2005;102(24):8692–8697.

75. Oh H, Bradfute SB, Gallardo TD, Nakamura T, Gaussin V, Mishina Y, *et al.* Cardiac progenitor cells from adult myocardium: homing, differentiation, and fusion after infarction. *Proc Natl Acad Sci USA* 2003;100(21):12313–12318.

76. Muller-Ehmsen J, Peterson KL, Kedes L, Whittaker P, Dow JS, Long TI, *et al.* Rebuilding a damaged heart: long-term survival of transplanted neonatal rat cardiomyocytes after myocardial infarction and effect on cardiac function. *Circulation* 2002;105(14):1720–1726.

77. Reffelmann T, Leor J, Muller-Ehmsen J, Kedes L, Kloner RA. Cardiomyocyte transplantation into the failing heart-new therapeutic approach for heart failure? *Heart Fail Rev* 2003;8(3):201–211.

78. Doetschman TC, Eistetter H, Katz M, Schmidt W, Kemler R. The *in vitro* development of blastocyst-derived embryonic stem cell lines: formation of visceral yolk sac, blood islands and myocardium. *J Embryol Exp Morphol* 1985;87:27–45.

79. Wobus AM, Boheler KR. Embryonic stem cells: prospects for developmental biology and cell therapy. *Physiol Rev* 2005;85(2):635–678.

80. Westfall MV, Pasyk KA, Yule DI, Samuelson LC, Metzger JM. Ultrastructure and cell-cell coupling of cardiac myocytes differentiating in embryonic stem cell cultures. *Cell Motil Cytoskeleton* 1997;36(1):43–54.

81. He JQ, Ma Y, Lee Y, Thomson JA, Kamp TJ. Human embryonic stem cells develop into multiple types of cardiac myocytes: action potential characterization. *Circ Res* 2003;93(1):32–39.

82. Maltsev VA, Rohwedel J, Hescheler J, Wobus AM. Embryonic stem cells differentiate *in vitro* into cardiomyocytes representing sinusnodal, atrial and ventricular cell types. *Mech Dev* 1993;44(1):41–50.

83. Lev S, Kehat I, Gepstein L. Differentiation pathways in human embryonic stem cell-derived cardiomyocytes. *Ann NY Acad Sci* 2005;1047:50–65.

84. Behfar A, Zingman LV, Hodgson DM, Rauzier JM, Kane GC, Terzic A, *et al.* Stem cell differentiation requires a paracrine pathway in the heart. *FASEB J* 2002;16(12):1558–1566.

85. Hodgson DM, Behfar A, Zingman LV, Kane GC, Perez-Terzic C, Alekseev AE, *et al.* Stable benefit of embryonic stem cell therapy in myocardial infarction. *Am J Physiol Heart Circ Physiol* 2004;287(2):H471–H479.

86. Swijnenburg RJ, Tanaka M, Vogel H, Baker J, Kofidis T, Gunawan F, *et al.* Embryonic stem cell immunogenicity increases upon differentiation after transplantation into ischemic myocardium. *Circulation* 2005;112(9 Suppl):I166–I172.

87. Klug MG, Soonpaa MH, Koh GY, Field LJ. Genetically selected cardiomyocytes from differentiating embronic stem cells form stable intracardiac grafts. *J Clin Invest* 1996;98(1):216–224.

88. Zandstra PW, Bauwens C, Yin T, Liu Q, Schiller H, Zweigerdt R, *et al.* Scalable production of embryonic stem cell-derived cardiomyocytes. *Tissue Eng* 2003;9(4):767–778.

89. Klug MG, Soonpaa MH, Field LJ. DNA synthesis and multinucleation in embryonic stem cell-derived cardiomyocytes. *Am J Physiol* 1995;269(6 Pt 2):H1913–H1921.

90. McDevitt TC, Laflamme MA, Murry CE. Proliferation of cardiomyocytes derived from human embryonic stem cells is mediated via the IGF/PI 3-kinase/Akt signaling pathway. *J Mol Cell Cardiol* 2005;39(6):865–873.

91. Snir M, Kehat I, Gepstein A, Coleman R, Itskovitz-Eldor J, Livne E, *et al.* Assessment of the ultrastructural and proliferative properties of human embryonic stem cell-derived cardiomyocytes. *Am J Physiol Heart Circ Physiol* 2003;285(6):H2355–H2363.

92. Xu C, Police S, Rao N, Carpenter MK. Characterization and enrichment of cardiomyocytes derived from human embryonic stem cells. *Circ Res* 2002;91(6):501–508.

93. Laflamme MA, Gold J, Xu C, Hassanipour M, Rosler E, Police S, *et al.* Formation of human myocardium in the rat heart from human embryonic stem cells. *Am J Pathol* 2005;167(3): 663–671.

94. Guan K, Nayernia K, Maier LS, Wagner S, Dressel R, Lee JH, *et al.* Pluripotency of spermatogonial stem cells from adult mouse testis. *Nature* 2006;440(7088):1199–1203.

10 Bone Marrow-Derived Stem Cells for Treatment of Ischaemic Heart Disease: Background and Experience from Clinical Trials

Ketil Lunde and Svend Aakhus

Introduction

Cardiovascular disease is a major global health problem, accounting for 17 million deaths worldwide each year, and the majority of these deaths are caused by ischaemic heart disease.[1] Interventions like acute revascularisation with thrombolysis and/or percutaneous coronary intervention (PCI) after acute occlusion of a coronary artery, as well as coronary artery bypass grafting for patients with extensive coronary artery disease, the use of implantable cardiac defibrillators, pharmacologic blockade of the sympathetic system with β-blockers and of the renin–angiotensin–aldosteron system with ACE inhibitors and/or AT-II receptor blockers, antithrombotic and/or anticoagulant therapy with aspirin, clopidogrel, warfarin and heparin and cholesterol-lowering drugs like statins with their multiple biologic effects have all improved prognosis for patients with ischaemic heart disease. Despite this progress, or maybe because of it since this progress for one thing has led to more patients surviving acute myocardial infarction, there is an increase in the population experiencing the problems caused by inadequate blood supply to the heart: angina pectoris and reduced cardiac pump function caused by hibernating or irreversibly damaged myocardium. Hypertension is generally considered the major risk factor for congestive heart disease.[2] However, Fox *et al.* identified ischaemic heart disease as the main cause of this disabling condition when coronary angiography was performed in all patients with a new diagnosis of heart failure.[3] Patients diagnosed with congestive heart failure (CHF) have a dismal prognosis with an expected one-year mortality of 20%.[2]

The above-mentioned interventions all limit the consequenses of impaired blood supply to the heart. However, none promote effects that could solve the problem, namely the proliferation of blood vessels and cardiomyocytes. Instead, the use of stem cells has been introduced as a novel strategy to achieve the optimistic therapeutic goal of regenerating the human heart. First, we will discuss the findings that led to the hypothesis that cells derived from bone marrow could be used for this purpose.

The long-standing paradigm that the human heart cannot regenerate has recently been challenged by several observations. Adult mature cardiomyocytes are terminally differentiated and have either no or limited regenerative capacity. However, there is growing evidence that there is a continuous low-rate turnover of cardiomyocytes throughout life.[4] Cardiomyocyte death by apoptosis is confirmed in pathological conditions in the human heart, and a low-grade degree of apoptosis is probably also a physiological phenomenon in the normal heart.[5] Some estimates indicate that the entire human heart would disappear in a few decades if there was no regeneration.[6] The evidence for regeneration of myocardium as a naturally physiological phenomenon is supported by observations of cells in cell cyclus within the myocardium. The number of such cells is increased in pathological processes as myocardial infarction,[7] in myocardial hypertrophia caused by aortic stenosis,[8] and in end-stage cardiomyopathy.[9] Recently, several cell populations with regenerative properties have been identified within the heart.[10–16] These cells typically lie in clusters and express markers that indicate cardiomyocyte differentiation. The origin of these cell populations is not clarified. They may represent intrinsic stem cell populations within the heart. However, some studies also indicate a bone marrow origin.

Evidence that cells derived from bone marrow are involved in myocardial regeneration

The bone marrow contains a hierarchy of several stem cell populations with varying potential for differentiation. The multiple adult progenitor cells (MAPC — primarily characterised by their lack of several surface markers) are primitive cells with the potential for *in vitro* differentiation to cell types of all three germ layers after being injected into early blastocysts.[17] These cells may be the origin of the haemangioblast that typically express CD 34 and CD 133,[18] and the mesenchymal stem cells (MSC). MSC are primarily characterised by their ability to adhere to plastic in culture and their ability for differentiation to tissues of mesodermal origin like bone, cartilage, and fat in culture. There are no uniformly accepted surface markers to identify MSC.[19] The haemangioblast is the precursor of the haematopoietic stem cells (HSC), which also typically express CD 34 and CD133,[20] and to the endothelial progenitor cells (EPC). Characterisation of EPC is not uniform, but some features include expression of CD 34 and the receptor for vascular endothelial growth factor, VEGFR2(KDR), the lack of markers seen on fully differentiated endothelial cells, and the potential to form mature endothelial cells in culture.[21] Circulating cells with similar properties are often referred to as circulating progenitor cells (CPC).

In addition, a stem cell population of side population (SP) cells, characterised by their ability to expel the Hoechst dye 33342 has been identified. SP-cells may have the ability to reconstitute the entire bone marrow after single cell grafting.[22] All these bone marrow cell populations are mononucleated and scarce (less than 1% of all cells).

Why cells derived for bone marrow could be useful for cardiac repair

The regenerative capacity of the bone marrow is well established in the clinic. Since the 1960s, autologous and allogenic bone marrow transplantation have reconstituted depleted bone marrow and saved the lives of thousands of patients with haematological and neoplastic diseases. In the end of the 1990s, several groups reported that bone marrow cells (BMC) were capable of generating a much wider spectrum of cells than previously thought by so-called transdifferentiation. This means that stem cells differentiate to cells outside their usual lineage. This ability is referred to as stem cell plasticity. *In vitro* experiments and animal models primarily in rodents claimed transdifferentiation of BMC to cells of such different organs as brain,[23] liver,[24] and skeletal muscle.[25] In 2001, these findings where extended to the heart by reports that transplanted BMC had transdifferentiated to endothelial cells, smooth muscle cells and cardiomyocytes, and partly regenerated the myocardium in experimental models of AMI in mice.[26,27] Another study found that transplanted cells reduced cardiomyocyte apoptosis by improved perfusion caused by formation of new blood vessels.[28]

Studies in patients after sex-mismatched bone marrow or cardiac transplantation have supported the notion that cells derived from bone marrow may contribute to regeneration of the heart. In male recipients of female hearts, varying degrees of Y-chromosome positive endothelial cells, Schwann cells, smooth muscle cells and cardiomyocytes have been found in the transplanted hearts.[29–32] Similar to these findings, in female recipients of male hearts, XX-chromosome positive cardiomyocytes were identified.[32] The origin of these cells is not clear, and it has been postulated that intrinsic cardiac stem cells from reminiscent recipient atrial tissue may have populated the transplanted heart.

However, studies in female recipients of male bone marrow support the notion that these cells derive from the bone marrow. In autopsy materials of such patients, Y-chromosome positive cardiomyocytes[33] and smooth muscle cells[34] have been identified. These findings suggest that circulating BMC have homed to the heart and transdifferentiated. Thus, regeneration of the heart by cells derived from bone marrow may be a naturally occurring phenomenon that handles the need for a continuous, low-grade turnover of cardiac cells. However, this mechanism is obviously insufficient in the setting of acute large cell loss, since myocardial infarction heals with scar formation.[35] Therefore, attempts to improve the effect of BMC mediated cardiac repair by direct transplantation of cells to the heart were tempting, and therefore tested promptly in the clinic after the publication of the initial studies in rodents.

After the first clinical studies were initiated, several groups failed to reproduce transdifferentiation of transplanted BMC in experimental myocardial infarction.[36–38] Thus, the scientific foundation for clinical trials with BMC has been challenged.[39] However, a vast number of experimental studies irrespective of the possible mechanism(s) involved have shown that transplantation of several different bone marrow cell populations cells have improved cardiac function after myocardial infarction. As a natural consequence of this discrepancy, alternative explanations for possible effects have been suggested, e.g. paracrine effects[40] and immune modulation.[41] A comprehensive discussion about the possible mechanisms involved is beyond the scope of this chapter.

Cell Populations

Since the cell population that may be responsible for a positive effect on cardiac function has not yet been identified, the pragmatic approach of using unfractioned cell populations has been chosen in the majority of clinical studies. However, since all progenitor/stem cells are mononuclear, most of the studies with unfractioned BMC have isolated mononuclear cells with the well-proven ficoll density gradient centrifugation before patient administration. Apheresis is another method for isolation of mononuclear cells from bone marrow or peripheral blood. Gelatine-polysuccinate density gradient sedimentation has also been used for preparation of bone marrow before cardiac application. By this method, only erythrocytes and platelets are depleted, and the final cell product contains a large number of polynucleated cells without proliferative capacity in addition to the mononuclear cells. Some groups have advocated the use of enriched progenitor cell populations due to the possible deleterious effects of injecting a large number of inflammatory cells. After isolation of mononuclear cells from peripheral blood, endothelial progenitor cells can be *ex vivo* cultivated within a few days. Bone marrow-derived MSC can be cultivated and greatly expanded within days to weeks out of mononuclear BMC. Other specific progenitor cell populations like CD 133+ or CD 34+ cells can be obtained from unfractioned bone marrow using techniques like FACS or MACS cell sorting. Mobilisation of bone marrow stem and progenitor cells to peripheral blood can be facilitated by the administration of cytokines like granulocyte-colony stimulating factor (G-CSF) and granulocyte-monocyte-colony stimulating factor (GM-CSF). Animal studies have suggested that BMC mediated cardiac repair can be achieved with this strategy.[42,43] In clinical trials, this approach has been used alone or as an adjunct to increase the number of stem/progenitor cells that can be obtained from peripheral blood prior to cardiac administration.

Selection of Patients

The possible positive effects of therapies with cells derived from bone marrow for treatment of patients with ischaemic heart disease will in theory be useful in three different clinical scenarios. (1) Acute myocardial infarction (AMI). Most of the experimental work was performed in the setting of AMI, and most clinical trials to date have studied this patient population. It has been presumed that despite an inflammatory environment that may be hostile for transplanted cells, there might still be signals within the infarcted myocardium that may target differentiation of transplanted cells towards constituents of functional myocardium. Patients with AMI would benefit from both recovery of contractile elements and improved blood supply. (2) The patients with end-stage ischaemic heart disease with no further revascularisation options, but without significant loss of viable myocardium. These patients would benefit mainly by improved blood supply. In this situation, an increase in myocardial muscle mass would potentially even be deleterious as this would increase the metabolic needs of the myocardium. (3) Patients with established CHF caused by ischaemic cardiomyopathy. Most of these patients have extensive myocardial scars after previous myocardial infarction, and need regeneration of all constituents of viable, functional myocardium. In clinical practice, the two latter populations will often be overlapping. This is illustrated in Fig. 1. Instead of a clear distinction between viable myocardium and a transmural, well-defined scar, MRI late enhancement show that this patient has mainly subendocardial infarction, but also areas with transmural scars. The SPECT perfusion scans show that the patient also has areas with exercise-induced ischaemia.

Cell transplantation therapies may be closest to clinical application for the patients with AMI. However, patient selection in this situation is crucial. Most AMI patients have an excellent prognosis if they receive prompt revascularisation and state-of-the art medical therapy. Only a small percentage of these patients are at risk of developing CHF. At present, we have limited established acute phase criteria to select patients with a dismal prognosis, patients where any possible effect of cell transplantation would be of use.[44] On the other hand, in the two latter patient populations with chronic heart disease, the patients that could benefit from cell therapy are easy to identify. Improved blood supply by administration of bone marrow cells to the patients with end-stage ischaemic heart disease may be a realistic option. However, while patients with established myocardial scars are those with the greatest need for a therapeutic option that could regenerate the myocardium, cell-based therapies in this situation are currently furthest from clinical application.

A less ambitious goal than regeneration of myocardium would be to improve the passive mechanical properties of the scar tissue. This approach could be useful for both patients with AMI and for those with established myocardial scars. This could theoretically be obtained by proliferation of fibroblasts and effects on the extracellular matrix via transdifferentiation and/or paracrine effects.

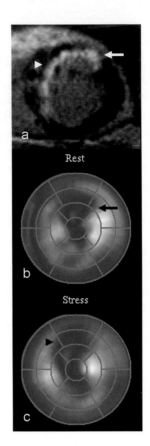

Fig. 1. MRI and SPECT images from a patient with previous anterior myocardial infarction. **(a)** MRI with gadolinium contrast. Bright areas (late enhancement) indicate myocardial scar. Note transmural infarction in free anterior wall (arrow), and subendocardial infarction in the interventricular septum (arrowhead). **(b)** SPECT "bulls-eye" obtained at rest shows anteroseptal perfusion defect. **(c)** During exercise, the perfusion defect extends compared to at rest, indicating reversible ischaemia in the peri-infarct zone.

Administration of the Cells

At present, three different techniques have been used for transplantation of cells to the human heart (Fig. 1). (1) Intracoronary injection through the lumen of an over-the-wire balloon catheter during intermittent balloon inflation for stop of flow during cell injections. This technique was assumed to be beneficial based on (a) the assumption that cells would be directed towards injured areas against a gradient of homing factors, (b) that the transplanted cells only would be directed to areas with sufficient blood supply, thereby attenuating cell death caused by ischaemia, (c) the assumption that stop of flow would reduce immediate wash-out of cells and therefore enhance extravasation and homing of cells to the infarcted areas, and (d) the less invasive approach than the alternative route of direct intramyocardial cell injections. (2) Percutaneous, catheter-based transendocardial intramyocardial injection from the lumen of the left ventricle. Different mapping systems are used to direct cell injections to the myocardial regions of interest. (3) Direct intramyocardial injections through the epicardium during open chest or alternatively minimal invasive approaches. This can be combined with coronary artery bypass grafting. It has been demonstrated that after intracoronary injection with the previously described stop-flow technique in patients with AMI, only 1% of radiolabeled mononuclear BMC remained in the heart 18 hours after injection of the cells.[45] In another study where BMC prepared by gelatine-polysuccinate density sedimentation were radiolabeled with 18F-FDG, 1.3%–2.6% of the cells were detected in the myocardium on PET-scans after 50–75 minutes.[46] However, when injecting cells expressing CD34, 14%–39% of the cells were detected within the heart using similar methods. Most cells were found in the infarct border zone. More surprisingly, also after intramyocardial injection of cells, most of the cells are drained from the heart. In an experimental swine model of AMI, Hou *et al.* found that one hour after administration of mononuclear cells from peripheral blood, 11% of the cells were found within the myocardium after intracoronary injection, versus 2.6% after intracoronary and 3.2% after interstitial retrograde coronary venous delivery.[47] In another study using an experimental swine AMI model, more MSC remained in the heart after intracoronary than transendocardial administration (both techniques were more efficient than intravenous delivery).[48] However, the superior retention of cells may have been caused by capillary plugging instead of homing to the myocardium, since cultivated MSC are relatively large cells. Indeed, intracoronary injection of MSC in dogs induced microinfarctions.[49] Cell harvesting, processing and administration is illustrated in Fig. 2.

Clinical Studies

AMI

The first studies were undertaken immediately after the reports claiming transdifferentiation of BMC to new myocardium in experimental myocardial infarction in mice. In 2002, Strauer *et al.* published results of ten patients treated with intracoronary injection of mononuclear BMC in the infarct-related coronary artery 7 ± 2 days after acute myocardial infarction reperfused with PCI in the acute phase.[50] They described the stop-flow PCI technique which has been used with only minor modifications in following studies, and most importantly, that the technique was feasible and seemed safe. Compared to the control group which comprised ten patients who refused cell therapy, the results indicated improved perfusion measured by SPECT, and regional left ventricular (LV) function by angiography after three months follow-up. This study was followed by the TOPCARE-AMI study.[51] Here, 59 patients with AMI treated with PCI received either mononuclear BMC ($n = 29$) or CPC ($n = 30$). Compared to historic controls, results indicated improved global and regional left ventricular function and reduction of infarct size. Similar results were obtained from several other unrandomised trials with small patient numbers.[52–54] While all these

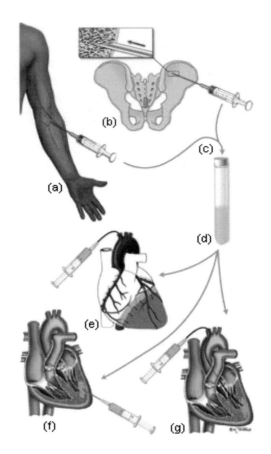

Fig. 2. Cell harvest, processing and administration. (**a**) Cell harvest from peripheral blood. (**b**) Bone marrow aspiration. (**c**) Cell processing. (**d**) Final cell product. (**e**) Intracoronary injection. (**f**) Direct intramyocardial injection. (**g**) Transendocardial intramyocardial injection.

trials confirmed that the treatment was feasible and seemed safe, and most of them indicated a beneficial effect of cell treatment, the lack of randomised control groups made it impossible to conclude about efficacy.

The natural course of AMI

For assessment of the results from randomised clinical trials in this field, it is worthwhile to compare the results with the expected outcome after AMI treated with state-of-the-art therapy. In the Capricorn echo substudy, treatment with the beta-blocker carvedilol in AMI patients resulted in an improvement in left ventricular ejection fraction (LVEF) of 5% after six months, compared to no change in the placebo group.[55] In a meta-analysis by Abdulla *et al.*, ACE inhibitors were found to increase LVEF with 3.75% in patients with myocardial infarction and LVEF below 45% at baseline.[56] Reperfusion with PCI in the acute phase has also been shown to increase the LVEF. In the Cadillac trial, Stone *et al.* found that patients treated with PCI with stenting increased LVEF with 3%–4% seven months after the acute event as measured by ventriculography.[57] In a population of 110 patients with AMI where 97 were treated with PCI, MRI was performed six days after the acute event and repeated after approximately nine months in 89 patients, LVEF increased with 6.7% and infarct size declined with 6.4% with a relative decline of 40% in patiens with microvascular obstruction.[58] In patients without microvascular obstruction, LVEF increased with 7.4% and infarct size declined with 2.0% with a relative decline of 25%. Baks *et al.* studied 22 patients treated with drug-eluting stents in the infarct-related coronary artery 2.5 hours after onset of symptoms.[59] All patients were treated with a beta-blocker and the majority also with an ACE inhibitor. Cardiac function was assessed with MRI after five days and repeated after five months. LVEF increased with 7%, and infarct size decreased with 31% (relative change). In addition, our own ASTAMI trial shows that patients with depressed ejection fraction at baseline have the greatest improvement (unpublished data). With the use of SPECT, the overall increase in LVEF from baseline (four days after the AMI) to six months follow-up was 11.2% for patients with LVEF below the median value at baseline (43%), and 4.7% for those with baseline LVEF above the median value. In the ASTAMI trial, all patients were given a beta-blocker and an ACE inhibitor or an ATII-receptor blocker.[60]

Thus, after AMI with conventional treatment, a marked improvement in left ventricular function should be expected, and this improvement should be greatest for those with a depressed ejection fraction at baseline. In addition, infarct size will diminish during follow-up. The magnitude of these changes will depend upon the timing of the baseline examination and the sensitivity of the imaging modality to detect serial changes. Due to recovery of myocardial stunning,[61] and overestimation of infarct size in the acute phase due to myocardial oedema,[62] the reported changes will be greater when baseline recordings are performed early after the acute event. If a statistical difference is reported in a clinical trial, one should assess if this difference was obtained by a positive effect of the intervention, or an unexpected poor outcome in the control group.

Randomised clinical trials with intracoronary injection of BMC in AMI

To our knowledge, seven randomised controlled trials including 20 or more patients with administration of bone-marrow derived cells in the acute phase of AMI have been published to date (February 2007). All used intracoronary administration of cells, with a similar technique as that first described by Strauer *et al.*[50] Results of these trials are summarised in Table 1. The primary end-point in the majority of these studies was the change in LVEF,[60,63–66] in one study it was cardiac death,[67] and one study defined a combined primary end-point of safety and efficacy.[68]

In the BOOST trial, 30 patients with AMI were randomised to intracoronary injection of BMC and 30 patients to a control group.[63] In the treatment group only,

Table 1. Randomised trials with 20 or more patients with intracoronary administration of BMC or CPC in AMI patients.

	N	Dose (10⁶)	Days after AMI	End-point evaluation	Follow-up (months)	Baseline LVEF (%)	ΔLVEF (%points)	P ΔLVEF	ΔLVEDV (ml)	ΔInfarct size
BOOST[63]	30 BMC 30 Control	2460 ± 940	4.8 ± 1.3	MRI	6	50.0 ± 10.0 51.3 ± 9.3	6.7 ± 6.5 0.7 ± 8.1	< 0.01	7.6 ± 20.0* 3.4 ± 11.1*	−14.1 ± 13.0 ml −10.5 ± 10.6 ml
BOOST[69]	30 BMC 30 Control			MRI	18		5.9 ± 8.9 3.1 ± 9.6	0.27	6.1 ± 20.3* 3.6 ± 15.1*	−12.8 ± 11.8 ml −10.1 ± 13.1 ml
Chen et al.[67]	34 MSC 35 Placebo	48 000–60 000	18	Echo LV Angio† PET	6	49 ± 9	18 48 ± 10	0.01 6	NA	NA
Janssens et al.[64]	34 BMC 33 Placebo	304 ± 128 (172 ± 72)	1	MRI	4	48.5 ± 7.2 46.9 ± 8.2	3.4 ± 6.9 2.2 ± 7.3	0.36	2.8 ± 15.2* 2.8 ± 15.0*	−10.2 ± 7.9 g −7.9 ± 8.5 g
ASTAMI[60]	50 mBMC 50 Control	87 ± 48	6.0 ± 1.3	SPECT	6	41.3 ± 10.4 42.6 ± 11.7	8.1 ± 11.2 7.0 ± 9.6	0.77	−11.2 ± 36.0 −1.8 ± 17.6	11.0 ± 12.7% points −7.8 ± 8.7% points
				Echo		45.7 ± 9.4 46.9 ± 9.6	3.1 ± 7.9 2.1 ± 9.2	0.70	8.9 ± 28.5 10.8 ± 29.1	NA
				MRI		54.8 ± 13.6 53.6 ± 11.6	1.2 ± 7.5 4.3 ± 7.1	0.054	−6.9 ± 34.3 −2.8 ± 20.0	−2.3 ± 11.2 ml −5.9 ± 13.6 ml
REPAIR-AMI[65]	101 mBMC 103 Placebo	236 ± 174	4.3 ± 1.3	LV Angio	4	47.5 ± 10.0 46.7 ± 10.3	5.5 ± 7.3 3.0 ± 6.5	0.01	12 ± 31 14 ± 33	NA

(Continued)

Table 1. (*Continued*)

	N	Dose (10^6)	Days after AMI	End-point evaluation	Follow-up (months)	Baseline LVEF (%)	ΔLVEF (%points)	p ΔLVEF	ΔLVEDV (ml)	ΔInfarct size
Meluzin et al.[68]	22 mBMC	100 (90–200)	7 ± 1.4	SPECT‡	3	41 ± 2	5 ± 1	0.04	1 ± 6	−10 ± 2
	22 mBMC	10 (9–20)				42 ± 2	3 ± 1	0.53	4 ± 7	−8 ± 2
	22 Control					42 ± 2	2 ± 1		16 ± 7	−8 ± 3
TCT-STAMI[66]	10 mBMC	39 ± 22	0.5	Echo	6	53.8 ± 9.2	4.8	NA	−0.4§	−7% points
	10 Placebo			SPECT: infarct size		58.2 ± 7.5	−1.9		4.8§	−2% points
Intracoronary application of CPC after G-CSF administration										
MAGIC Cell-3-DES[77]	25 CPC	1400 ± 500	4	MRI	6	52.0 ± 9.9	5.1 ± 9.1	0.046	3.4	−12.5 ± 13.3 ml
	25 Control					53.2 ± 13.3	−0.2 ± 8.6		10.1	0.8 ± 14.3 ml
Li et al.[78]	35 CPC	73 ± 73 CD34+	6	Echo	6	50.0 ± 8.2	7.1	0.04	−25	NA
	23 Control					51.0 ± 8.1	1.6		8.6	

Values are mean ± SD or median (range) unless otherwise stated. BMC: Bone marrow cells, CPC: circulating progenitor cell, mBMC: mononuclear BMC, LVEF: left ventricular ejection fraction, LVEDV: left ventricular end-diastolic volume, LV: angio left ventricular angiography, NA: not applicable, G-CSF: granulocyte colony-stimulating factor. Δ = change (follow-up − baseline). *LVEDV index ($LVEDV/m^2$). †Uncertainity about which modality used for LVEF. Change values for LVEDV and infarct size not reported in both groups. ‡Values for LVEF, LVEDV and infarct size are mean ± SEM. §Left ventricular end-diastolic diameter.

128 ± 33 ml of bone marrow was aspirated from the posterior iliac crest under a brief, general anaesthesia. Unique to this trial, gelatine-density gradient sedimentation was used for preparation of the aspirated bone marrow, and since this method includes polynucleated cells in the final cell product, cell numbers were higher than in other studies. Intracoronary BMC administration was performed after approximately five days. Assessment of cardiac function with MRI was performed 3.5 ± 1.5 days after the acute event, and repeated after six months. LVEF increased with 6.7% in the BMC group and with 0.7% in the control group, $p < 0.01$. There were no differences between groups for the change in left ventricular end-diastolic volume (LVEDV) and infarct size. However, MRI was repeated after 18 months follow-up and at this time-point, the significant difference between groups for the change in LVEF had vanished.[69] The increase in LVEF from baseline to 18 months was 5.9% in the BMC group and 3.1% in the control group. The authors concluded that the effect of BMC therapy was sustained, while the control group had experienced an unexpected improvement. On the basis of the available data on the natural course after AMI, we believe the conclusion should be reversed, i.e. the statistically significant difference at six months was more likely due to the unexpected lack of improvement in the control group. In this study, cardiac function was also assessed by echocardiography.[70] A significant difference between groups for the change in E/A ratio was interpretated as an improvement in diastolic function after BMC therapy. However, no differences were found for the E/E ratio which better predict left ventricular filling pressure. Interestingly, LVEF measured by echocardiography was similar in both groups at all time-points.

In the study by Janssens *et al.*, 33 patients were randomised to intracoronary injection of BMC and 34 to a placebo group.[64] In local anaesthesia, 130 ± 22 ml bone marrow was aspirated from the hip bone in all patients one day after successful PCI for STEMI. The bone marrow was processed using ficoll density gradient centrifugation, and four to six hours later, patients received either BMC or placebo solution. The placebo medium was 0.9% sodium chloride with 5% autologous serum. MRI was performed four days after the acute event and repeated after four months. LVEF increased with 3.4% in the BMC group and 2.2% in the control group ($p = 0.36$). However, there was a significantly greater decline in infarct size determined by late enhanced MRI in the BMC than in the control group. In addition, the authors argue for a transient effect of BMC therapy on regional myocardial function assessed by strain Doppler echocardiography. However, the reported differences for these parameters are not uniformly consistent.

In our own ASTAMI trial, we studied patients with anterior wall AMI reperfused with PCI with stent on culprit lesion in the LAD between two and 12 hours after onset of symptoms.[60] Patients were randomised either to BMC therapy ($n = 50$, of whom 47 received intracoronary BMC therapy) or to a control groups ($n = 50$) where neither bone marrow aspiration nor sham intracoronary injection were performed. We chose to study anterior wall infarctions since this infarct localisation has the greatest impact on left ventricular function, and most imaging modalities have best precision for assessment of function in anterior segments of the left ventricle. We aspirated 50 ml bone marrow from the hip bone in local anaesthesia four to seven days after the infarction, and mononuclear cells were obtained with the use of ficoll density gradient centrifugation. Intracoronary administration of BMC with the stop-flow PCI technique was performed either the same ($n = 4$) or the following day ($n = 43$). Baseline recordings for ECG-gated SPECT (primary imaging modality) and echocardiography were performed before the time for BMC therapy in both groups (four days after the AMI for SPECT and 4.5 days for echocardiography). Baseline MRI recordings were performed two to three weeks after the infarction. After six months follow-up, LVEF increased with 8.1% in the BMC group and with 7.0% in the control group ($p = 0.77$) as measured by SPECT. With the use of echocardiography and MRI, the absolute increase in LVEF was smaller. The reason for this may have been that we used a three-dimensional approach by SPECT (4D-MSPECT software), which is probably a more sensitive technique to discover serial changes of LV volumes than the two-dimensional approaches we used by echocardiography (Simpsons method) and MRI (area-length method). Another aspect is that the potential for improvement of LV function caused by recovery after stunning and reduction of infarct size caused by decline of tissue oedema was greatest for SPECT and least by MRI, according to the time-points for baseline recordings. Results were consistent since neither of the imaging modalities demonstrated any differences between groups for the change in (a) LVEF, (b) remodeling as measured by the LVEDV, nor (c) infarct size. The increase in LVEF was greater for those with low LVEF at baseline, regardless of treatment allocation. A secondary end-point in the ASTAMI trial was the influence of BMC therapy on exercise capacity. We found that patients in the treatment group significantly improved time to exhaustion on bicycle ergometer tests, accompanied by an improvement in heart rate response to exercise.[71] These results indicate a favourable effect of BMC therapy. However, the ASTAMI study was an open-labeled study, and since there was no difference between the groups for more effort-independent measures like peak VO_2, these differences may reflect a placebo-effect only.

The largest study to date was the REPAIR-AMI trial which included 204 AMI patients.[65] Also in this study, patients were treated with PCI with stent in the acute phase. A hundred and one patients were randomised to intracoronary injections of BMC and 103 patients to intracoronary injections of placebo medium. In both groups, 50 ml of bone marrow was aspirated from the hip bone in local anaesthesia. Also in this study, mononuclear cells were obtained by ficoll density gradient centrifugation. The placebo medium was 10 ml of the cell culture medium X-vivo10 with 2 ml of autologous serum. BMC/placebo injections were performed three to seven days after the infarction. After four months follow-up, LVEF increased with 5.5% in the BMC group and with 3.0% in the control group ($p = 0.01$). Pre-specified sub-group analyses showed that the benefit was confined to: (a) patients with depressed ejection fraction at baseline. Among the patients with a baseline LVEF at or below the median value (48.9%), the change in LVEF was 7.5% in the

BMC group and 2.5% in the control group, compared to a similar increase in LVEF for those with a baseline LVEF above the median value (4.0% in the BMC group and 3.7% in the placebo group). (b) When BMC were administered five days or more after the AMI (7.0% increase in LVEF in the BMC group and 1.9% in the control group compared to 4.5% in the BMC group and 3.9% in the control group when BMC/placebo were administered earlier).

In the REPAIR-AMI study, single-plane ventriculography was used for assessment of LV function. It is well known that there are several limitations related to this method. A recent comparison with MRI has shown that this method is particularly inaccurate in patients with regional wall motion abnormalities.[72]

There are also data from the REPAIR-AMI trial that indicate improved clinical outcome following BMC therapy.[73] After 12 months follow-up, there were significantly fewer reinfarctions ($p = 0.029$) and revascularisations ($p = 0.026$) in the BMC group compared to the control group. Given these differences, there was also a significant difference in favour of the BMC group for the pre-specified cumulative end-point of death, reinfarctions, and revascularisations ($p = 0.009$). These results are of course encouraging for the future of BMC research, but since the REPAIR-AMI was not powered to detect improvements in clinical outcome, the issue of clinical effects remains unresolved. Compared to other studies in this field, it appears that the differences between groups may have been influenced by an unexpected poor outcome in the placebo group.

Meluzin et al. randomised AMI patients of whom the majority were reperfused with acute PCI to one of three groups:[68] (1) high-dose mononuclear BMC (10^8 cells), (2) low-dose mononuclear BMC (10^7 cells), or (3) control group were neither bone marrow aspiration nor placebo injections were performed. While 73 patients were included in the study, the final analysis included only 22 patients in each group. The patient population was also somewhat heterogeneous, caused by the inclusion of eight patients with delayed PCI. Acknowledging these limitations, the authors reported that after three months follow-up, the LVEF measured by SPECT increased with 2% in the control group, with 3% in the low-dose group and with 5% in the high-dose group. There was a significant difference between the high-dose and the control group only ($p = 0.041$). There were also significant differences in favour of the cell treatment groups for the measurement of tissue Doppler systolic velocities from basal ventricular segments. There were no significant differences between the groups for the LVEDV or the infarct size measured as the perfusion defect.

A special feature in the randomised TCT-STAMI study was the administration of cells or placebo in the hyper-acute phase of the AMI.[66] Immediately after revascularisation of the infarct-related coronary artery with successful PCI, 40 ml of bone marrow was aspirated from the hip bone in local anaesthesia. Only three hours after the PCI, patients in the cell group ($n = 10$) received intracoronary injections of mononuclear BMC, and the control group received a similar volume of the bone marrow supernatant. LVEF and LVEDV was assessed with echocardiography, and left ventricular perfusion with SPECT one week and six months after the infarction.

There was a significant increase in LVEF and a reduction in left ventricular end-diastolic diameter and the perfusion defect at six months compared to baseline in the BMC group. Rather unexpected, there were no significant changes for these values at six months follow-up in the control group. It is difficult to interpret treatment effects in the TCT-STAMI study, since results of statistical comparisons between groups were only reported for baseline values.[74]

Chen *et al.* randomised 69 patients treated with acute PCI for AMI to intra-coronary injection of autologous MSC ($n = 34$) or placebo ($n = 35$).[67] Bone marrow (60 ml) was aspirated from the hip bone in local anaesthesia eight days after the infarction. MSC were cultivated *ex vivo* for ten days before administration to the cell treatment group. The placebo medium was 0.9% sodium chloride. There are two remarkable features of this trial. First, a very high number of MSC were obtained during the ten-day cultivation period, and $8-10 \times 10^9$ cells were injected. Second, the improvement in LVEF in the treatment group was dramatic. With the use of either LV angiography or echocardiography (which method was used for which values is not accurately reported in their publication), LVEF increased with 18% points in the MSC group and 6% points in the control group. In addition, the authors found higher values for perfusion and reduced LVEDV in the MSC compared to the control group at three months follow-up. While these results are indeed encouraging, they need to be confirmed in other studies. To date, this is the only published study with MSC in the clinical setting.

Intracoronary administration of bone-marrow derived cells after cytokine mobilisation

The combined approach of subcutaneous G-CSF injections followed by collection of mononuclear cells by apheresis of peripheral blood and subsequent intracoronary injectons with the PCI stop-flow technique in patients with myocardial infarction has been tested in the MAGIC Cell-1.[75,76] and the MAGIC Cell-3-DES.[77] trials. The first was a small trial which randomised patients with acute or old myocardial infarction to either: (1) G-CSF followed by apheresis and intracoronary injection of mononuclear BMC ($n = 10$), (2) only G-CSF ($n = 10$), and (3) control where neither placebo subcutaneous nor intracoronary cell injections were performed. Results from this study are somewhat difficult to interpret due to the heterogeneous patient population and since the study protocol was changed during the study due to safety concerns about administration of G-CSF upfront of PCI. After two years follow-up, the change in LVEF was significantly greater in the G-CSF/cell infusion group than in the G-CSF group, but not compared to the control group.[76] In the randomised MAGIC Cell-3-DES trial which evaluated the combined G-CSF/cell infusion approach after PCI with a drug eluting stent, there were four groups: (1) AMI patients with cell injection ($n = 25$), (2) AMI control ($n = 25$), (3) old myocardial infarction and cell injection ($n = 16$), and (4) old myocardial infarction control ($n = 16$).[77] MRI was performed at baseline and repeated after six months. LVEF increased with 5.1%

in the cell treated AMI patients, compared to no change in the AMI control group ($p = 0.046$). There was also a significant reduction in infarct size in favour of the cell treatment group, but also for the infarct size, this was achieved compared to a control group where no change was observed. The combined G-CSF/cell infusion approach was also used by Li et al., who studied 70 patients with AMI treated wtih PCI.[78] After six months, LVEF measured with echocardiography increased with 7.1% in the treatment group and 1.6% in the control group. It is however unclear whether the treatment allocation was randomized. In addition, 12 of the 35 patients allocated to the control group dropped out during the follow-up period, compared to no dropouts in the treatment group. Results of the study by Li et al. and the AMI patients in the MAGIC Cell-3-DES trial are summarised in the lower section of Table 1.

G-CSF only

The singular application of subcutaneous G-CSF injections has been tested in AMI patients after successful revascularisation with PCI in a few randomised trials. Recent experimental data suggest that effects of G-CSF may be mediated through direct, anti-apoptotic action on cardiomyocytes rather than via increased homing of BMC to the infarcted heart.[79]

Valgimigli et al. randomised patients to G-CSF ($n = 10$) or placebo ($n = 10$) 1.5 days after AMI, of whom 14 had been treated with PCI in the acute phase, and found a trend in favour of the G-CSF group, but no significant differences between groups for the change in LVEF ($p = 0.068$) or LVEDV ($p = 0.054$).[80] In the open-labeled FIRSTLINE-AMI study, patients were randomised to G-CSF ($n = 25$) or control ($n = 25$) after AMI treated with primary PCI.[81] With the use of echocardiography, significant differences in favour of the G-CSF group were found at four months follow-up for LVEF ($p < 0.001$), LVEDV ($p = 0.002$), regional wall thickening ($p < 0.001$) and the wall motion score index (WMSI) ($p < 0.001$). It is noteworthy that these differences were obtained when compared to a rather unexpected outcome in the control group, since LVEF decreased and only a slight improvement in the WMSI was found in control group patients. Similar results were reported after 12 months follow-up.[82] More recently, three larger double-blind, placebo-controlled trials could not confirm any beneficial effect of G-CSF treatment in patients with AMI.[83-85] In a recent review, Kastrup et al. conclude that at present, there is no indication for the use of G-CSF mobilised stem cells for clinical regenerative treatment of infarcted myocardium.[86]

End-stage ischaemic heart disease with no further revascularisation options

As previously discussed, the main therapeutic goal for these patients is to improve blood supply. However, most of these patients also have areas of myocardial scarring caused by either acute or chronic ischaemia. In this patient scenario, most trials have been small phase 1 studies either without or with small non-randomised control groups. Transendomyocardial injection of different BMC populations via percutaneous cell injection devices have been performed in several studies. Mononuclear BMC obtained by density gradient centrifugation were injected by Tse *et al.* ($n = 8$).[87] and Beeres *et al.* ($n = 25$),[88] while Fuchs *et al.* ($n = 27$)[89] and Briguori *et al.* ($n = 10$)[90] simply used freshly aspirated, filtered bone marrow. In these studies, intramyocardial cell injections were guided by electromechanical mapping to identify ischaemic, but viable areas of the myocardium. Results obtained by this method indicate improved perfusion and myocardial function, and it has clearly been demonstrated that cell injection by this method is feasible and seems safe as long as small volumes are injected and areas with thin myocardium are avoided. Preliminary data from the randomised, placebo-controlled PROTECT-CAD study show increased exercise treadmill time (primary end-point), increased LVEF and perfusion after endomyocardial injections of mononuclear BMC ($n = 19$) compared to a placebo group ($n = 19$).[91] These promising results need to be confirmed in larger, randomised trials.

The combined G-CSF/cell infusion approach was used by Boyle *et al.* in five patients.[92] Following aphereresis of peripheral blood, CD34$^+$ cells were enriched by MACS cell sorting and administered via the intracoronary route. Perfusion improved, but concerns were raised by the observation that one patient developed an acute coronary syndrome, and another developed lentigo maligna, events that potentially may have been caused by CD34+cell induced plaque angiogenesis, instability and rupture and tumor angiogenesis, respectively. Erbs *et al.* performed a larger, randomised trial including 26 patients with chronic coronary artery occlusion with signs of myocardial ischaemia after successful recanalisation of the occluded coronary artery by PCI.[93] G-CSF was applied in all patients for four days prior to collection of 400 ml peripheral blood. Mononuclear cells were isolated by ficoll density gradient centrifugation and cultivated for four days. CPC were then injected intracoronarily ($n = 13$). The control group had intracoronary injection of cell-free serum ($n = 13$). After the three-month follow-up period, LVEF measured with MRI increased with 7.1% in the treatment group compared to no change in the control group ($p < 0.05$). In addition, perfusion increased and infarct size declined in the cell treatment group.

The sole use of cytokine-induced mobilisation of stem and progenitor cells for improvement of myocardial perfusion in patients with chronic, stable coronary artery disease has been tested in a few, randomised studies. Seiler *et al.* injected GM-CSF ($n = 10$) or placebo ($n = 11$) first directly into the narrowed coronary artery followed by two weeks of subcutaneous administration, and found that perfusion improved in the GM-CSF group.[94] However, safety concerns arose from results of a following study from the same group. In this randomised study, only systemic administration of GM-CSF or placebo was performed.[95] During the study period, two out of the seven patients in the treatment group and none of the placebo patients experienced an acute coronary syndrome. Similar findings were reported by Hill *et al.* who gave G-CSF to 16 patients with stable, coronary disease.[96] One patient experienced an acute coronary syndrome in the study drug administration period, and one patient died from myocardial infarction 17 days after treatment. In this study, no changes were observed for wall motion score and perfusion, and there was a trend towards an increased number of myocardial segments with inducible ischaemia after G-CSF treatment.

Congestive heart failure caused by ischaemic cardiomyopathy

BMC-based therapies have been tested also in this clinical scenario, in patients where the main problems are related to myocardial scarring caused by old myocardial infarction, and to a minor degree to persistent myocardial ischaemia. The scarce number of trials that have been performed in this patient population is probably a reflection of the huge challenge represented by the need to replace scarred with viable, functional myocardium. Theoretically, BMC-induced improvements in cardiac function could be possible also via other mechanisms than replacing the scar. Cell-induced angiogenesis may lead to improved perfusion and a decline in cardiomyocyte apoptosis, and by improved passive properties of the scar. In a small study comprising six patients, G-CSF was given for four days prior to collection of peripheral blood mononuclear cells by apheresis and direct intramyocardial cell injections.[97] Efficacy of this approach was unanswered by this study, due to the small sample size and the confounding effects of the CABG performed in conjunction with cell injections. The intracoronary administration of BMC in these patients has been tested in three trials. In the IACT study, 18 patients with old myocardial infarction and an open infarct-related artery were treated with intracoronary injection of mononuclear BMC.[98] Eighteen patients who refused cell therapy were used as the control group. After three months, LVEF measured with biplane ventriculography increased significantly with 8% in the BMC group compared to a non-significant increase of 1% in the control group. The authors also reported that the BMC treatment group improved maximal oxygen uptake and metabolism by 18-FDG scans, and infarct size declined. Similar to most of the studies mentioned, interpretation of the results is hampered by small sample size, a non-randomised control group and a short observation period. However, similar results were found in a recent randomised study with a crossover design.[99] Here, 75 patients who suffered an AMI at least three months previous to inclusion, were randomised to either intracoronary injection of BMC, ($n = 28$), CPC ($n = 24$) or control ($n = 23$). After three months, control group patients were randomised to either intracoronary BMC or CPC administration, and the BMC and CPC group patients crossed over to receive either CPC or BMC, respectively. A favourable outcome was only reported after BMC treatment, with a 2.9% increase in LVEF three months after treatment compared to a decrease in LVEF of 0.4% after CPC treatment and 1.2% in controls. Both in this study and in the IACT-study, LV ventriculography was used for assessment of LVEF.

As mentioned earlier, a significant proportion of patients have problems caused both by an established myocardial scars, and ischaemia in viable parts of the myocardium. Several groups have injected BMC as an adjunctive therapy to CABG in such patients. Galinanes *et al.* used unmanipulated, filtered bone marrow harvested from the sternum, and cells were injected into the scars in 14 patients.[100] Only myocardial segments which both were bypass grafted and injected with cells improved function. Stamm *et al.* injected CD133+ cells, and found indications of improved myocardial perfusion.[101] Hendrikx *et al.* randomised patients to either CABG and intramyocardial injections of mononuclear BMC ($n = 10$), or CABG only ($n = 10$).[102] Four months later, there were no significant differences between groups for the increase in LVEF. The transendocardial approach was tested by Perin *et al.*, who treated 14 patients with previous myocardial infarction, LVEF below 40% and reversible ischaemia.[103,104] Viable, but ischaemic segments were identified with electromechanic mapping, and selected for intramyocardial injections of mononuclear BMC. Compared to a non-randomised control group ($n = 9$), LVEF, perfusion and maximal oxygen uptake increased. Results of published randomised studies with 20 or more patients with chronic ischaemic heart disease are summarised in Table 2.

Safety

When introducing a new therapeutic option, patent safety is paramount. Data from experimental research can provide important information about possible risks. One example is the formation of teratomas in the heart after transplantation of undifferentiated embryonic stem cells in mice.[105] However, several other examples show that only clinical experience can elucidate the possible risks of a new therapy. One important example related to stem cell therapy is the occurrence of ventricular arrhythmias after intramyocardial injection of skeletal myoblasts in patients with heart failure which was quite unexpected,[106] since arrhythmias were not observed in studies with myoblast transplantation after experimental infarction. The first clinical trials with bone marrow derived cells were performed only four to five years ago, and limited data exist about long-term safety. Here, we will discuss safety concerns that have arisen so far.

Cell harvest

Bone marrow aspiration is obviously an uncomfortable procedure for the patient, and may trigger physiological responses. Activation of the sympathetic nerve system may be deleterious, in particular in the setting of an AMI. Widimsky *et al.* reported one case of fatal stent thrombosis in direct relation to bone marrow aspiration.[107] Our experience from the ASTAMI trial is that patients tolerated the procedure well when performed in local anaesthesia after premedication with analgetics and tranquilisers.

Table 2.　Randomised trials with BMC or CPC with 20 or more patients with chronic ischaemic heart disease.

	N	Dose (10⁶)	Days after AMI	End-point evaluation	Follow-up (months)	Baseline LVEF (%)	ΔLVEF (%points)	P ΔLVEF	ΔLVEDV (ml)	ΔInfarct size
Assmus et al.[99]	35 BMC	205 ± 110	I.C.	LV Angio	3	41 ± 11	2.9 ± 3.6	< 0.001*	0 ± 10†	NA
	34 CPC	22 ± 11				39 ± 10	−0.4 ± 2.2		−3 ± 18†	
	23 Control					43 ± 13	−1.2 ± 3.0		−3 ± 17†	
Erbs et al.[93]	13 CPC (G-CSF)	22–200	I.C.	MRI	3	51.7 ± 3.7‡	7.2	< 0.01	−0.2	3.0 ml
	13 Placebo					55.8 ± 2.8‡	0		−7.6	1.1 ml
MAGIC Cell-3-DES[77]	16 CPC (G-CSF)	1400 ± 500	I.C.	MRI	6	48.5 ± 12.9	0	NS	5.2	−5.5 ml
	16 Control					45.1 ± 10.2	0.2		2.1	1.5 ml
Hendrikx et al.[102]	10 BMC	60 ± 31	I.M	MRI	4	42.9 ± 10.3	6.1 ± 8.6	0.41	0.2 ± 35.0	NA
	10 Control					39.5 ± 5.5	3.6 ± 9.1		3.4 ± 28.2	
	CABG in all patients									

Values are mean(±SD) or range unless otherwise stated. BMC: Bone marrow cells, CPC: circulating progenitor cell, LVEF: left ventricular ejection fraction, LVEDV: left ventricular end-diastolic volume, NA: not applicable, G-CSF: granulocyte colony-stimulating factor, I.C.: intracoronary, NS: not significant, I.M.: intramyocardial, CABG: coronary artery bypass grafting. Δ = change (follow-up − baseline). *vs. control, †LVEDV index (LVEDV/m²), ‡Mean ± SEM.

Another aspect is the possibility for contamination of the cell product. Even though all cell processing was performed according to GMP conditions, and an aseptic procedure was used for bone marrow aspiration in the ASTAMI trial, bacterial testing revealed contamination of skin bacteria in two of the aspirates. In one of these patients, BMC were administered to the patient immediately after cell processing, and contamination of the cell product was discovered the next day. The patient was treated with antibiotics, and no clinical events occurred. After this event, we decided that BMC should not be administered to the patient before overnight bacterial testing was negative. Administration of a contaminated BMC product was also reported by Meluzin *et al.*[68]

Cell processing

The techniques for processing of bone marrow-derived cells are based on more than 40 years experience from bone marrow transplantation for treatment of haemato-logical and neoplastic diseases. Thus, the use of autologous, unmanipulated cell products is by all matter relatively safe. In a non-randomised study in AMI patients, Bartunek *et al.* used an irreversibly attached mouse antibody for MACS cell sorting of CD133+ cells prior to intracoronary administration.[54] They observed an unex-pected high rate of restenosis, and it has been suggested that the mouse antibody may have provoked a deleterious inflammatory reaction. Manipulation of the cells may be advantageous in some respects. Mangi *et al.* found that genetic manipu-lated MSC overexpressing Akt were superior to the wild-type MSC for myocardial regeneration.[108] However, such manipulation may also increase the risk for uncon-trolled growth, differentiation and tumor formation.[109] In some studies, artificial cell culture mediums have been used for cell processing and storage. Such substances may provide some theoretical advantages compared to alternative autologous serum and plasma with regard to, for example, buffering capacity. However, artificial cell culture mediums are primarily manufactured for *in vitro* research purposes, and few cell culture media are approved for clinical use. The safety after cardiac application of such substances is unknown.

Administration

The intracoronary administration of cells expose the patient for the small, but not neg-ligible risk of a left-sided heart catherisation with an overall mortality of 0.14%.[110] The stop-flow intracoronary PCI cell delivery technique have inherent risks as well. In the ASTAMI study, 47 patients received intracoronary cell injections. Of these, 34 (72%) had chest pain, and 36 (77%) had ischaemic changes on the electro-cardiogram during balloon inflations. No patients had significant increase of tro-ponins after the procedure and we did not observe any complications related to the procedure. Serious adverse events have however been reported. In the study by Meluzin *et al.* where 44 patients received intracoronary cell injections, one patient had coronary artery dissection and another stent thrombosis as direct complications to the procedure.[68] Widimsky and Penicka reported severe haemodynamic deterioration from intracoronary injection of BMC during coronary occlusion in two patients with idiopathic dilated cardiomyopathy.[111] Hypotension followed by troponin elevation was reported after intracoronary infusion of mononuclear BMC in a patient with ischaemic cardiomyopathy.[112] Intracoronary injection of MSC produced microinfarctions in dogs,[49] but this was not confirmed in the clinical study where this approach was tested.[67] Uncertainty about safety with intracoronary administration of MSC remains, as the authors of the clinical trial did not report troponin values after cell administration. The percutaneous transendocardial intramyocardial cell delivery technique possess obvious risks for intramyocardial damage during injections. In a study where vascular endothelial growth factor was injected into the myocardium with a similar technique,[113] pericardial tamponade, high-degree atrioventricular block, ST-sement elevation, myocardial infarction, embolic events and sepsis were procedure-related adverse events. Still, few adverse events have been reported after cell injections with this approach. However, this procedure certainly warrants high operator technical skills, careful patient selection and the use of accurate imaging/mapping devices to guide intramyocardial injections to the myocardial regions of interest. Intramyocardial cell injections during open-heart surgery have not been reported to provoke complications directly related to the procedure.

Cell related adverse events

Transplantation of skeletal myoblasts provoked ventricular arrhythmias. So far, there has not been an excess of arrhythmias after treatment with bone marrow-derived cells, at least after intracoronary administration. This was extensively studied in the BOOST trial, where an electrophysiologic examination was performed in more than 90% of the patients at six months follow-up.[63] There were no differences between the BMC and the control group for the occurrence of ventricular arrhythmias. An electrophysiological study was performed in nine patients after intramyocardial injections of mononuclear BMC, and six patients had inducible ventricular arrhythmias.[102] Since this was observed in a small study with patients at risk for this complication, a causal relation to the BMC therapy cannot be established.

Cells derived from bone marrow may improve neovascularisation, but some reports have highlighted that these cells may also be involved in the development of atherosclerotic lesions.[34,114] Thus, the intracoronary administration of these cells could possibly aggravate coronary artery atherosclerosis.[115] As mentioned earlier, increased rates for in-stent restenosis was found after intracoronary injection of CD133+ cells.[54] The authors argue that restenosis was caused by the cells *per se*, and not the possible deleterious immune reaction caused by the mouse antibody used for cell selection. Increased rates for in-stent restenosis was also found in the MAGIC Cell-1 trial.[75] However, since G-CSF was administered in advance of the PCI, this observation may have been caused by a high level of circulating inflammatory cells at the time of the revascularisation procedure. In contrast, we did not find evidence for increased rates of restenosis by quantitative coronary angiography after the intra-coronary administration of unfractioned bone marrow cells,[116] in agreement with other studies.[63,117]

With the exception of the single case with lentigo maligna after administration of CD34+ cells,[92] there is no evidence that autotransplantation of cells derived from bone marrow may provoke neoplastic disease. Long-term results are needed to further assess this possibility.

Discussion

The experiences obtained from clinical trials of BMC therapy for patients with ischaemic heart disease have so far gained more questions than answers. First, it is in our opinion not yet established that any therapy based on cells derived from bone marrow improves cardiac function to a clinically significant degree, irrespective of the patient population studied, the cell population applied and the mode of cell administration. No large-scale randomised trials have been performed, and most randomised trials have studied effects of intracoronary administration of BMC after AMI. In the trials with unfractioned bone marrow cells, either no effect,[60] a short-term effect,[63,69] or a small effect was found.[64–66,68] Only the study using MSC showed a marked improvement, but these results need to be confirmed.[67] The mechanism(s) that may be responsible for the possible advantageous effect of BMC therapy has not yet been identified. Acknowledging these fundamental limitations, it is at present difficult to identify how cell therapy strategies may be improved. Here, we will mention a few central unresolved issues.

(1) *Cell type.* We do not know which is the preferred cell type for any of the described clinical scenarios, or perhaps what mixture of cell types that may be optimal. Recent experimental data demonstrating a synergistic neovascularisation effect by a combination of different CPC populations support the pragmatic approach of using unfractioned cells.[118,119]

(2) *Cell preparation.* Standardised techniques for preparation of viable, functional cells have been developed in the field of haematology during years of experience. These methods, proven to be efficacious for bone marrow transplantation, have now been used also for preparation of BMC intended for cardiac application. Cell preparation must be performed according to GMP quality criteria. It has been suggested that subtle differences in these established protocols is of importance for the efficacy of BMC treatment in AMI patients.[120] These results obtained from *in vitro* studies and a mouse hindlimb ischaemia model with questionable clinical relevance, need to be confirmed.

(3) *Cell numbers.* We do not know the number of different cells that may be necessary to improve cardiac function. Some simple calculations are worthwhile[121]

in this regard. A myocardial infarction comprising 30% of the left ventricle leads to loss of approximately 1700×10^6 cardiomyocytes. In the trials that have been performed so far, the number of stem/ progenitor cells that possibly remained in the heart after administration were far lower. Even though Meluzin *et al.* suggested a dose–response effect, they did not find a correlation between the number of CD34+ progenitor cells and improvement in cardiac function.[68] The lack of a significant correlation between cell numbers and the change in LVEF was confirmed in the TOPCARE-AMI study,[122] the BOOST study[63] and the ASTAMI study.[60]

(4) *Timing of cell administration.* This issue is relevant in the setting of an AMI. Data from the REPAIR-AMI study suggest that BMC therapy was only advantageous when applied five days or later after the acute event. This was not supported by results of the ASTAMI study, where cells were delivered at a mean time of six days after the AMI and no effect on LVEF was observed compared to the control group. It is not known whether repeated cell administration will be beneficial.

(5) *Mode of administration.* It is not established which cell delivery technique gives best engraftment of cells within the heart, and if different techniques should be used for different cell types and patient populations. The intracoronary injection of cells during no-flow was assumed to result in more cells administered to the infarcted area.[50] A recent study in experimental AMI in pigs has shown that the stop-flow intracoronary cell injection technique was not superior to continuous intracoronary infusion for engraftment of mononuclear BMC.[123]

Furthermore, what is the ideal study design? A double-blind study design is considered the gold standard for clinical trials. If the goal is to assess the effects of an intervention compared to standard, state-of-the-art therapy, it is important to eliminate confounding effects of the placebo procedure. In the placebo controlled AMI trials, bone marrow aspiration, left sided heart catherisation and sham intracoronary injection were performed.[64–66] Cytokine release can be induced by bone marrow aspiration[124] and the PCI procedure for cell administration.[125] Intermittent balloon occlusion for stop-flow can also affect the infarct process,[126] and introduce several confounders. In addition, patient safety is paramount, and severe complications related to the bone marrow aspiration[107] and the stop-flow PCI technique has been reported.[68,111,112] Ethical concerns arise in particular for placebo controlled trials, since placebo group patients are only exposed to the risk of the sham procedure and not to any possible benefit of the therapy. Possibly, an open-labeled study design is preferable as long as end-point analyses are performed by investigators unaware of treatment allocation.

Conclusion

At present, there is no clinical indication for therapies with bone marrow cells for patients with ischaemic heart disease. The low incidence of serious adverse events after treatment with BMC and CPC justifies further clinical research. In our opinion, these studies should be performed in the absence of a confounding placebo procedure. The clinical studies should be parallelled with further basic research that hopefully will improve the potential for therapies with cells derived from bone marrow or peripheral blood. Results of future adequately powered, randomised, clinical trials using accurate methods to detect a benefit are awaited.

References

1. Callow AD. Cardiovascular disease 2005 — the global picture. *Vasc Pharmacol* 2006;45(5):302–307.
2. Rosamond W, Flegal K, Friday G, Furie K, Go A, Greenlund K, *et al*. Heart disease and stroke statistics — 2007 update: a report from the American Heart Association Statistics Committee and Stroke Statistics Subcommittee. *Circulation* 2007; 115(5):e69–171.
3. Fox KF, Cowie MR, Wood DA, Coats AJS, Gibbs JSR, Underwood SR, *et al*. Coronary artery disease as the cause of incident heart failure in the population. *Eur Heart J* 2001;22(3):228–236.
4. Ellison GM, Torella D, Karakikes I, Nadal-Ginard B. Myocyte death and renewal: modern concepts of cardiac cellular homeostasis. *Nat Clin Pract Cardiovasc Med* 2007;(Suppl 1): S52–S59.
5. Kang PM, Izumo S. Apoptosis and heart failure: a critical review of the literature. *Circ Res* 2000;86(11):1107–1113.
6. Nadal-Ginard B, Kajstura J, Leri A, Anversa P. Myocyte death, growth, and regeneration in cardiac hypertrophy and failure. *Circ Res* 2003;92(2):139–150.
7. Beltrami AP, Urbanek K, Kajstura J, Yan SM, Finato N, Bussani R, *et al*. Evidence that human cardiac myocytes divide after myocardial infarction. *N Engl J Med* 2001;344(23):1750–1757.
8. Urbanek K, Quaini F, Tasca G, Torella D, Castaldo C, Nadal-Ginard B, *et al*. Intense myocyte formation from cardiac stem cells in human cardiac hypertrophy. *Proc Natl Acad Sci USA* 2003;100(18):10440–10445.
9. Kajstura J, Leri A, Finato N, Di Loreto C, Beltrami CA, Anversa P. Myocyte proliferation in end-stage cardiac failure in humans. *Proc Natl Acad Sci USA* 1998;95(15):8801–8805.
10. Beltrami AP, Barlucchi L, Torella D, Baker M, Limana F, Chimenti S, *et al*. Adult cardiac stem cells are multipotent and support myocardial regeneration. *Cell* 2003;114(6):763–776.
11. Oh H, Bradfute SB, Gallardo TD, Nakamura T, Gaussin V, Mishina Y, *et al*. Cardiac progenitor cells from adult myocardium: homing, differentiation, and fusion after infarction. *Proc Natl Acad Sci USA* 2003;100(21):12313 – 12318.
12. Matsuura K, Nagai T, Nishigaki N, Oyama T, Nishi J, Wada H, *et al*. Adult cardiac Sca-1-positive cells differentiate into beating cardiomyocytes. *J Biol Chem* 2004;279(12):11384–11391.
13. Messina E, De AL, Frati G, Morrone S, Chimenti S, Fiordaliso F, *et al*. Isolation and expansion of adult cardiac stem cells from human and murine heart. *Circ Res* 2004;95(9):911–921.
14. Laugwitz KL, Moretti A, Lam J, Gruber P, Chen Y, Woodard S, *et al*. Postnatal isl1+ cardioblasts enter fully differentiated cardiomyocyte lineages. *Nature* 2005;433(7026):647–653.
15. Pfister O, Mouquet F, Jain M, Summer R, Helmes M, Fine A, *et al*. CD31 — but not CD31+ cardiac side population cells exhibit functional cardiomyogenic differentiation. *Circ Res* 2005;97(1):52–61.
16. Tomita Y, Matsumura K, Wakamatsu Y, Matsuzaki Y, Shibuya I, Kawaguchi H, *et al*. Cardiac neural crest cells contribute to the dormant multipotent stem cell in the mammalian heart. *J Cell Biol* 2005;170(7):1135–1146.

17. Jiang Y, Jahagirdar BN, Reinhardt RL, Schwartz RE, Keene CD, Ortiz-Gonzalez XR, et al. Pluripotency of mesenchymal stem cells derived from adult marrow. *Nature* 2002;418(6893): 41–49.

18. Choi K, Kennedy M, Kazarov A, Papadimitriou JC, Keller G. A common precursor for hematopoietic and endothelial cells. *Development* 1998;125(4):725–732.

19. Roufosse CA, Direkze NC, Otto WR, Wright NA. Circulating mesenchymal stem cells. *Int J Biochem Cell Biol* 2004;36(4):585–597.

20. Wognum AW, Eaves AC, Thomas TE. Identification and isolation of hematopoietic stem cells. *Arch Med Res* 2003;34(6):461–475.

21. Leor J, Marber M. Endothelial progenitors: a new Tower of Babel? *J Am Coll Cardiol* 2006;48(8):1588–1590.

22. Matsuzaki Y, Kinjo K, Mulligan RC, Okano H. Unexpectedly efficient homing capacity of purified murine hematopoietic stem cells. *Immunity* 2004;20(1):87–93.

23. Brazelton TR, Rossi FM, Keshet GI, Blau HM. From marrow to brain: expression of neuronal phenotypes in adult mice. *Science* 2000;290(5497):1775–1779.

24. Theise ND, Nimmakayalu M, Gardner R, Illei PB, Morgan G, Teperman L, et al. Liver from bone marrow in humans. *Hepatology* 2000;32(1):11–16.

25. Ferrari G, Cusella-De Angelis G, Coletta M, Paolucci E, Stornaiuolo A, Cossu G, et al. Muscle regeneration by bone marrow-derived myogenic progenitors. *Science* 1998;279(5356): 1528–1530.

26. Orlic D, Kajstura J, Chimenti S, Jakoniuk I, Anderson SM, Li B, et al. Bone marrow cells regenerate infarcted myocardium. *Nature* 2001;410(6829):701–705.

27. Jackson KA, Majka SM, Wang H, Pocius J, Hartley CJ, Majesky MW, et al. Regeneration of ischemic cardiac muscle and vascular endothelium by adult stem cells. *J Clin Invest* 2001;107(11):1395–1402.

28. Kocher AA, Schuster MD, Szabolcs MJ, Takuma S, Burkhoff D, Wang J, et al. Neo-vascularization of ischemic myocardium by human bone-marrow-derived angioblasts prevents cardiomyocyte apoptosis, reduces remodeling and improves cardiac function. *Nat Med* 2001;7(4):430–436.

29. Quaini F, Urbanek K, Beltrami AP, Finato N, Beltrami CA, Nadal-Ginard B, et al. Chimerism of the transplanted heart. *N Engl J Med* 2002;346(1):5–15.

30. Laflamme MA, Myerson D, Saffitz JE, Murry CE. Evidence for cardiomyocyte repopulation by extracardiac progenitors in transplanted human hearts. *Circ Res* 2002;90(6):634–640.

31. Muller P, Pfeiffer P, Koglin J, Schafers HJ, Seeland U, Janzen I, et al. Cardiomyocytes of noncardiac origin in myocardial biopsies of human transplanted hearts. *Circulation* 2002;106(1): 31–35.

32. Bayes-Genis A, Salido M, Sole RF, Puig M, Brossa V, Camprecios M, et al. Host cell-derived cardiomyocytes in sex-mismatch cardiac allografts. *Cardiovasc Res* 2002;56(3):404–410.

33. Deb A, Wang S, Skelding KA, Miller D, Simper D, Caplice NM. Bone marrow-derived cardiomyocytes are present in adult human heart: a study of gender-mismatched bone marrow transplantation patients. *Circulation* 2003;107(9):1247–1249.

34. Caplice NM, Bunch TJ, Stalboerger PG, Wang S, Simper D, Miller DV, et al. Smooth muscle cells in human coronary atherosclerosis can originate from cells administered at marrow transplantation. *Proc Natl Acad Sci USA* 2003;100(8):4754–4759.

35. Mallory GK, White PD, Salcedo-Salgar J. The speed of healing of myocardial infarction: a study of the pathologic anatomy in 72 cases. *Am Heart J* 1939;18:647–671.

36. Balsam LB, Wagers AJ, Christensen JL, Kofidis T, Weissman IL, Robbins RC. Haematopoietic stem cells adopt mature haematopoietic fates in ischaemic myocardium. *Nature* 2004;428(6983):668–673.

37. Murry CE, Soonpaa MH, Reinecke H, Nakajima H, Nakajima HO, Rubart M, *et al.* Haematopoietic stem cells do not transdifferentiate into cardiac myocytes in myocardial infarcts. *Nature* 2004;428(6983):664–668.

38. Nygren JM, Jovinge S, Breitbach M, Sawen P, Roll W, Hescheler J, *et al.* Bone marrow-derived hematopoietic cells generate cardiomyocytes at a low frequency through cell fusion, but not transdifferentiation. *Nat Med* 2004;10(5):494–501.

39. Chien KR. Stem cells: lost in translation. *Nature* 2004;428(6983):607–608.

40. Heil M, Ziegelhoeffer T, Mees B, Schaper W. A different outlook on the role of bone marrow stem cells in vascular growth: bone marrow delivers software not hardware. *Circ Res* 2004;94(5): 573–574.

41. Thum T, Bauersachs J, Poole-Wilson PA, Volk HD, Anker SD. The dying stem cell hypothesis: immune modulation as a novel mechanism for progenitor cell therapy in cardiac muscle. *J Am Coll Cardiol* 2005;46(10):1799–1802.

42. Orlic D, Kajstura J, Chimenti S, Limana F, Jakoniuk I, Quaini F, *et al.* Mobilized bone marrow cells repair the infarcted heart, improving function and survival. *Proc Natl Acad Sci USA* 2001;98(18):10344–10349.

43. Minatoguchi S, Takemura G, Chen XH, Wang N, Uno Y, Koda M, *et al.* Acceleration of the healing process and myocardial regeneration may be important as a mechanism of improvement of cardiac function and remodeling by postinfarction granulocyte colony-stimulating factor treatment. *Circulation* 2004;109(21):2572–2580.

44. Lipiecki J, Durel N, Ponsonnaille J. Which patients with ischaemic heart disease could benefit from cell replacement therapy? *Eur Heart J* 2006;8(Suppl H):H3–H7.

45. Penicka M, Widimsky P, Kobylka P, Kozak T, Lang O. Early tissue distribution of bone marrow mononuclear cells after transcoronary transplantation in a patient with acute myocardial infarction. *Circulation* 2005;112(4):e63–e65.

46. Hofmann M, Wollert KC, Meyer GP, Menke A, Arseniev L, Hertenstein B, *et al.* Monitoring of bone marrow cell homing into the infarcted human myocardium. *Circulation* 2005;111(17):2198–2202.

47. Hou D, Youssef EA-S, Brinton TJ, Zhang P, Rogers P, Price ET, *et al.* Radiolabeled cell distribution after intramyocardial, intracoronary, and interstitial retrograde coronary venous delivery: implications for current clinical trials. *Circulation* 2005;112(Suppl 9):I-150.

48. Freyman T, Polin G, Osman H, Crary J, Lu M, Cheng L, *et al.* A quantitative, randomized study evaluating three methods of mesenchymal stem cell delivery following myocardial infarction. *Eur Heart J* 2006;27(9):1114–1122.

49. Vulliet PR, Greeley M, Halloran SM, MacDonald KA, Kittleson MD. Intra-coronary arterial injection of mesenchymal stromal cells and microinfarction in dogs. *Lancet* 2004;363(9411):783–784.

50. Strauer BE, Brehm M, Zeus T, Kostering M, Hernandez A, Sorg RV, *et al.* Repair of infarcted myocardium by autologous intracoronary mononuclear bone marrow cell transplantation in humans. *Circulation* 2002;106(15):1913–1918.

51. Schächinger V, Assmus B, Britten MB, Honold J, Lehmann R, Teupe C, *et al.* Transplantation of progenitor cells and regeneration enhancement in acute myocardial infarction: final one-year results of the TOPCARE-AMI Trial. *J Am Coll Cardiol* 2004;44(8):1690–1699.

52. Fernandez-Aviles F, San Roman JA, Garcia-Frade J, Fernandez ME, Penarrubia MJ, de la Fuente L, *et al.* Experimental and clinical regenerative capability of human bone marrow cells after myocardial infarction. *Circ Res* 2004;95(7):742–748.

53. Kuethe F, Richartz BM, Sayer HG, Kasper C, Werner GS, Hoffken K, *et al.* Lack of regeneration of myocardium by autologous intracoronary mononuclear bone marrow cell transplantation in humans with large anterior myocardial infarctions. *Int J Cardiol* 2004;97(1):123–127.

54. Bartunek J, Vanderheyden M, Vandekerckhove B, Mansour S, De Bruyne B, De Bondt P, et al. Intracoronary injection of CD133-positive enriched bone marrow progenitor cells promotes cardiac recovery after recent myocardial infarction: feasibility and safety. *Circulation* 2005;112(Suppl 9):I-178.
55. Doughty RN, Whalley GA, Walsh HA, Gamble GD, Lopez-Sendon J, Sharpe N. Effects of carvedilol on left ventricular remodeling after acute myocardial infarction: the CAPRICORN Echo Substudy. *Circulation* 2004;109(2):201–206.
56. Abdulla J, Barlera S, Latini R, Kjoller-Hansen L, Sogaard P, Christensen E, et al. A systematic review: effect of angiotensin converting enzyme inhibition on left ventricular volumes and ejection fraction in patients with a myocardial infarction and in patients with left ventricular dysfunction. *Eur J Heart Fail* 2007;9(2):129–135.
57. Stone GW, Grines CL, Cox DA, Garcia E, Tcheng JE, Griffin JJ, et al. Comparison of angioplasty with stenting, with or without abciximab, in acute myocardial infarction. *N Engl J Med* 2002;346(13):957–966.
58. Hombach V, Grebe O, Merkle N, Waldenmaier S, Hoher M, Kochs M, et al. Sequelae of acute myocardial infarction regarding cardiac structure and function and their prognostic significance as assessed by magnetic resonance imaging. *Eur Heart J* 2005;26(6):549–557.
59. Baks T, van Geuns RJ, Biagini E, Wielopolski P, Mollet NR, Cademartiri F, et al. Recovery of left ventricular function after primary angioplasty for acute myocardial infarction. *Eur Heart J* 2005;26(11):1070–1077.
60. Lunde K, Solheim S, Aakhus S, Arnesen H, Abdelnoor M, Egeland T, et al. Intracoronary injection of mononuclear bone marrow cells in acute myocardial infarction. *N Engl J Med* 2006;355(12):1199–1209.
61. Kloner RA, Jennings RB. Consequences of brief ischemia: stunning, preconditioning, and their clinical implications: part 1. *Circulation* 2001;104(24):2981–2989.
62. Schulz-Menger J, Gross M, Messroghli D, Uhlich F, Dietz R, Friedrich MG. Cardiovascular magnetic resonance of acute myocardial infarction at a very early stage. *J Am Coll Cardiol* 2003;42(3):513–518.
63. Wollert KC, Meyer GP, Lotz J, Ringes-Lichtenberg S, Lippolt P, Breidenbach C, et al. Intracoronary autologous bone-marrow cell transfer after myocardial infarction: the BOOST randomized controlled clinical trial. *Lancet* 2004;364(9429):141–148.
64. Janssens S, Dubois C, Bogaert J, Theunissen K, Deroose C, Desmet W, et al. Autologous bone marrow-derived stem-cell transfer in patients with ST-segment elevation myocardial infarction: double-blind, randomized controlled trial. *Lancet* 2006;367(9505):113–121.
65. Schächinger V, Erbs S, Elsässer A, Haberbosch W, Hambrecht R, Hölschermann H, et al. Intracoronary bone marrow-derived progenitor cells in acute myocardial infarction. *N Engl J Med* 2006;355(12):1210–1221.
66. Ge J, Li Y, Qian J, Shi J, Wang Q, Niu Y, et al. Efficacy of emergent transcatheter transplantation of stem cells for treatment of acute myocardial infarction (TCT-STAMI). *Heart* 2006;92(12):1764–1767.
67. Chen Sl, Fang Ww, Ye F, Liu YH, Qian J, Shan SJ, et al. Effect on left ventricular function of intracoronary transplantation of autologous bone marrow mesenchymal stem cell in patients with acute myocardial infarction. *Am J Cardiol* 2004;94(1):92–95.
68. Meluzin J, Mayer J, Groch L, Janousek S, Hornacek I, Hlinomaz O, et al. Autologous transplantation of mononuclear bone marrow cells in patients with acute myocardial infarction: the effect of the dose of transplanted cells on myocardial function. *Am Heart J* 2006;152(5):975–915.
69. Meyer GP, Wollert KC, Lotz J, Steffens J, Lippolt P, Fichtner S, et al. Intracoronary bone marrow cell transfer after myocardial infarction: eighteen months' follow-up data from the randomized, controlled BOOST (Bone marrow transfer to enhance ST-elevation infarct regeneration) trial. *Circulation* 2006;113(10):1287–1294.

70. Schaefer A, Meyer GP, Fuchs M, Klein G, Kaplan M, Wollert KC, *et al.* Impact of intracoronary bone marrow cell transfer on diastolic function in patients after acute myocardial infarction: results from the BOOST trial. *Eur Heart J* 2006;27(8):929–935.
71. Lunde K, Solheim S, Aakhus S, Arnesen H, Moum T, Abdelnoor M, Egeland T, Endresen K, Ilebekk A, Mangschau A, Forfang K. Exercise capacity and quality of life after intracoronary injection of autologous mononuclear bone marrow cells in acute myocardial infarction: results from the autologous stem cell transplantation in acute myocardial infarction (ASTAMI) randomized controlled trial. *Am Heart J* 2007;154:710.e1–710.e8.
72. Grebe O, Kestler HA, Merkle N, Wohrle J, Kochs M, Hoher M, *et al.* Assessment of left ventricular function with steady-state-free-precession magnetic resonance imaging. Reference values and a comparison to left ventriculography. *Z Kardiol* 2004;93(9):686–695.
73. Schächinger V, Erbs S, Elsasser A, Haberbosch W, Hambrecht R, Holschermann H, *et al.* Improved clinical outcome after intracoronary administration of bone-marrow-derived progenitor cells in acute myocardial infarction: final 1-year results of the REPAIR-AMI trial. *Eur Heart J* 2006;27(23):2775–2783.
74. Vickers AJ, Altman DG. Statistics notes: analysing controlled trials with baseline and follow up measurements. *Br Med J* 2001;323(7321):1123–1124.
75. Kang HJ, Kim HS, Zhang SY, Park KW, Cho HJ, Koo BK, *et al.* Effects of intracoronary infusion of peripheral blood stem-cells mobilised with granulocyte-colony stimulating factor on left ventricular systolic function and restenosis after coronary stenting in myocardial infarction: the MAGIC cell randomized clinical trial. *Lancet* 2004;363(9411):751–756.
76. Kang HJ, Kim HS, Koo BK, Kim YJ, Lee D, Sohn DW, *et al.* Intracoronary infusion of the mobilized peripheral blood stem cell by G-CSF is better than mobilization alone by G-CSF for improvement of cardiac function and remodeling: 2-year follow-up results of the Myocardial Regeneration and Angiogenesis in Myocardial Infarction with G-CSF and Intra-Coronary Stem Cell Infusion (MAGIC Cell) 1 trial. *Am Heart J* 2007;153(2):237–238.
77. Kang HJ, Lee Hy, Na SH, Chang SA, Park KW, Kim HK, *et al.* Differential effect of intracoronary infusion of mobilized peripheral blood stem cells by granulocyte colony-stimulating factor on left ventricular function and remodeling in patients with acute myocardial infarction versus old myocardial infarction: the MAGIC Cell-3-DES randomized, controlled trial. *Circulation* 2006;114(Suppl 1):I-145.
78. Li Z-Q, Zhang M, Jing Y-Z, Wei-wei Z, Ying L, Li-jie C, *et al.* The clinical study of autologous peripheral blood stem cell transplantation by intracoronory infusion in patients with acute myocardial infarction (AMI). *Int J Cardiol* 2007;115(1):52–56.
79. Harada M, Qin Y, Takano H, Minamino T, Zou Y, Toko H, *et al.* G-CSF prevents cardiac remodeling after myocardial infarction by activating the Jak-Stat pathway in cardiomyocytes. *Nat Med* 2005;11(3):305–311.
80. Valgimigli M, Rigolin GM, Cittanti C, Malagutti P, Curello S, Percoco G, *et al.* Use of granulocyte-colony stimulating factor during acute myocardial infarction to enhance bone marrow stem cell mobilization in humans: clinical and angiographic safety profile. *Eur Heart J* 2005;26(18):1838–1845.
81. Ince H, Petzsch M, Kleine HD, Schmidt H, Rehders T, Korber T, *et al.* Preservation from left ventricular remodeling by front-integrated revascularization and stem cell liberation in evolving acute myocardial infarction by use of granulocyte-colony-stimulating factor (FIRSTLINE-AMI). *Circulation* 2005;112(20):3097–3106.
82. Ince H, Petzsch M, Kleine HD, Eckard H, Rehders T, Burska, T, *et al.* Prevention of left ventricular remodeling with granulocyte colony-stimulating factor after acute myocardial infarction: final 1-year results of the front-integrated revascularization and stem cell liberation in evolving acute myocardial infarction by granulocyte colony-stimulating factor (FIRSTLINE-AMI) trial. *Circulation* 2005;112(Suppl 9):I-73.

83. Zohlnhofer D, Ott I, Mehilli J, Schomig K, Michalk F, Ibrahim T, *et al.* Stem cell mobilization by granulocyte colony-stimulating factor in patients with acute myocardial infarction: a randomized controlled trial. *J Am Med Assoc* 2006;295(9):1003–1010.

84. Ripa RS, Jorgensen E, Wang Y, Thune JJ, Nilsson JC, Sondergaard L, *et al.* Stem cell mobilization induced by subcutaneous granulocyte-colony stimulating factor to improve cardiac regeneration after acute st-elevation myocardial infarction: result of the double-blind, randomized, placebo-controlled stem cells in myocardial infarction (STEMMI) trial. *Circulation* 2006;113(16): 1983–1992.

85. Engelmann MG, Theiss HD, Hennig-Theiss C, Huber A, Wintersperger BJ, Werle-Ruedinger AE, *et al.* Autologous bone marrow stem cell mobilization induced by granulocyte colony-stimulating factor after subacute ST-segment elevation myocardial infarction undergoing late revascularization: final results from the G-CSF-STEMI (granulocyte colony-stimulating factor ST-segment elevation myocardial infarction) trial. *J Am Coll Cardiol* 2006;48(8):1712–1721.

86. Kastrup J, Ripa RS, Wang Y, Jorgensen E. Myocardial regeneration induced by granulocyte-colony-stimulating factor mobilization of stem cells in patients with acute or chronic ischaemic heart disease: a non-invasive alternative for clinical stem cell therapy? *Eur Heart J* 2006;27(23):2748–2754.

87. Tse HF, Kwong YL, Chan JK, Lo G, Ho CL, Lau CP. Angiogenesis in ischaemic myocardium by intramyocardial autologous bone marrow mononuclear cell implantation. *Lancet* 2003;361(9351):47–49.

88. Beeres SL, Bax JJ, Bbets-Schneider P, Stokkel MP, Fibbe WE, van der Wall EE, *et al.* Sustained effect of autologous bone marrow mononuclear cell injection in patients with refractory angina pectoris and chronic myocardial ischemia: twelve-month follow-up results. *Am Heart J* 2006;152(4):684–686.

89. Fuchs S, Kornowski R, Weisz G, Satler LF, Smits PC, Okubagzi P, *et al.* Safety and feasibility of transendocardial autologous bone marrow cell transplantation in patients with advanced heart disease. *Am J Cardiol* 2006;97(6):823–829.

90. Briguori C, Reimers B, Sarais C, Napodano M, Pascotto P, Azzarello G, *et al.* Direct intramyocardial percutaneous delivery of autologous bone marrow in patients with refractory myocardial angina. *Am Heart J* 2006;151(3):674–680.

91. Unlisted authors. Progress in clinical trials. *Clin Cardiol* 2007;29:230.

92. Boyle AJ, Whitbourn R, Schlicht S, Krum H, Kocher A, Nandurkar H, *et al.* Intra-coronary high-dose CD34+ stem cells in patients with chronic ischemic heart disease: a 12-month follow-up. *Int J Cardiol* 2006;109(1):21–27.

93. Erbs S, Linke A, Adams V, Lenk K, Thiele H, Diederich KW, *et al.* Transplantation of blood-derived progenitor cells after recanalization of chronic coronary artery occlusion: first randomized and placebo-controlled study. *Circ Res* 2005;97(8):756–762.

94. Seiler C, Pohl T, Wustmann K, Hutter D, Nicolet PA, Windecker S, *et al.* Promotion of collateral growth by granulocyte-macrophage colony-stimulating factor in patients with coronary artery disease: a randomized, double-blind, placebo-controlled study. *Circulation* 2001;104(17):2012–2017.

95. Zbinden S, Zbinden R, Meier P, Windecker S, Seiler C. Safety and efficacy of subcutaneous-only granulocyte-macrophage colony-stimulating factor for collateral growth promotion in patients with coronary artery disease. *J Am Coll Cardiol* 2005;46(9):1636–1642.

96. Hill JM, Syed MA, Arai AE, Powell TM, Paul JD, Zalos G, *et al.* Outcomes and risks of granulocyte colony-stimulating factor in patients with coronary artery disease. *J Am Coll Cardiol* 2005;46(9):1643–1648.

97. Ozbaran M, Omay SB, Nalbantgil S, Kultursay H, Kumanlioglu K, Nart D, *et al.* Autologous peripheral stem cell transplantation in patients with congestive heart failure due to ischemic heart disease. *Eur J Cardiothorac Surg* 2004;25(3):342–350.

98. Strauer BE, Brehm M, Zeus T, Bartsch T, Schannwell C, Antke C, *et al.* Regeneration of human infarcted heart muscle by intracoronary autologous bone marrow cell transplantation in chronic coronary artery disease: the IACT study. *J Am Coll Cardiol* 2005;46(9):1651–1658.

99. Assmus B, Honold J, Schachinger V, Britten MB, Fischer-Rasokat U, Lehmann R, *et al.* Transcoronary transplantation of progenitor cells after myocardial infarction. *N Engl J Med* 2006;355(12):1222–1232.

100. Galinanes M, Loubani M, Davies J, Chin D, Pasi J, Bell PR. Autotransplantation of unmanipulated bone marrow into scarred myocardium is safe and enhances cardiac function in humans. *Cell Transplant* 2004;13(1):7–13.

101. Stamm C, Westphal B, Kleine HD, Petzsch M, Kittner C, Klinge H, *et al.* Autologous bone-marrow stem-cell transplantation for myocardial regeneration. *Lancet* 2003;361(9351):45–46.

102. Hendrikx M, Hensen K, Clijsters C, Jongen H, Koninckx R, Bijnens E, *et al.* Recovery of regional but not global contractile function by the direct intramyocardial autologous bone marrow transplantation: results from a randomized controlled clinical trial. *Circulation* 2006;114(suppl 1):I-101.

103. Perin EC, Dohmann HF, Borojevic R, Silva SA, Sousa AL, Mesquita CT, *et al.* Transendocardial, autologous bone marrow cell transplantation for severe, chronic ischemic heart failure. *Circulation* 2003;107(18):2294–2302.

104. Perin EC, Dohmann HFR, Borojevic R, Silva SA, Sousa ALS, Silva GV, *et al.* Improved exercise capacity and ischemia 6 and 12 months after transendocardial injection of autologous bone marrow mononuclear cells for ischemic cardiomyopathy. *Circulation* 2004;110(Suppl 11): II-213.

105. Nussbaum J, Minami E, Laflamme MA, Virag JA, Ware CB, Masino A, *et al.* Transplantation of undifferentiated murine embryonic stem cells in the heart: teratoma formation and immune response. *FASEB J* 2007;21:1345–1357.

106. Menasche P, Hagege AA, Vilquin JT, Desnos M, Abergel E, Pouzet B, *et al.* Autologous skeletal myoblast transplantation for severe postinfarction left ventricular dysfunction. *J Am Coll Cardiol* 2003;41(7):1078–1083.

107. Widimsky P, Penicka M, Lang O, Kozak T, Motovska Z, Jirmar R, *et al.* Intracoronary transplantation of bone marrow stem cells: background, techniques, and limitations. *Eur Heart J Suppl* 2006;8(Suppl H):H16–H22.

108. Mangi AA, Noiseux N, Kong D, He H, Rezvani M, Ingwall JS, *et al.* Mesenchymal stem cells modified with Akt prevent remodeling and restore performance of infarcted hearts. *Nat Med* 2003;9(9):1195–1201.

109. Koc ON, Gerson SL. Akt helps stem cells heal the heart. *Nat Med* 2003;9(9):1109–1110.

110. Braunwald E, Zipes DP, Libby P. *Braunwald's Heart Disease: A Textbook of Cardiovascular Medicine.* W.B. Saunders, Philadelphia, 2005.

111. Widimsky P, Penicka M. Complications after intracoronary stem cell transplantation in idiopathic dilated cardiomyopathy. *Int J Cardiol* 2006;111(1):178–179.

112. Blatt A, Cotter G, Leitman M, Krakover R, Kaluski E, Milo-Cotter O, *et al.* Intracoronary administration of autologous bone marrow mononuclear cells after induction of short ischemia is safe and may improve hibernation and ischemia in patients with ischemic cardiomyopathy. *Am Heart J* 2005;150(5):986.

113. Kastrup J, Jorgensen E, Ruck A, Tagil K, Glogar D, Ruzyllo W, *et al.* Direct intramyocardial plasmid vascular endothelial growth factor-A165 gene therapy in patients with stable severe angina pectoris: a randomized double-blind placebo-controlled study — the Euroinject One trial. *J Am Coll Cardiol* 2005;45(7):982–988.

114. Hu Y, Davison F, Zhang Z, Xu Q. Endothelial replacement and angiogenesis in arteriosclerotic lesions of allografts are contributed by circulating progenitor cells. *Circulation* 2003;108(25):3122–3127.

115. Epstein SE, Stabile E, Kinnaird T, Lee CW, Clavijo L, Burnett MS. Janus phenomenon: the interrelated tradeoffs inherent in therapies designed to enhance collateral formation and those designed to inhibit atherogenesis. *Circulation* 2004;109(23):2826–2831.

116. Lunde K, Solheim S, Forfang K, Arnesen H, Brinch L, Bjørnerheim R, Ragnarsson A, Egeland T, Endresen K, Ilebekk A, Mangschau A, Aakhus S. Anterior myocardial infarction with acute percutaneous coronary intervention and intracoronary injection of autologous mononuclear bone marrow cells. Safety, clinical outcome, and serial changes in left ventricular function during 12 months follow-up. *J Am Coll Cardiol* 2008;51:674–676.

117. Assmus B, Walter DH, Lehmann R, Honold J, Martin H, Dimmeler S, *et al.* Intracoronary infusion of progenitor cells is not associated with aggravated restenosis development or atherosclerotic disease progression in patients with acute myocardial infarction. *Eur Heart J* 2006;27(24):2989–2995.

118. Yoon CH, Hur J, Park KW, Kim JH, Lee CS, Oh IY, *et al.* Synergistic neovascularization by mixed transplantation of early endothelial progenitor cells and late outgrowth endothelial cells: role of angiogenic cytokines and matrix metalloproteinases. *Circulation* 2005;112(11):1618–1627.

119. Rookmaaker MB, Verhaar MC, Loomans CJ, Verloop R, Peters E, Westerweel PE, *et al.* CD34+ cells home, proliferate, and participate in capillary formation, and in combination with. *Arterioscler Thromb Vasc Biol* 2005;25(9):1843–1850.

120. Seeger F, Tonn T, Krzossok N, Aicher A, Zeiher AM. Cell isolation procedures matter: a comparison of different isolation protocols of bone marrow mononuclear cells used for cell therapy in patients with myocardial infarction. *Circulation* 2006;114(Abstract Suppl):II-51.

121. Beltrami CA, Finato N, Rocco M, Feruglio GA, Puricelli C, Cigola E, *et al.* Structural basis of end-stage failure in ischemic cardiomyopathy in humans. *Circulation* 1994;89(1):151–163.

122. Britten MB, Abolmaali ND, Assmus B, Lehmann R, Honold J, Schmitt J, *et al.* Infarct remodeling after intracoronary progenitor cell treatment in patients with acute myocardial infarction (TOPCARE-AMI): mechanistic insights from serial contrast-enhanced magnetic resonance imaging. *Circulation* 2003;108(18):2212–2218.

123. Moelker AD, Baks T, van den Bos EJ, van Geuns RJ, de Feyter PJ, Duncker DJ, *et al.* Reduction in infarct size, but no functional improvement after bone marrow cell administration in a porcine model of reperfused myocardial infarction. *Eur Heart J* 2006;27(24):3057–3064.

124. Cluitmans FH, Esendam BH, Veenhof WF, Landegent JE, Willemze R, Falkenburg JH. The role of cytokines and hematopoietic growth factors in the autocrine/paracrine regulation of inducible hematopoiesis. *Ann Hematol* 1997;75(1–2):27–31.

125. Paroli M, Mariani P, Accapezzato D, D'Alessandro M, Di Russo C, Bifolco M, *et al.* Modulation of tachykinin and cytokine release in patients with coronary disease undergoing percutaneous revascularization. *Clin Immunol* 2004;112(1):78–84.

126. Staat P, Rioufol G, Piot C, Cottin Y, Cung TT, L'Huillier I, *et al.* Postconditioning the human heart. *Circulation* 2005;112(14):2143–2148.

11 Skeletal Myoblast Transplantation for Ischemic Heart Failure

Philippe Menasché

Cell therapy is currently emerging as a potentially new treatment of heart failure. It is primarily based on the assumption that this condition develops when a critical number of cardiomyocytes has been irreversibly lost, and that, consequently, function could be improved by re-colonization of the areas of akinetic myocardium with new contractile cells. Even if self-repair endogenous mechanisms have been described recently in the adult human heart, they are likely overwhelmed in patients with extensive postinfarction scars (or diffusively fibrotic non-ischemic cardiomyopathies), thereby supporting the concept of engrafting the damaged areas with an exogenous supply of donor cells.

Theoretical and Experimental Basis for the Use of Skeletal Myoblasts

So far, multiple cell types have been tested experimentally for cardiac repair. Indeed, fetal cardiomyocytes have been the first to provide the "proof-of-concept" as these cells were shown to successfully engraft in scar tissue, establish connexions with host cardiomyocytes through gap junctions and improve left ventricular function.[1,2] However, the issues associated with their use (ethics, availability, sensitivity to ischemia, immunogenicity) have rapidly led, in a clinical perspective, to switch to alternate cell type best suited for human applications.

In this context, skeletal myoblasts (also called satellite cells) which normally lie in a quiescent state under the basal membrane of skeletal muscular fibers but are rapidly mobilized, proliferate and fuse to effect repair following muscle injury, have generated a great interest motivated by their clinically relevant attractive characteristics: (1) an autologous origin which makes accessibility easy and avoids any form of immune rejection, (2) a high degree of scalability in culture (one billion cells can be yielded from an initial small biopsy over a two- to three-week time frame), (3) a myogenic lineage restriction which provides a safeguard against tumor formation, and (4) a high resistance to ischemia, which is a major advantage given the poorly vascularized environment in which they are to be implanted. A large number of experimental studies, both in small and large animal models of myocardial injury, have established that myoblasts injected into postinfarction scars differentiated into typical multinucleated myotubes and that this engraftment was associated with a sustained improvement in left ventricular function.[3] More recently, the beneficial effects of myoblast transplantation have been further supported by sonomicrometry data in a dog model of heart failure[4] although the most consistent effect of the procedure seems to be a limitation of remodeling, as shown by magnetic resonance imaging in rabbit hearts[5] and pressure–volume loops in sheep.[6] These observations have contributed to shed some light on the possible mechanisms of action of implanted myoblasts. Thus, currently, a prevailing hypothesis is that cells exert a buttressing effect (which may actually be not specific for myoblasts) on the infarcted segment which improves scar elasticity and ultimately translates into a limitation of ventricular dilatation. A reduction of wall stress placed on remote myocardium subsequent to reduced remodeling could, in turn, contribute to improve heart function.[6] An alternate possibility is that cells affect the surrounding myocardium by releasing cytokines and growth factors that may paracrinally increase local angiogenesis, change the composition of the extracellular matrix or even recruit the putative resident cardiac stem cells. The role of these paracrine factors has been primarily established following bone marrow cell transplantation[7,8] but microarray studies conducted in collaboration with Professor Felipe Prosper, University of Navarra, Spain, suggest that similar mechanisms might be in play following myoblast transplantation. However, a more straightforward demonstration of this paracrine hypothesis would require to show that the effects of transplanted myoblasts can be reproduced by injecting the conditioned medium recovered from the myogenic cell cultures and it is fair to acknowledge that such a proof-of-principle experiment is still missing. Finally, although myoblasts could feature stretch-activated contractions, their lack of connexions with the neighboring cardiomyocytes prevents them to beat in synchrony with the host heart through a functional syncytium[9,10] and makes thus dubious that they can noticeably contribute to augment pump function directly.

Update on Clinical Studies

Despite the unsettled mechanistic issues associated with myoblast transplantation, the bulk of animal data has been deemed convincing enough to justify a move towards clinical applications that actually started in June 2000, when we performed the first human transplantation of autologous myoblasts in a patient with severe ischemic heart failure.[11] This case was the first of a series of ten which included patients with a low left ventricular ejection fraction (LVEF $\leq$ 35%), a history of myocardial infarction with a residual scar without evidence for viability and an indication for CABG which was consistently performed in areas different from the transplanted ones. Overall, 871 million cells, of which 87% were myoblasts, were injected.

This cohort has now been followed for an average of 52 months (18 to 58) after transplantation.[12] Overall, the patients have been symptomatically improved with a low incidence of hospitalizations for heart failure (0.13/patient-years). Ejection fraction initially increased after surgery; it then levelled off and is now comparable with the preoperative value; left ventricular dimensions have remained equally stable over time. Early ventricular arrhythmias prompted the implantation of an internal cardioverter-defibrillator (ICD) in five patients, of whom three have demonstrated new arrhythmic events at later time points (six, seven, and 18 months after implantation). One patient died 17.5 months after his surgery from a presumed stroke and the pathologic examination of his heart showed clusters of myotubes in scar tissue.[13] Overall, these findings have been corroborated by the three other adjunct-to-CABG trials. In the study of Gavira and co-workers,[14] 12 patients who had received an average of 221 million myoblasts were found, at the one-year postoperative follow-up, to have incurred a significant increase in both ejection fraction (from 35.5% to 55.1%, $p < 0.01$) and regional contractility (as evidenced by a decrease in the wall motion score index from 3.02 to 1.36, $p < 0.0001$). No cardiac arrhythmias were reported. On the basis of positron emission tomography (PET)-based measurements, viability and perfusion were also found to have improved in the myoblast-transplanted segments. The study of Siminiak *et al.*[15] included ten patients who received doses of cells ranging from 4 ± 10^5 to 5 ± 10^7. The mean ejection fraction increased from 35.2% to 42.0% ($p < 0.05$) at four months and this effect was maintained throughout the 12-month follow-up period. Four episodes of sustained ventricular tachycardia were documented (two during the early postoperative period and two others after two weeks). Finally, Dib and co-workers[16] have reported on 30 patients, of whom 24 were treated with skeletal myoblasts at the time of CABG according to a dose-escalating protocol (12 were divided into four three-patient groups receiving 1, 3, 10, and 30×10^7 while 12 received a fixed dose of 3×10^8). Like in the previous studies, the echocardiographic ejection fraction increased from 28% to 35% at one year ($p = 0.02$) and 36% at two years ($p = 0.01$). In some patients, PET and magnetic resonance imaging also documented improved viability in the transplanted areas. Non-sustained ventricular tachycardia were reported in three cases. The remaining six patients of this series received the cells during implantation of a left ventricular assist device (LVAD) as a bridge to heart transplantation. The histopathological studies of the hearts after device explantation or death confirmed engraftment of myofibers in four out of the six patients. Of note, whereas these three studies shared in common with our trial the profile of the patient population and the technique of cell injections, they basically differed by the fact that the cell-implanted area was consistently revascularized.

In parallel to these surgical trials, three catheter-based studies have been reported. One has entailed administration of myoblasts through a trans-coronary sinus route. This catheter features an extendable needle through which a microcatheter is advanced directly into the target area for cell injections under endovascular ultrasound guidance. Experimentally, this system has been shown to result in a successful myoblast engraftment[17] and the ten-patient clinical study has confirmed both the feasibility and procedural safety of this approach.[18] The other two percutaneous trials have entailed an endoventricular transfer of myoblasts under electromechanical guidance. One of the studies (ten patients) reported a one-year improvement in systolic velocity of the cell-injected segments and an increase in global ejection fraction during low-dose dobutamine infusion.[19] The second study also demonstrated an improved function in the six treated patients compared with six case-matched controls.[20]

Put together, these studies yet allow to draw some conclusions pertaining to feasibility and safety. Thus, it is well demonstrated that a small muscular biopsy can be upscaled under Good Manufacturing Practice (GMP) conditions in a two- to three-week period to yield several hundreds million cells with a high degree of viability and myogenic purity. Likewise, the safety of multiple cell injections in and around the postinfarction scar has been well established and no intraoperative procedural complications have been described so far. The postoperative adverse events that have been reported are those commonly encountered in high-risk patients undergoing CABG and have never been specifically ascribed to cell implantations *per se*. Indeed, the only serious safety concern relates to the occurrence of ventricular arrhythmias potentially induced by engrafted myoblasts. This problem was raised after initial reports of sustained ventricular tachycardias that usually occurred early after myoblast transplantation.[15,21] The prevailing hypothesis to explain these events is that injected myoblasts engraft as electrically insulated clusters and co-culture experiments have indeed documented that this pattern results in a slowing of conduction velocity that may set the stage for re-entry circuits.[22] The pathogenetic role of this lack of coupling between donor myoblasts and recipient cardiomyocytes is further supported by the finding that genetically-induced overexpression of connexin-43 decreases arrhythmogenesis in the co-culture experiments. However, alternate mechanisms cannot be ruled out and include delayed cardiac repolarization (an effect attributed to the secretion of insulin growth factor-1 by myoblasts and a subsequent myocyte hypertrophy causing action potential prolongation) and myoblast automaticity which could trigger stretch-mediated fibrillation-like contractions of the neighboring cardiomyocytes.[23] The *in vivo* translation of these observations is not straightforward since the increased incidence of sustained ventricular tachycardia reported in a rat model of myoblast transplantation[24] contrasts with the observations made in the canine left ventricular wedge preparation assessed by optimal mapping, that abnormal impulse propagation and arrhythmia inducibility are common features of all preparations with myocardial infarction regardless of whether myoblasts have been transplanted or not.[25] These discrepant data may, to some extent, be reconciled by taking into account the following two considerations. First, the graft size seems to influence the incidence of arrhythmias. Thus, reentry was not induced in cocultures containing few myoblasts (1%–5%) but consistently occurred when this percentage exceeded 20%.[22] Likewise, in a rabbit model of doxorubicin-induced cardiomyopathy, a positive correlation was found between the number of transplanted cells (skeletal myoblasts or bone marrow-derived mesenchymal stem cells) in the pacing site and the dispersion of activation time.[26] This size effect could explain why bone marrow cells which share in common with myoblasts the lack of connexin-43 and similar *in vitro* patterns of reduced conduction velocity[27] but differ by their low-engraftment rate, do not seem to cause clinically detectable arrhythmias. A second critical determinant of arrhythmogenesis could be the exact location of the cell injections as, experimentally, myoblasts delivered in the core of the scar seem less arrhythmogenic than those injected in the border zone.[28] This would be consistent with the expectedly greater electrical instability when slowing of the conduction velocity due to myoblast clusters superimposes on that already present in the border zone. Indeed, in the canine left ventricular wedge preparation used by Fouts *et al.*[25] skeletal myoblasts did not influence arrhythmia inducibility when they were injected in scar tissue whereas they occasionally caused abnormal impulse propagation when injected in normal myocardium. The situation of the human infarcted heart is complex in that scars usually feature a patchy pattern with islands of near-normal myocardium interspersed with areas of fibrosis; the occurrence of arrhythmic episodes occasionnally reported after clinical myoblast transplantation might then be dependent on whether cells have been injected in the core of the infarct or along its margins. The magnitude of cell delivery-induced local inflammation could further modulate the patterns of local heterogeneity of impulse propagation increasing the susceptibility to ventricular arrhythmias.

However, despite the theoretical and experimental data that may rationalize the pro-arrhythmic risk of skeletal myoblast transplantation, it is critical to remind that the interpretation of the clinical events is somewhat clouded by the 'background noise' due to the arrhythmogenic nature of the underlying heart failure disease. This has been well exemplified in the PATCH trial, a study that randomized high-risk patients (LVEF $\leq$ 36%) undergoing CABG to prophylactic implantation of an implantable cardioverter-defibrillator (ICD) or no additional treatment, where the one-year actuarial incidence of first discharge was 50% in the defibrillator group.[29] These considerations have prompted us to implant an ICD in all patients enrolled in the randomized MAGIC trial that we have implemented (see below), not only for obvious safety reasons (most of the included patients actually matched the MADIT II criteria) but also for collecting ICD read-outs allowing an objective comparison of the prevalence of arrhythmias between the treated patients and those allocated to the placebo-injected arm of the study. In contrast to some speculations, the six-month results do not indicate a statistically increased risk of arrhythmias in the myoblast-transplanted patients but these reassuring safety data need to be interpreted cautiously because of the still small sample size. More generally, the not unexpectedly high prevalence of arrhythmic events in the placebo-injected arm confirms the survival benefit conferred by ICDs in this high-risk patient population.

Whereas the phase I studies have been important for assessing both feasibility and safety, the efficacy data they have generated should be viewed much more cautiously because interpretation of patient outcomes is confounded by several major methodologic issues such as the small sample sizes, the open-label design, the lack of true control groups, and the common concomitant revascularization of the cell-transplanted segments. This is the reason why we have undertaken the multicenter Myoblast Autologous Grafting in Ischemic Cardiomyopathy (MAGIC) trial which featured a randomized, double blind, placebo-controlled dose-ranging design. All included patients have undergone a muscular biopsy followed, three weeks later, by injection of cells (at two doses: 400 and 800 million) or a placebo solution in the core and at the margins of akinetic non-viable scars. The consistency of the final cell therapy product has been ensured by a centralized production in two-core manufacturing facilities working under similar procedural guidelines. As previously mentioned, all patients have been implanted with an ICD prior to hospital discharge. The primary end points of the study have thus been safety (occurrence of major adverse cardiac events and ventricular arrhythmias over the first six postoperative months) and efficacy (changes in regional wall motion and global ejection fraction, as assessed by echocardiograms read in a core lab by blinded investigators). The results obtained in the 97 patients enrolled in the trial show that although myoblast transplantation failed to increase contractility of the injected segments as well as global left ventricular function, the highest dose of cells resulted in a significant reversal of left ventricular remodeling.[30] This finding further supports the concept that myoblasts, whether delivered surgically or by catheter, can exert therapeutic effects through decreased fibrosis and changes in extracellar matrix remodeling[31] and, if sustained over time, might have a favorable impact on clinically meaningful patient outcomes. In parallel to the MAGIC trial, a catheter-based open-label single-center study[32] has included 23 patients randomized to myoblast injections by an endoventricular approach according to a dose-escalating protocol (12 patients allocated to four three-patient blocks ranging from 30 to 600 million cells) or optimal medical management alone (11 patients). All these patients had ejection fractions below 40% and old (>10 years) infarcts. The six-month interim results, as assessed by echocardiography and SPECT, provide reassuring safety outcomes (only one arrhythmic episode in the 300-million cell subgroup) and suggest a trend towards smaller left ventricular dimensions without changes in ejection fraction in the myoblast-treated patients but the still limited amount of available data mandates to wait for the completion of the trial before drawing any definite conclusion.

Remaining Issues

Cell transfer

With the current techniques, both direct intramyocardial and catheter-based injections are still associated with a substantial degree of leakage that decreases the amount of cell engraftment in the target area and leads to an unwanted systemic dissemination of the cells. These events have been well illustrated by the finding that injections of male myoblasts in open-chest female recipients results in the identification of the Y chromosome in several extracardiac tissues (regardless of whether there is reperfusion or not).[33] This leakage issue thus needs to be addressed by improvements in techniques and modalities of cell transfer. Indeed, the importance of the technique is illustrated by the mere observation that multiple small deposits of myoblasts are more effective in improving engraftment and function than "macrodepots" i.e. a smaller number of punctures delivering larger amounts of cells per spot.[34] However, a substantial progress could likely be made if separate needle injections could be replaced by a multiple-shot device allowing a time-saving and reproducible cell delivery procedure. Methods not based on injections are also considered and exemplified by cell-seeded temperature-sensitive sheets that are laid over the infarct area to which they stick owing to their physical properties. This approach has yielded encouraging experimental results[35] but its potential clinical applications are likely to remain confined to surgical cell transplantation whereas one can easily anticipate that the development of myoblast cell therapy and its large-scale implementation necessarily require percutaneous routes of application. This is particularly important in view of recent laboratory data suggesting that the benefits of these cells might extend to non-ischemic globally dilated cardiomyopathies, thereby potentially opening new therapeutic perspectives for these patients who cannot currently be offered any other option than cardiac transplantation.[36,37] It is also noteworthy that only catheter-based techniques could allow repeated myoblast injections which have recently been shown to be more effective for improving engraftment and function that a single bolus of the same number of cells.[38]

Cell survival

Up to 90% of the injected myoblasts die within a few days after transplantation and although the surviving fraction proliferates, it cannot catch up the initial cell loss.[39] This high rate of graft attrition results from the interplay of multiple factors including inflammation, apoptosis, hypoxia and loss of the normal relationships between transplanted cells and the extracellular matrix. These events are clinically relevant because the functional benefit of myoblast transplantation seems to be tightly dependent to the number of engrafted cells.[40] One can therefore predict that the expected benefits of the procedure will remain suboptimal as long as cell survival is not substantially increased. Among the different strategies that can be considered for achieving this objective, two have already proven to be effective. The first consists of increasing the blood supply to the grafted area to address the ischemic component of cell death. This can be accomplished through a variety of methods including direct revascularization by a coronary artery bypass graft or balloon angioplasty, co-injection of angiogenic growth factors,[41] transfection of the grafted myoblasts by genes encoding some of these factors[42] or co-transplantation of angiogenic bone marrow-derived cells.[43] The second strategy consists of embedding cells in a bioinjectable scaffold designed to restore a three-dimensional microenvironment. In this setting, the first proof-of-concept was brought by fibrin glue used as a matrix for the injected myoblasts and which actually resulted in a better cell engraftment[44]; however, both the viscosity-related poor injectability and possible arrhythmogenicity of fibrin glue[45] would probably preclude its clinical use and warrant the search for alternate scaffolds among which self-assembling nanopeptides are particularly interesting, partly because of their capacity to bind growth factors whose controlled release over time may favorably affect the trophicity of the graft and of the surrounding host myocardium.[46] Although most of the above-mentioned approaches are probably the easiest to implement clinically because they look efficacious, relatively user-friendly and safe, other cell-survival enhancing strategies also need to be considered and particularly include transfection of the cells to be grafted with genes encoding pro-survival factors[47,48] or their preconditioning by physical[39] or pharmacological (potassium channel agonists) methods.[49,50]

Obviously, the rigorous assessment of the efficacy of these various strategies requires a means of noninvasively monitoring the fate of the engrafted cells. A great deal of interest is paid to labeling of the cells with iron particles allowing their subsequent visualization by magnetic resonance imaging[51] but the reliability of this technique is still debated because of the possibility to detect signals whereas the grafted cells are actually dead and the potential for an iron-mediated impairment of cell functionality.[52]

Cell functionality

We have previously stressed that the lack of physical connexions between skeletal myoblasts and host cardiomyocytes makes unlikely that the grafted cells may form a functional syncytium and thus contribute to augment the heart's contractile function. Theoretically, this hurdle can be overcome by transfecting myoblasts by the gene encoding connexin-43 but although this approach results in an increased conduction velocity and a limitation of the arrhythmic events in co-culture systems,[22] it has not yet been shown that this forced expression of the gap junction proteins translated into an effective electromechanical integration of the graft and the attendant improvement in function. Indeed, the key conceptual question here is to determine whether the contractile properties of transplanted cells are mandatory for cell therapy to be effective. The increased recognition (primarily based on bone marrow cell transplantation studies) that a minimal degree of engraftment does not preclude a functional improvement[53] currently leads some investigators to consider that cells are beneficial by mechanisms independent from their intrinsic contractile properties, i.e. changes in scar elasticity and paracrine effects, in which case attempts at achieving their coupling with host cardiomyocytes may become irrelevant. Conversely, if one sticks to the original paradigm of repopulating the scar tissue with a sufficient number of newly formed donor-derived cardiomyocytes, other cell types than myoblasts have to be considered and currently, three main sources have been putatively identified. The first source comprises cardiac progenitors residing in extracardiac tissues like skeletal muscle, bone marrow, fat tissue or umbilical cord; however, so far, the cardiogenic potential of these progenitors has primarily been established in murine models and the clinical relevance of these findings still remains elusive. In the setting of myogenic cells, two cell sources named skeletal precursors of adult cardiomyocytes (SPOC)[54] and muscle-derived stem cells,[55] respectively, have been reported to yield better engraftment rates and functional outcomes than "conventional" myoblasts in mouse models of myocardial infarction but the applicability of these approaches to the human situation remains to be established. The second source consists of cardiac stem cells purportedly harbored in small intramyocardial niches[56]; indeed, although these cells have raised the theoretically appealing idea that they could be harvested through an endomyocardial biopsy, grown *in vitro* and reinjected, several questions still need to be answered, particularly their exact phenotype and their persistence in adulthood. Recent data from our department indicate that the few c-*kit*-positive cells that can be found in either right ventricular biopsies taken in heart transplant recipients or right appendage tissue specimens collected during coronary artery bypass surgery, express hematopoietic markers and, more specifically, those of mast cells without evidence for expression of markers of the cardiac lineage like Nkx2.5.[57] Put together, these data raise a cautionary note about the existence of these cardiac stem cells in ischemic patients who would need them most. The third potential source of cardiac precursors is represented by embryonic stem cells. Not unexpectedly, the pluripotentiality of these cells makes possible to direct their fate towards the cardiomyogenic lineage by appropriate *in vitro* specification procedures. There is no doubt that the therapeutic use of embryonic stem cells still raises important concerns primarily related to availability, ethics, propagation under Good Manufacturing Practice conditions and immunogenicity (while teratoma formation does not seem to be an issue as long as the injected cells have been appropriately lineage-specified). Despite these shortcomings, these cells possibly hold a real therapeutic promise in that they remain so far the only ones which have been convincingly shown in both small and large animal models of myocardial infarction[58–60] and atrioventricular block[61] to differentiate into cardiomyocytes and achieve a successful electromechanical integration within the transplanted heart.

Economic Issues

As mentioned at the onset of this review, one of the historical arguments favouring the use of skeletal myoblasts has been their autologous origin. As experience has accumulated, it has become evident that patient-specific products also raise some practical concerns. They include (1) a delay in treatment as at least two weeks are required for upscaling the muscular biopsy, a time lag that may not always been well tolerated by unstable patients or even accepted by patients who increasingly behave as consumers eager to have their health problem fixed as rapidly as possible; (2) the logistical complexity related to shipment of the biopsy from the transplant centre to the cell manufacturing facility followed by the back trip of the final cell product; (3) the interpatient variability which makes difficult to reproducibly end up with an homogeneously characterized cell therapy product; and (4) the cost of customized quality controls which need to be repeated for each patient-specific batch. Thus, in the perspective of a treatment which would extend beyond the boundaries of a well-defined clinical trial to become readily applicable to a large number of patients, it is clear that an "off-the-shelf" product derived from a cell bank would offer significant advantages in terms of practicality, reproducibility and cost savings. Such an allogeneic product raises, in turn, the major concern of immuno-genicity which could possibly be addressed by an immunosuppressive treatment (with its attendant risks), knock-out of some major histocompatibility genes or even co-transplantation of allogenic mesenchymal stem cells if their tolerogenicity was unequivocally demonstrated.[62] Clearly, there is a need for studies designed to thoughtfully compare the risk-benefit and cost-effectiveness ratios between autol-ogous and allogeneic myoblast-based cell therapy products.

Methodologic Issues

Regardless of its results, the MAGIC trial should hopefully be helpful in orienting future cell therapy studies with regard to the patient profile, type and dose of cells, delivery method and outcome assessment. Thus, the decreased referrals for coronary artery bypass surgery make probably realistic to rather consider a catheter-based method of cell transfer to broaden the number of eligible patients and speed the rate of inclusions. Because the size of the myoblasts precludes intracoronary injections, an electromechanical-guided endoventricular approach is likely to be the most suitable for upcoming trials. It remains, however, unsettled whether injections should be targeted at the core of the scar area, the border zone or both. Although the inclusion criteria also have to be fine-tuned, the already demonstrated effects of high doses of cells on remodeling may justify an earlier timing of treatment in postinfarct patients whose left ventricular dimensions are not yet too enlarged. Furthermore, the lack of effects in the low-dose cell group of the MAGIC trial leads to recommend delivery of a high dose of myoblasts in conjunction with a scaffold and growth factors targeted at enhancing graft survival. In parallel, preclinical studies should be conducted to assess whether it remains appropriate to use an unfractionated skeletal myoblast population or if it is more effective to rather sort it to yield a purified subpopulation of more immature cells featuring a greater plasticity. The above-mentioned critical choice between autologous patient-specific and allogeneic bank-derived cells will also have to be investigated. Finally, outcome measures should rather concentrate on robust clinical end points like major adverse cardiovascular events, particularly rehospitalizations for heart failure, and no longer on the exclusive surrogates used so far such as ventricular function or brain natriuretic peptide values. Namely, although stem cells cannot be fully assimilated to drugs,[63] one can anticipate that in the current era of cost-containment, a high level of evidence for efficacy will be required by the health authorities for registration and reimbursement. It is, therefore, obvious that only sufficiently powered multicenter trials can allow to generate this kind of data. In turn, these trials raise several issues that have been well identified during the MAGIC study, particularly interinstitutional variability in patient management and recruitment and cost. The former factor can probably be more easily handled in a catheter-based study than in a surgical one while the lessons drawn from the MAGIC experience strongly suggest that a careful selection of centers combining a large case load, real commitment to cardiac cell therapy and solid experience in clinical research is a mandatory factor for making the trial outcome successful. However, funding remains a primary concern and, at this stage, it can only be hoped that the intellectual property associated with myoblast processing and the subsequent possibility of ultimately commercializing a patented product will motivate biotech or larger pharmaceutical companies to get involved in this field and contribute to move it forward.

References

1. Leor J, Patterson M, Quinones MJ, *et al.* Transplantation of fetal myocardial tissue into the infarcted myocardium of rat. A potential method for repair of infarcted myocardium? *Circulation* 1996;94(9 Suppl):II332–II336.
2. Scorsin M, Hagege AA, Marotte F, *et al.* Does transplantation of cardiomyocytes improve function of infarcted myocardium? *Circulation* 1997;96(9 Suppl):II188–II193.
3. Dowell JD, Rubart M, Pasumarthi KB, *et al.* Myocyte and myogenic stem cell transplantation in the heart. *Cardiovasc Res* 2003;58:336–350.
4. He KL, Yi GH, Sherman W, *et al.* Autologous skeletal myoblast transplantation improved hemo-dynamics and left ventricular function in chronic heart failure dogs. *J Heart Lung Transplant* 2005;24:1940–1949.
5. Van den Bos EJ, Thompson RB, Wagner A, *et al.* Functional assessment of myoblast transplan-tation for cardiac repair with magnetic resonance imaging. *Eur J Heart Fail* 2005;7:435–443.
6. McConnell PI, del Rio CL, Jacoby DB, *et al.* Correlation of autologous skeletal myoblast survival with changes in left ventricular remodeling in dilated ischemic heart failure. *J Thorac Cardiovasc Surg* 2005;130:1001–1009.
7. Kinnaird T, Stabile E, Burnett MS, *et al.* Local delivery of marrow-derived stromal cells augments collateral perfusion through paracrine mechanisms. *Circulation* 2004;109:1543–1549; Erratum 2005;112:e73.
8. Gnecchi M, He H, Liang OD, *et al.* Paracrine action accounts for marked protection of ischemic heart by Akt-modified mesenchymal stem cells. *Nat Med* 2005;11:367–368.
9. Leobon B, Garcin I, Menasche P, Vilquin JT, Audinat E, Charpak S. Myoblasts transplanted into rat infarcted myocardium are functionally isolated from their host. *Proc Natl Acad Sci USA* 2003;100:7808–7811.
10. Rubart M, Soonpaa MH, Nakajima H, *et al.* Spontaneous and evoked intracellular calcium tran-sients in donor-derived myocytes following intracardiac myoblast transplantation. *J Clin Invest* 2004;114:775–783.

11. Menasche P, Hagege AA, Scorsin M, et al. Myoblast transplantation for heart failure. *Lancet* 2001;357:279–280.

12. Hagege AA, Marolleau JP, Vilquin JT, et al. Skeletal myoblast transplantation in ischemic heart failure: long-term follow-up of the first phase I cohort of patients. *Circulation* 2006;114(Suppl 1): I108–I113.

13. Hagege AA, Carrion C, Menasche P, et al. Viability and differentiation of autologous skeletal myoblast grafts in ischaemic cardiomyopathy. *Lancet* 2003;361:491–492.

14. Gavira JJ, Herreros J, Perez A, et al. Autologous skeletal myoblast transplantation in patients with nonacute myocardial infarction: 1-year follow-up. *J Thorac Cardiovasc Surg* 2006;131:799–804.

15. Siminiak T, Kalawski R, Fiszer D, et al. Autologous skeletal myoblast transplantation for the treatment of postinfarction myocardial injury: phase I clinical study with 12 months of follow-up. *Am Heart J* 2004;148:531–537.

16. Dib N, Michler RE, Pagani FD, et al. Safety and feasibility of autologous myoblast transplantation in patients with ischemic cardiomyopathy: four-year follow-up. *Circulation* 2005;112: 1748–1755.

17. Brasselet C, Morichetti MC, Messas E, et al. Skeletal myoblast transplantation through a catheter-based coronary sinus approach: an effective means of improving function of infarcted myocardium. *Eur Heart J* 2005;26:1551–1556.

18. Siminiak T, Fiszer D, Jerzykowska O, et al. Percutaneous trans-coronary-venous transplantation of autologous skeletal myoblasts in the treatment of post-infarction myocardial contractility impairment: the POZNAN trial. *Eur Heart J* 2005;26:1188–1195.

19. Biagini E, Valgimigli M, Smits PC, et al. Stress and tissue Doppler echocardiographic evidence of effectiveness of myoblast transplantation in patients with ischaemic heart failure. *Eur J Heart Fail* 2006;8:641–648.

20. Ince H, Petzsch M, Rehders TC, Chatterjee T, Nienaber CA. Transcatheter transplantation of autologous skeletal myoblasts in postinfarction patients with severe left ventricular dysfunction. *J Endovasc Ther* 2004;11:695–704.

21. Menasche P, Hagege AA, Vilquin JT, et al. Autologous skeletal myoblast transplantation for severe postinfarction left ventricular dysfunction. *J Am Coll Cardiol* 2003;41:1078–1083.

22. Abraham MR, Henrikson CA, Tung L, et al. Antiarrhythmic engineering of skeletal myoblasts for cardiac transplantation. *Circ Res* 2005;97:159–167.

23. Itabashi Y, Miyoshi S, Yuasa S, et al. Analysis of the electrophysiological properties and arrhythmias in directly contacted skeletal and cardiac muscle cell sheets. *Cardiovasc Res* 2005;6:561–570.

24. Fernandes S, Amirault JC, Lande G, et al. Autologous myoblast transplantation after myocardial infarction increases the inducibility of ventricular arrhythmias. *Cardiovasc Res* 2006;69:348–358.

25. Fouts K, Fernandes B, Mal N, et al. Electrophysiological consequence of skeletal myoblast transplantation in normal and infarcted canine myocardium. *Heart Rhythm* 2006;3:452–461.

26. Chen M, Fan ZC, Liu XJ, et al. Effects of autologous stem cell transplantation on ventricular electrophysiology in doxorubicin-induced heart failure. *Cell Biol Int* 2006;30:576–582.

27. Chang MG, Tung L, Sekar RB, et al. Proarrhythmic potential of mesenchymal stem cell transplantation revealed in an *in vitro* coculture model. *Circulation* 2006;113:1832–1841.

28. Soliman AM, Krucoff MW, Crater S, et al. Cell location may be a primary determinant of safety after myoblast transplantation into the infarcted heart. *J Am Coll Cardiol* 2004;43:15A.

29. Bigger JT Jr. Prophylactic use of implanted cardiac defibrillators in patients at high risk for ventricular arrhythmias after coronary-artery bypass graft surgery. Coronary Artery Bypass Graft (CABG) Patch Trial Investigators. *N Engl J Med* 1997;337:1569–1575.

30. Menasché P, Alfieri O, Janssens S, et al. The myoblast autologous grafting in ischemic cardiomyopathy (MAGIC) trial. First randomized placebo-controlled study of myoblast transplantation. *Circulation* 2008;117:1189–1200.

31. Gavira JJ, Perez-Ilzarbe M, Abizanda G, *et al.* A comparison between percutaneous and surgical transplantation of autologous skeletal myoblasts in a swine model of chronic myocardial infarction. *Cardiovasc Res* 2006;71:744–753.
32. Dib N, Dinsmore J, Mozak R, White B, Moravec S, Diethrich EB. Safety and feasability of percutaneous autologous skeletal myoblast transplantation for ischemic cardiomyopathy: six-month interim analysis. *Circulation* 2006;114(Suppl II):II88.
33. Dow J, Simkhovich BZ, Kedes L, Kloner RA. Washout of transplanted cells from the heart: a potential new hurdle for cell transplantation therapy. *Cardiovasc Res* 2005;67:301–307.
34. Ott HC, Kroess R, Bonaros N, *et al.* Intramyocardial microdepot injection increases the efficacy of skeletal myoblast transplantation. *Eur J Cardiothorac Surg* 2005;27:1017–1021.
35. Memon IA, Sawa Y, Fukushima N, *et al.* Repair of impaired myocardium by means of implantation of engineered autologous myoblast sheets. *J Thorac Cardiovasc Surg* 2005;130:1333–1341.
36. Pouly J, Hagege AA, Vilquin JT, *et al.* Does the functional efficacy of skeletal myoblast transplantation extend to nonischemic cardiomyopathy? *Circulation* 2004;110:1626–1631.
37. Guarita-Souza LC, Carvalho KA, Woitowicz V, *et al.* Simultaneous autologous transplantation of cocultured mesenchymal stem cells and skeletal myoblasts improves ventricular function in a murine model of Chagas disease. *Circulation* 2006;114(Suppl 1):I120–I124.
38. Premaratne GU, Tambara K, Fujita M, *et al.* Repeated implantation is a more effective cell delivery method in skeletal myoblast transplantation for rat myocardial infarction. *Circ J* 2006;70: 1184–1189.
39. Maurel A, Azarnoush K, Sabbah L, *et al.* Can cold or heat shock improve skeletal myoblast engraftment in infarcted myocardium? *Transplantation* 2005;80:660–665.
40. Tambara K, Sakakibara Y, Sakaguchi G, *et al.* Transplanted skeletal myoblasts can fully replace the infarcted myocardium when they survive in the host in large numbers. *Circulation* 2003;108(Suppl 1):II259–II263.
41. Azarnoush K, Maurel A, Sebbah L, *et al.* Enhancement of the functional benefits of skeletal myoblast transplantation by means of co-administration of hypoxia-inducible factor 1-alpha. *J Thorac Cardiovasc Surg* 2005;130:173–179.
42. Yau TM, Kim C, Ng D, *et al.* Increasing transplanted cell survival with cell-based angiogenic gene therapy. *Ann Thorac Surg* 2005;80:1779–1786.
43. Memon IA, Sawa Y, Miyagawa S, *et al.* Combined autologous cellular cardiomyoplasty with skeletal myoblasts and bone marrow cells in canine hearts for ischemic cardiomyopathy. *J Thorac Cardiovasc Surg* 2005;130:646–653.
44. Christman KL, Vardanian AJ, Fang Q, *et al.* Injectable fibrin scaffold improves cell transplant survival, reduces infarct expansion, and induces neovasculature formation in ischemic myocardium. *J Am Coll Cardiol* 2004;44:654–660.
45. Bunch TJ, Mahapatra S, Johnson SB, Packer DL. Epicardial wave break and arrhythmogenicity after fibrin sealant injections: a novel cell delivery scaffold. *Circulation* 2006;114(Suppl II):II265.
46. Hsieh PC, Davis ME, Gannon J, MacGillivray C, Lee RT. Controlled delivery of PDGF-BB for myocardial protection using injectable self-assembling peptide nanofibers. *J Clin Invest* 2006;116:237–248.
47. Kutschka I, Kofidis T, Chen IY, *et al.* Adenoviral human BCL-2 transgene expression attenuates early donor cell death after cardiomyoblast transplantation into ischemic rat hearts. *Circulation* 2006;114(Suppl 1):I174–I180.
48. Lim SY, Kim YS, Ahn Y, *et al.* The effects of mesenchymal stem cells transduced with Akt in a porcine myocardial infarction model. *Cardiovasc Res* 2006;70:530–542.
49. Zhang M, Methot D, Poppa V, *et al.* Cardiomyocyte grafting for cardiac repair: graft cell death and anti-death strategies. *J Mol Cell Cardiol* 2001;33:907–921.
50. Pasha Z, Wang Y, Zhang D, Zhao T, Xu M, Ashraf M. Pharmacological preconditioning enhances survival of transplanted mesenchymal stem cells in the ischemic myocardium. *Circulation* 2006;114(Suppl II):II19.

51. Stuckey DJ, Carr CA, Martin-Rendon E, *et al.* Iron particles for noninvasive monitoring of bone marrow stromal cell engraftment into, and isolation of viable engrafted donor cells from, the heart. *Stem Cells* 2006;24:1968–1975.

52. Amsalem Y, Mardor Y, Daniels D, *et al.* Reparative capacity of mesenchymal stem cells is impaired by iron-oxide labeling after extensive myocardial infarction. *Circulation* 2006;114(Suppl II): II661.

53. Limbourg FP, Ringes-Lichtenberg S, Schaefer A, *et al.* Haematopoietic stem cells improve cardiac function after infarction without permanent cardiac engraftment. *Eur J Heart Fail* 2005;7: 722–729.

54. Winitsky SO, Gopal TV, Hassanzadeh S, *et al.* Adult murine skeletal muscle contains cells that can differentiate into beating cardiomyocytes *in vivo*. *PLoS Biol* 2005;3:e87.

55. Oshima H, Payne TR, Urish KL, *et al.* Differential myocardial infarct repair with muscle stem cells compared to myoblasts. *Mol Ther* 2005;12:1130–1141.

56. Parmacek MS, Epstein JA. Pursuing cardiac progenitors: regeneration redux. *Cell* 2005;120: 295–298.

57. Pouly J, Bruneval P, Mandet C, *et al.* Cardiac stem cells in the real world. *J Thorac Cardiovasc Surg* 2008;135:673–678.

58. Hodgson DM, Behfar A, Zingman LV, *et al.* Stable benefit of embryonic stem cell therapy in myocardial infarction. *Am J Physiol Heart Circ Physiol* 2004;287:H471–H479.

59. Menard C, Hagege AA, Agbulut O, *et al.* Transplantation of cardiac-committed mouse embryonic stem cells to infarcted sheep myocardium: a preclinical study. *Lancet* 2005;366:1005–1012.

60. Tomescot A, Leschik J, Bellamy V, *et al.* Differentiation *in vivo* of cardiac committed human embryonic stem cells in post-myocardial infarcted rats. *Stem Cells* 2007;25:2200-2205.

61. Kehat I, Khimovich L, Caspi O, *et al.* Electromechanical integration of cardiomyocytes derived from human embryonic stem cells. *Nat Biotechnol* 2004;22:1282–1289.

62. Ryan JM, Barry FP, Murphy JM, Mahon BP. Mesenchymal stem cells avoid allogeneic rejection. *J Inflamm (Lond)* 2005;2:8.

63. Rosen MR. Are stem cells drugs? The regulation of stem cell research and development. *Circulation* 2006;114:1992–2000.

12 Myocardial Tissue Regeneration Observed in Stem-Cell Seeded Bioengineered Scaffolds

Yen Chang, Chun-Hung Chen, Hao-Ji Wei, Wei-Wen Lin,
Shiaw-Min Hwang, Huihua Kenny Chiang,
Sung-Ching Chen, Po-Hong Lai and Hsing-Wen Sung

Introduction

A patch is often mandatory to repair failing myocardium or congenital cardiac malformations.[1] Currently, the patches clinically used are made of Dacron polyester fabric, polytetrafluoroethylene (PTFE), glutaraldehyde-treated bovine pericardium, or antibiotic-preserved or cryopreserved homografts.[2] However, cardiac repair by currently available patches is limited by the inability to promote local tissue regeneration within the implanted materials.[3]

In our previous study, it was found that acellular bovine pericardial tissues fixed with genipin can provide a natural microenvironment for host cell migration and may be used as a tissue-engineering extracellular matrix (ECM) to accelerate tissue regeneration.[4] Genipin, a naturally occurring crosslinking agent, can be obtained from its parent compound – geniposide, which can be isolated from the fruits of *Gardenia jasminoides* ELLIS.[5] It has been used to fix biological tissues or amino-group containing biomaterials for biomedical applications.[5] It was found that genipin is significantly less cytotoxic than glutaraldehyde.[6,7] Moreover, the genotoxicity of genipin was tested *in vitro* using Chinese hamster ovary (CHO-K1) cells. The results suggested that glutaraldehyde may produce a weakly clastogenic response in CHO-K1 cells. In contrast, genipin does not cause clastogenic response in CHO-K1 cells.[8]

In the study, the acellular bovine pericardial tissue was further treated with acetic acid and subsequently with collagenase to increase its pore size and interconnectivity. It is generally accepted that a tissue-engineering ECM must be highly porous for a sufficient cell density to be seeded *in vitro*, for blood invasion to occur *in vivo* and for oxygen and nutrients to be supplied to cells.[9] It is known that the human heart cannot regenerate significantly because adult cardiomyocytes are terminally differentiated and cannot replicate after injury.[10] In an attempt to overcome this limitation, the porous acellular bovine pericardium was additionally seeded with bone marrow-derived mesenchymal stem cells (MSCs) as a patch to repair a surgically created myocardial defect in the right ventricle of a syngeneic rat model.

Recently, the identification of stem cells capable of contributing to tissue regeneration has raised the possibility that cell therapy could be employed for repair of damaged myocardium.[11-13] Bone marrow-derived MSCs retain the ability to differentiate into various types of tissue cells and contribute to the regeneration of a variety of mesenchymal tissues including bone, cartilage, muscle, and adipose.[14-16] Several studies performed on rodents, pigs and human demonstrated that MSCs own the potency to differentiate into a cardiomyocyte phenotype in the heart.[17-18]

Materials and Methods

Preparation and characterization of test samples

The procedures used to remove the cellular components from bovine pericardia were based on a method previously reported by Courtman et al.[19] To increase the pore size and porosity within test samples, the acellular tissues were treated additionally with acetic acid and subsequently with collagenase. Details of the methodology used to prepare porous acellular bovine pericardia were previously reported.[20,21] Finally, the acellular tissues with a porous structure were fixed in a 0.05% genipin (Challenge Bioproducts, Taiwan) aqueous solution (phosphate buffered saline, PBS, pH 7.4) at 37°C for 3 days.

The degree of crosslinking of the genipin-fixed porous acellular pericardium (the acellular patch) was determined by measuring its fixation index and denaturation temperature ($n = 5$).[22] The fixation index, determined by the ninhydrin assay, was defined as the percentage of free amino groups in test tissues reacted with genipin subsequent to fixation. The denaturation temperature of test samples was measured by a Perkin-Elmer differential scanning calorimeter (model DSC-7, Norwalk, Connecticut, USA). The pore size of the acellular patch, stained with hematoxylin and eosin (H&E), was determined under a microscope with a computer-based image analysis system (Image-Pro® Plus, Media Cybernetics, Silver Spring, MD, USA) and its porosity was measured by helium pycnometery ($n = 5$).[23] Mechanical testing of the acellular patch was conducted by an Instron material testing machine (Mini 44, Canton, MA, USA, $n = 5$).[24]

To facilitate cell infiltration and repopulation, the dense layer on each side of the fixed porous acellular tissues was further sliced off using a cryostat microtome (Leica CM3050S, Leica Microsystems Nussloch GmbH, Nussloch, Germany). After preparation of test samples, the porous acellular tissues were processed for scanning electron microscopic (SEM, S-4200, Hitachi, Japan) examinations to investigate their ultrastructures.[25] The obtained samples were then sterilized in a graded series of ethanol solutions for populating MSCs in vitro and for the subsequent animal study. A commercially available e-PTFE patch (Gore-Tex, W.L. Gore & Associates, Inc., Flagstaff, AZ, USA) was used as a control.

Isolation and culture of bone marrow MSCs

Bone marrow MSCs were isolated as described previously.[26,27] Lewis rats were anesthetized with intramuscular administration of ketamine hydrochloride (22 mg/kg) followed by an intraperitoneal injection of sodium pentobarbital (30 mg/kg). Femora and tibia of rats were collected and the adherent soft tissues were carefully removed. Bone marrow was aspirated with a syringe containing 1 ml heparin with a 26-gauge needle and disaggregated into a single-cell suspension by sequential passage through a 26-gauge needle. Mononuclear cells were separated by density-gradient centrifugation over Ficoll-Paque ($\rho = 1.077$, Amersham, Uppsala, Sweden) at $1100 \times g$ for 30 min. The cells were rinsed twice with PBS to remove Ficoll–Paque and then seeded onto plastic tissue culture plates at 1×10^5 cells/cm^2 with culture medium (alpha-Modified Minimum Essential Medium, alpha-MEM, supplemented with 20% fetal bovine serum, 100 U/ml penicillin and 100 μg/ml streptomycin, GIBCO Laboratory, Life Technologies, Grand Island, NY, USA).

Three days later, non-adherent cells were removed by changing the medium. After 14 days in culture, adherent cells formed homogenous spindle-shaped colonies. Adherent cells were passed with 0.05% trypsin-EDTA (GIBCO Laboratory) and split at a ratio of 1:3. The culture medium was changed every 3–4 days and the passages 2–4 MSCs were used for the following studies.

In vitro graft preparation, seeding MSCs onto porous acellular tissues

To induce myogenic differentiation, the DNA-demethylating agent 5-azacytidine (5-aza) was added on the third day and incubated with MSCs for 24 hours.[27] Subsequently, the induced MSCs were labeled *in vitro* for later identification by adding 100 μg/ml 5-bromo-2'-deoxyuridine (BrdU) containing media to 50% confluent cultures for 24 h.[28] The cultured cells were immunohistochemically stained for cardiac-specific troponin T and BrdU.[14] Samples ($n = 5$) were analyzed using a FACScan flow cytometer (Becton Dickinson, San Jose, CA, USA).

At the completion of the incubation period, the cells were washed, harvested from culture dishes using 0.05% trypsin solution and resuspended to a concentration of 1×10^6 bone marrow MSCs in 50 μl of culture medium, and seeded onto the prepared porous acellular tissues (in a disk shape with a diameter of 5 mm and a thickness of 1.6 mm, the acellular/MSC patch). Thirty minutes after incubation, drops of culture medium were added at the circumference of the acellular/MSC patch. After another 90 minutes of incubation, 20 ml of culture medium were added to the culture dishes.

After incubation for 20 hours, the acellular/MSC patch was removed from the incubator. To investigate the cell attachment and distribution on the acellular/MSC patch, samples were fixed in 10% phosphate-buffered formalin for two days and subsequently processed for the H&E stain and SEM examinations.

Animal study

Animal care and use was performed in compliance with the "Guide for the Care and Use of Laboratory Animals prepared by the Institute of Laboratory Animal" Resources, National Research Council, and published by the National Academy Press, revised 1996. Adult syngeneic Lewis rats weighing 250–275 g were used for the right ventricular wall replacement. The prepared patches were used to repair a transmural defect surgically created in the right ventricle of syngeneic rat hearts based on a method reported by Ozawa et al.[29] Briefly, rats were anesthetized with isoflurane (3.0%) and intubated for continuous ventilation with room air supplemented with oxygen and isoflurane (1.5%–3.0%). The rat heart was exposed through a median sternotomy. A purse-string suture (5–6 mm in diameter) was placed in the free wall of right ventricle. Each end of the suture was passed through a 24-gauge plastic vascular cannula (Insyte, Becton Dickinson Vascular Access, Sandy, UT, USA), which was used as a tourniquet. The tourniquet was tightened and the bulging part of the ventricular wall inside the purse-string stitch was lopped off. The tourniquet was transiently loosened to confirm a complete transmural resection.

Subsequently, the e-PTFE, acellular or acellular/MSC patch was sutured along the margin of the purse-string stitch with an over-and-over method with 7-0 polypropylene (Prolene, Ethicon, Inc., Somerville, NJ, USA) to cover the defect created in the right ventricle. After completion of suturing, the tourniquet was released and the purse-string stitch was removed. The chest incision was closed in layers with running sutures of 3-0 silk. Finally, rats were removed from ventilation and recovered under a warming lamp. The implanted samples were retrieved at four- and 12-week post-operatively ($n = 5$ at each time point) and the appearance of each retrieved sample was grossly examined and photographed and then used for histological examinations.

Epicardial electrograms

A 12-lead electrocardiographic system (PC-ECG 1200M, Norav Medical, Kiryat Bialik, Israel) was used to acquire the epicardial electrograms of the implanted patch and its adjacent rat native myocardium immediately after implantation and at retrieval.[30]

Echocardiography (assessment of the cardiac function)

Echocardiography was performed before patch implantation and at retrieval for each studied group. Rats were anesthetized with sodium pentobarbital (30 mg/kg) and isoflurane was used as a supplement to maintain mild anesthesia. Cardiac ultrasonography was performed with a commercially available echocardiographic system (SONOS 5500, Agilent Technologies, Andover, MA, USA) equipped with a 12-MHz broadband sector transducer. The heart was imaged in the two-dimensional mode in short-axis views at the mid-papillary level of left ventricle to evaluate contractions of each patch-implanted myocardium at the right ventricle.[31]

Histological examinations

The samples used for light microscopy were fixed in 10% phosphate buffered formalin and prepared for histological examinations. In the histological examinations, the fixed samples were embedded in paraffin and sectioned into a thickness of 5 μm and then stained with H&E. Also, sections of test samples were stained with Masson's trichrome and elastic van Gieson (EVG) for the detection of collagen fibrils and muscle fibers and stained with safranin-O to visualize glycosaminoglycans. Additional sections were stained with a van Gieson solution to visualize mesothelial cells.[32]

Immunohistochemical staining with a monoclonal antibody against BrdU (Caltag Laboratories, Burlingame, CA) was used to identify the cells populated on the acellular/MSC patch and revealed by a peroxidase–antiperoxidase technique.[33] Additionally, sections of the retrieved samples were stained with a monoclonal antibody against α-sacromeric actin (clone 5C5, Serotec, Kidlington, Oxford, UK).[34]

A monoclonal antibody against α-smooth muscle actin (α-SMA, DAKO, DAKO Corp., Carpinteria, CA, USA) was used to identify smooth muscle cells (SMCs). Additional sections were stained for factor VIII with immunohistological technique with a monoclonal anti-factor VIII antibody (DAKO).[35] The density of neomicrovessels in each studied sample was quantified with the computer-based image analysis system mentioned above and converted to vessels/mm^2.[36] Five different microscopic fields ($\times$200 by ECLIPSE-E800, Nikon, Tokyo, Japan) of each patch portion of the right ventricular wall were randomly selected.

Statistical analysis

Statistical analysis for the determination of differences in the measured properties between groups was accomplished using one-way analysis of variance and determination of CIs, which was performed with a computer statistical program (Statistical Analysis System, Version 6.08, SAS Institute Inc., Cary, NC, USA). All data are presented as mean values $\pm$SD. Differences were considered to be statistically significant when the p values were less than 0.05.

Results

Characteristics of the acellular patch

After the genipin fixation, the color of bovine pericardial tissues became bluish. The pore size and porosity of the acellular patch were 159.8$\pm$26.7 μm and 94.9$\pm$1.7%,

respectively, while its denaturation temperature and fixation index measured were
$75.5 \pm 0.3°C$ and $57.8 \pm 5.4\%$, respectively. The ultimate tensile strength of the
acellular patch was 2.1 ± 0.3 MPa. As shown in Figs. 1a, 1c, and 1e, there was a
dense layer on each side of the porous acellular bovine pericardium. After slicing
off these dense layers with a cryostat microtome, a porous structure beneath was
revealed (Figs. 1b, 1d, and 1f).

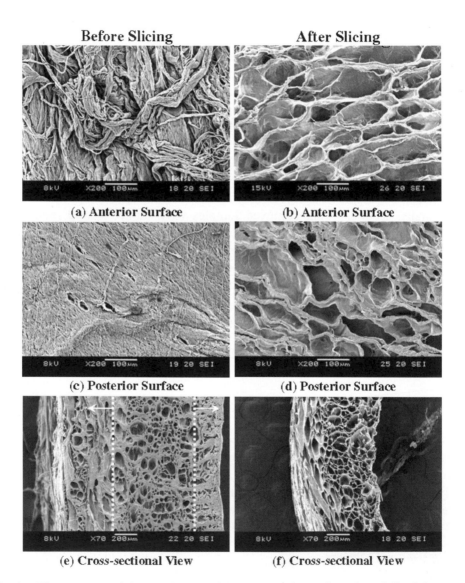

Fig. 1. Ultrastructures of the anterior (**a and b**) and posterior surfaces (**c and d**) of the porous
acellular bovine pericardium, before and after slicing off the dense layers on its both sides using a
cryostat microtome, together with their cross-sectional views (**e and f**) examined by SEM. (From
Wei HJ, *et al.*, *Biomaterials* 2006;27:5412.)

Phenotype of MSCs before implantation

Morphology of the isolated cells initially appeared small and rounded with a tendency to grow in clusters. Non-adherent cells were removed by medium change at 24 hours and every four days thereafter. Elongated cells with a spindle-shaped morphology appeared after 72 hours and reached confluence after ten to 14 days. Seven days after 5-aza treatment, flow cytometric analysis of the 5-aza-induced MSCs demonstrated a significant difference in the expression of troponin T as compared to the control group (Fig. 2a). The nuclei of the cells induced by 5-aza were labeled with BrdU for 24 hours pre-transplantation (Fig. 2b). It was found that $91.6 \pm 3.2\%$ of the culture cells stained positively.

Characteristics of the acellular/MSC patch

After cell populating, MSCs were mostly present in the surface layers of porous acellular tissues and had a uniform, viable spindle-shaped morphology (the acellular/MSC patch, Figs. 3a to 3c). BrdU incorporation was evident in >90% of the populated MSCs (Fig. 3d).

Epicardial electrograms

Immediately after implantation, no epicardial electrogram signals were observed on all studied groups (Fig. 4). There was no electrogram signal present on the e-PTFE patch throughout the entire course of the study. In contrast, local electrograms appeared on the epicardial surfaces of the acellular and acellular/MSC patches at four-week post-operatively and their amplitudes increased significantly with time (at 12-week post-operatively). However, the amplitude of the electrogram observed on the acellular/MSC patch was significantly stronger than the acellular patch.

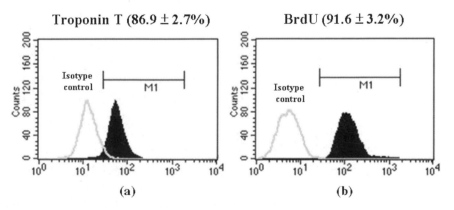

Fig. 2. Representative results from five independently performed experiments of flow cytometric analysis of **(a)** 5-aza-induced MSCs and **(b)** BrdU-labeled MSCs.

SEM

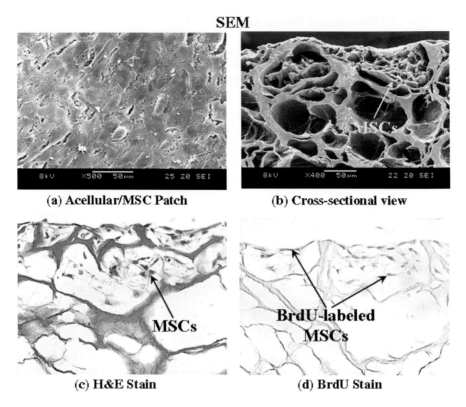

(a) Acellular/MSC Patch   (b) Cross-sectional view

(c) H&E Stain   (d) BrdU Stain

Fig. 3.   SEM micrographs of the acellular/MSC patch (**a**) and its cross-sectional view (**b**); photomicrographs of the acellular/MSC patch stained with H&E ($\times 400$ magnification, **c**) and that stained with a monoclonal antibody against BrdU ($\times 400$ magnification, **d**). (From Wei HJ, *et al.*, *Biomaterials* 2006;27:5413.)

Echocardiography

Short-axis two-dimensional images of the normal heart, the e-PTFE-patch-implanted heart, the acellular-patch-implanted heart and the acellular/MSC-patch-implanted heart at retrieval were obtained using a commercially available echocardiographic system. The results indicated that all studied patches were akinetic. It was noted that the differences in global contraction of the native myocardium and all patch-implanted myocardia at the right ventricle were difficult to differentiate.

Gross examination

At retrieval, on the outer surface, no apparent tissue adhesion was observed on the acellular and acellular/MSC patches for all the animals studied, while a mild-moderate adhesion to the chest wall was seen for the e-PTFE patch in two of the animals studied at each time point (Fig. 5). On the inner surface, intimal thickening was observed for all studied groups; however, no thrombus formation was found.

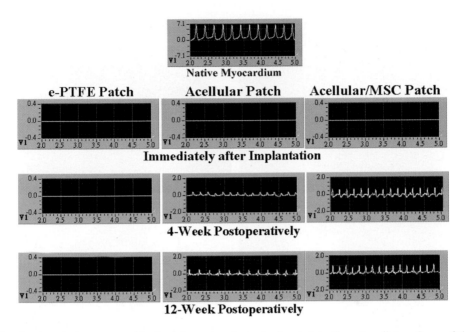

Fig. 4. Local electrograms obtained from the electrodes located on the epicardial surfaces of the e-PTFE, acellular and acellular/MSC patches and their adjacent rat native myocardia immediately after implantation and those observed at four- and 12-week post-operatively.

No aneurysmal dilation of the implanted patches was seen for all studied groups throughout the entire course of the study.

Histological findings

At four-week post-operatively, host cells together with neo-tissue fibrils and neo-capillaries were clearly observed in the inner and outer layers of the acellular and acellular/MSC patches, an indication of tissue regeneration (Fig. 6). In contrast, there were only a few infiltrated cells (mostly inflammatory cells) found in the most inner and outer layers of the e-PTFE patch. An intact layer of endothelial cells was found on the intimal thickening generated on the inner surfaces of the acellular and acellular/MSC patches identified by the factor VIII staining (Fig. 6). In contrast, endothelial cells did not universally and totally cover the entire inner surface of the e-PTFE patch (Fig. 6).

The outer surfaces of the acellular and acellular/MSC patches were positively stained with van Gieson (Fig. 6), indicating the presence of mesothelial cells. However, no mesothelial cells were found on the outer surface of the e-PTFE patch (Fig. 6). Instead, a thick layer of fibrous tissue was firmly attached onto the outer surface of the e-PTFE patch (Fig. 6).

In the middle layer of the acellular and acellular/MSC patches, host cells together with neo-tissue fibrils were found to fill most of their inside pores (Fig. 7).

Endocardial Surface   **Epicardial Surface**

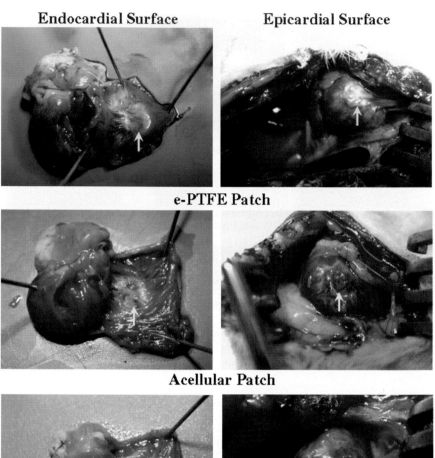

e-PTFE Patch

Acellular Patch

Acellular/MSC Patch

Fig. 5. Photographs of the inner (endocardial) and outer (epicardial) surfaces of the e-PTFE, acellular and acellular/MSC patches retrieved at four-week post-operatively. The implanted patch is indicated by →.

The neo-tissue fibrils regenerated in these patches were identified to be neo-muscle fibers (stained red) with a few neo-collagen fibrils (stained blue) by the Masson's trichrome stain (Fig. 7). The neo-muscle fibers seen in the patches were further confirmed by the EVG stain (stained brown, Fig. 7). Also, there were some neo-glycosaminoglycans regenerated within the pores of the patches recognized by the safranin-O stain (stained pink, Fig. 7). In contrast, no apparent tissue regeneration was observed inside the e-PTFE patch (Fig. 7).

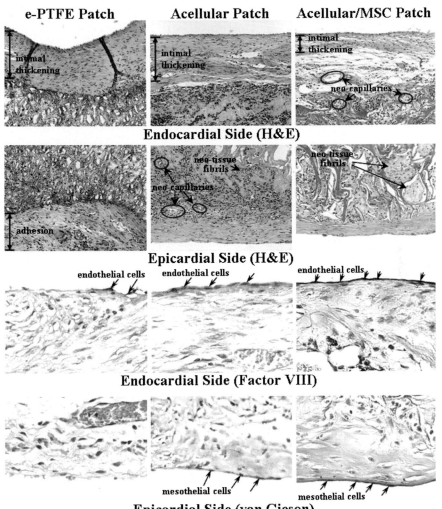

Fig. 6. Photomicrographs of the inner (endocardial) and outer (epicardial) layers of the e-PTFE, acellular and acellular/MSC patches stained with H&E retrieved at four-week post-operatively (×200 magnification); and those stained with an antibody against factor VIII and van Gieson (×800 magnification). (From Chang Y, *et al.*, *J Thorac Cardiovasc Surg* 2005;130:705.e9; Wei HJ, *et al.*, *Biomaterials* 2006;27:5415.)

α-SMA positively stained cells were observed in the acellular and acellular/MSC patches (Fig. 8), indicating the presence of smooth muscle cells. In contrast, no such stained cells were observed in the e-PTFE patch. Additionally, BrdU-labeled cells were clearly identified in the acellular/MSC patch (Fig. 8). Some of the identified BrdU-labeled cells were further stained positively for α-sacromeric actin (Fig. 8), indicating that a portion of the implanted bone marrow MSCs had been differentiated towards the myocytic lineage and expressed cytoplasmic α-sacromeric

e-PTFE Patch   Acellular Patch   Acellular/MSC Patch

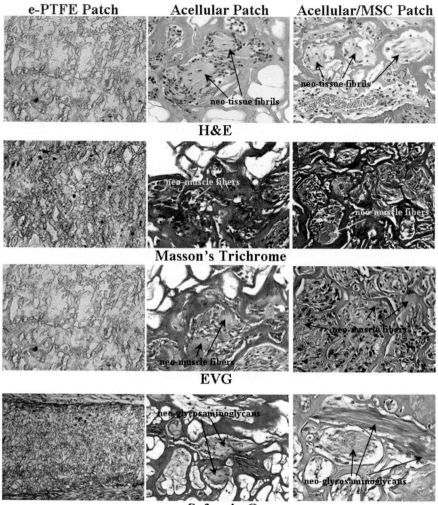

H&E

Masson's Trichrome

EVG

Safranin O

Fig. 7. Photomicrographs of the middle layers of the e-PTFE, acellular and acellular/MSC patches stained with H&E ($\times$400 magnification), Masson's trichrome ($\times$400 magnification), EVG ($\times$400 magnification) and safranin-O ($\times$400 magnification) retrieved at four-week post-operatively. (From Chang Y, *et al.*, *J Thorac Cardiovasc Surg* 2005;130:705.e10; Wei HJ, *et al.*, *Biomaterials* 2006;27:5416.)

actin. Moreover, a few capillary walls composed of BrdU-labeled endothelial cells (Fig. 8) and some BrdU-labeled smooth muscle cells (Fig. 8) were recognized in the acellular/MSC patch. The densities of neo-microvessels found in the acellular and acellular/MSC patches were 189 ± 37 and 327 ± 41 vessels/mm^2, respectively. However, it was still much less than that seen in the native myocardium (2118 ± 320 vessels/mm^2).

At 12-week post-operatively, there were still no signs of tissue regeneration (no host cells, no capillaries, no neo-tissue fibrils and no cells stained positively for

α-Smooth Muscle Actin

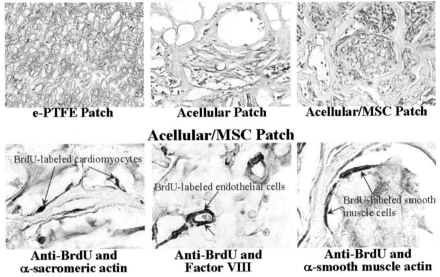

Fig. 8. Photomicrographs of immunohistochemistry staining of a monoclonal antibody against α-smooth muscle actin for the e-PTFE, acellular and acellular/MSC patches retrieved at four-week post-operatively (×400 magnification); a monoclonal antibody against BrdU (stained blue) and α-sacromeric actin (stained brown, ×2000 magnification), a monoclonal antibody against BrdU (stained blue) and factor VIII (stained brown, ×2000 magnification) and a monoclonal antibody against BrdU (stained blue) and α-smooth muscle actin (stained brown, ×2000 magnification) for the acellular/MSC patch. (From Chang Y, et al., *J Thorac Cardiovasc Surg* 2005;130:705,e5; Wei HJ, et al., *Biomaterials* 2006;27:5417.)

α-sacromeric actin or α-SMA) observed inside the e-PTFE patch. The neo-tissue fibrils seen in the acellular and acellular/MSC patches appeared to be more compact and organized than their counterparts observed at four-week post-operatively. BrdU-, α-sacromeric-actin- and α-SMA-positively stained cells were still observed in the acellular/MSC patch. The densities of neo-microvessels observed in the acellular (187 ± 23 vessels/mm^2) and acellular/MSC (335 ± 20 vessels/mm^2) did not significantly increase with time.

Discussion

Cell transplantation is a promising therapy for patients with ischemia and ventricle dysfunction.[27,37,38] However, the engraftment process is limited and a large mature scar does not benefit from cell transplantation alone. Resection of the aneurysm and surgical remodeling of the ventricle to restore chamber size and shape may improve the cardiac function under ideal circumstances.[39] Nevertheless, the benefits of procedures of surgical scar resection and ventricular remodeling may be limited by the use of non-viable synthetic patch materials.[39]

In the study, a porous acellular bovine pericardial tissue populated with MSCs was used to repair a myocardial defect created in the right ventricle of a syngeneic rat model. Acellular biological tissues have been proposed to be used as natural biomaterials for tissue repair and tissue engineering.[4,19] Natural biomaterials are composed of extracellular matrix proteins that are conserved among different species and that can serve as scaffolds for cell attachment, migration, and proliferation.[4] This can be a large advantage over synthetic materials. Tissue engineering is aimed to manipulate regeneration of host tissues in the implanted ECMs to repair the defects in human body.

Recently, bone marrow-derived MSCs have gained attention for their therapeutic potential. Under differentiation-inducing culture conditions, MSCs have the capacity to differentiate into other cell types.[26] During myogenic differentiation, MSCs continued to proliferate and readily formed myotubes. Additionally, the results obtained in the flow cytometric analysis showed that the induced MSCs stained positively for cardiac-specific troponin T.

At retrieval, we found that all studied patches did not thin and dilate throughout the entire course of the study. This indicated that the mechanical strengths of all studied groups were strong enough to tolerate the right ventricular pressure without aneurysmal dilation during the long-term implantation. The e-PTFE patch has been accepted worldwide as one of the most reliable non-degradable synthetic materials in terms of durability, low thrombogenicity, and comfortable handling for surgery intervention.[40] Additionally, an advantage of the e-PTFE patch is that it does not thin and dilate in response to the intraventricular pressure.[29] Unfortunately, the e-PTFE patch is not viable and does not grow as a child's heart or blood vessel grows.[41]

At four-week post-operatively, host cells together with neo-tissue fibrils and neo-capillaries were clearly observed in the outer (epicardial) layers of the acellular and acellular/MSC patches, indicating that the outer layers of these two patches became well integrated with their host tissues (Fig. 6). Additionally, an intact layer of neo-mesothelial cells (i.e. completed remesothelialization, identified by the van Gieson stain) was present on top of the neo-tissue fibrils regenerated in the outer layer of the acellular and acellular/MSC patches (Fig. 6). However, such observations did not occur for the e-PTFE patch (Fig. 6). Instead, a thick layer of fibrous tissue was firmly attached onto the outer surface of the e-PTFE patch.

It is known that the epicardium forms the outer covering of the heart and has an external layer of flat mesothelial cells.[42] Whitaker *et al.* reported that a pure culture of mesothelial cells was able to induce fibrinolysis.[43] Another study suggested that the mesothelial fibrinolytic properties are associated with the secretion of tissue plasminogen activator.[44] These results likely explained the observation that once the surfaces of the implanted acellular and acellular/MSC patches were populated with mesothelial cells, they remained resistant to adhesion formation.

An intimal thickening covered with endothelial cells was found on the inner (endocardial) surfaces of the e-PTFE, acellular and acellular/MSC patches (Fig. 6). This finding suggested that host endocardial endothelial cells or endothelial progenitor cells were involved in the endothelialization on the inner surfaces of the implanted patches.[4,45] No thrombus formation was observed on all studied groups. It is known that the most thrombo-resistant graft surface is that provided by normal endothelial cells.[46] The basis for the thrombo-resistance of endothelial cells may possibly be related to their role as a source of plasminogen activator, prostacyclin (PGI_2), or both.[47] Plasminogen is a key proenzyme in fibrinolysis and thrombolytic system.[48]

Muscle fibers (with some collagen fibrils), glycosaminoglycans and microvessels were regenerated in the middle layers of the acellular and acellular/MSC patches, indicated by the H&E, Masson's trichrome, EVG, and safranin-O stains (Fig. 7), an indication of tissue regeneration. Additionally, α-SMA positively stained cells were seen in the acellular and acellular/MSC patches, indicating the presence of smooth muscle cells (Fig. 8). It is known that smooth muscle cells permit formation of a muscular tissue (observed in Fig. 7, stained red with Massson's trichrome and stained brown with EVG) in addition to collagen formation (stained blue with Massson's trichrome).[49] These findings suggested that host progenitor cells from the systemic circulation or the surrounding tissue might be relevant to the presence of smooth muscle cells in the implanted acellular and acellular/MSC patches.[45] Previous studies reported that smooth-muscle-cell transplantation into myocardial infarct scar tissues improves heart function and prevents ventricle dilation.[37]

However, cardiomyocytes were not found within the acellular patch. The main benefits of the cardiomyocytes are thought to be an increase in myocardial wall tension and elasticity which minimize ventricular dilation.[37] In contrast, some cardiomyocytes, which stained positively for BrdU and α-sarcomeric actin, were observed in the acellular/MSC patch (Fig. 8), indicating that the implanted MSCs, transferred within the patch, can engraft and differentiate into cardiomyocytes. It was reported that the MSCs transplanted into a myocardium environment can express myogenic-specific proteins such as α-sarcomeric actin, sarcomeric MHC, desmin, troponin T, and phospholamban.[27,50] Our finding confirmed that the adult heart may provide an environment for the cardiomyogenic differentiation of the implanted MSCs.

Additionally, it was reported that the MSCs transplanted in the myocardium can be differentiated into endothelial cells and smooth muscle cells.[50–52] This fact was also confirmed in our study. As shown, a few capillary walls composed of BrdU-labeled endothelial cells (Fig. 8) together with some BrdU-labeled smooth muscle cells (Fig. 8) were identified in the acellular/MSC patch. Since only a few neo-capillaries and smooth muscle cells were positively labeled with BrdU, the majority of the neo-capillaries and smooth muscle cells observed in the acellular/MSC patch must be originated from the host progenitor cells in blood or the surrounding tissues.

In contrast, only a few infiltrated inflammatory cells were present in the most outer and inner layers of the e-PTFE patch (Fig. 6). Also, no apparent tissue regeneration was observed in the middle layer of the e-PTFE patch (Fig. 7) and no electrogram signals were observed on the epicardial surface of the e-PTFE patch throughout the entire course of the study (Fig. 4). On the contrary, the amplitude of the local electrograms on the acellular and acellular/MSC patches increased significantly with increasing the implantation duration (Fig. 4), an indication of a better electrical conductance. The increase in the amplitude of the local electrograms on the acellular and acellular/MSC patches might be due to the regenerated tissues observed in its pores.

At 12-week post-operatively, the neo-tissue fibrils (neo-muscle fibers, neo-glycosaminoglycans, and neo-capillaries) observed in the acellular and acellular/MSC patches were more compact and organized than their counterparts observed at fourth-week post-operatively. BrdU-labeled cardiomyocytes, endothelial cells and smooth muscle cells were still observed in the acellular/MSC patch. The densities of neo-capillaries found in the acellular and acellular/MSC patches throughout the entire course of the study were still significantly lower than that observed in the native myocardium.

The aforementioned results indicated that the acellular/MSC patch might preserve the structure of the ventricular wall while providing the potential for growth. However, the echocardiographic results showed that the acellular/MSC patch was akinetic, in spite of the presence of cardiomyocytes and a normality of the local electrograms. This may be attributed to the fact that there were not enough cardiomyocytes present in the acellular/MSC patch. Additionally, these cardiomyocytes were not in contact with their host counterparts and therefore were not able to beat synchronously with the host myocardium.

References

1. Akhyari P, Fedak PWM, Weisel RD, Lee TYJ, Verma S, Mickle DAG, *et al.* Mechanical stretch regimen enhances the formation of bioengineered autologous cardiac muscle grafts. *Circulation* 2002;106(Suppl I):I137–I142.
2. Ozawa T, Mickle DAG, Weisel RD, Koyama N, Wong H, Ozawa S, *et al.* Histologic changes of nonbiodegradable and biodegradable biomaterials used to repair right ventricular heart defects in rats. *J Thorac Cardiovasc Surg* 2002;124:1157–1164.
3. Robinson KA, Li J, Mathison M, Redkar A, Cui J, Chronos NAF, *et al.* Extracellular matrix scaffold for cardiac repair. *Circulation* 2005;112(Suppl I):I135–I143.
4. Chang Y, Tsai CC, Liang HC, Sung HW. *In vivo* evaluation of cellular and acellular bovine pericardia fixed with a naturally occurring crosslinking agent (genipin). *Biomaterials* 2002;23:2447–2457.
5. Sung HW, Chen CN, Huang RN, Hsu JC, Chang WH. *In vitro* surface characterization of a biological patch fixed with a naturally occurring crosslinking agent. *Biomaterials* 2000;21:1353–1362.
6. Sung HW, Huang RN, Huang LLH, Tsai CC. *In vitro* evaluation of cytotoxicity of a naturally occurring cross-linking reagent for biological tissue fixation. *J Biomater Sci Polym Ed* 1999;10:63–78.

7. Chang Y, Tsai CC, Liang HC, Sung HW. Reconstruction of the right ventricular outflow tract with a bovine jugular vein graft fixed with a naturally occurring crosslinking agent (genipin) in a canine model. *J Thorac Cardiovasc Surg* 2001;122:1208–1218.

8. Tsai CC, Huang RN, Sung HW. *In vitro* evaluation of the genotoxicity of a naturally occurring crosslinking agent (genipin) for biological tissue fixation. *J Biomed Mater Res* 2000;52:58–67.

9. Matsuda T, Nakayama Y. Surface microarchitectural design in biomedical applications: *in vitro* transmural endothelialization on microporous segmented polyurethane films fabricated using an excimer laser. *J Biomed Mater Res* 1996;31:235–242.

10. Pasumarthi KB, Field LJ. Cardiomyocyte cell cycle regulation. *Circ Res* 2002;90:1044–1054.

11. Yoon YS, Lee N, Scadova H. Myocardial regeneration with bone-marrow-derived stem cells. *Biol Cell* 2005;97:253–263.

12. Laflamme MA, Gold J, Xu C, Hassanipour M, Rosler E, Police S, et al. Formation of human myocardium in the rat heart from human embryonic stem cells. *Am J Pathol* 2005;167:663–671.

13. Malouf NN, Coleman WB, Grisham JW, Lininger RA, Madden VJ, Sproul M, et al. Adult-derived stem cells from the liver become myocytes in the heart *in vivo*. *Am J Pathol* 2001;158:1929–1935.

14. Pittenger MF, Mackay AM, Beck SC, Jaiswal RK, Douglas R, Mosca JD, et al. Multilineage potential of adult human mesenchymal stem cells. *Science* 1999;284:143–147.

15. Minguell JJ, Erices A, Conget P. Mesenchymal stem cells. *Exp Biol Med* 2001;226:507–520.

16. Soares MB, Lima RS, Rocha LL, Takyia CM, Pontes-de-Carvalho L, Campos de Carvalho AC, et al. Transplanted bone marrow cells repair heart tissue and reduce myocarditis in chronic chagasic mice. *Am J Pathol* 2004;164:441–447.

17. Orlic D, Kajstura J, Chimenti S, Jakoniuk I, Anderson SM, Li B, et al. Bone marrow cells regenerate infarcted myocardium. *Nature* 2001;410:701–705.

18. Ryu JH, Kim IK, Cho SW, Cho MC, Hwang KK, Piao H. Implantation of bone marrow mononuclear cells using injectable fibrin matrix enhances neovascularization in infarcted myocardium. *Biomaterials* 2005;26:319–326.

19. Courtman DW, Pereira CA, Kashef V, McComb D, Lee JM, Wilson GJ. Development of a pericardial acellular matrix biomaterial: biochemical and mechanical effects of cell extraction. *J Biomed Mater Res* 1994;28:655–702.

20. Wei HJ, Chang Y, Linag HC, Huang YC, Chang Y, Sung HW. Construction of varying porous structures in acellular bovine pericardia as a tissue-engineering extracellular matrix. *Biomaterials* 2005;26:1905–1913.

21. Chang Y, Lee MH, Liang HC, Hsu CK, Sung HW. Acellular bovine pericardia with distinct porous structures fixed with genipin as an extracellular matrix. *Tissue Eng* 2004;10:881–892.

22. Sung HW, Chang Y, Chiu CT, Chen CN, Liang HC. Crosslinking characteristics and mechanical properties of a bovine pericardium fixed with a naturally occurring crosslinking agent. *J Biomed Mater Res* 1999;47:116–126.

23. Martin I, Shastri VP, Padera RF, Yang J, Mackay AJ, Langer R, et al. Selective differentiation of mammalian bone marrow stromal cells cultured on three-dimensional polymer foams. *J Biomed Mater Res* 2001;55:229–235.

24. Lee JM, Haberer SA, Boughner DR. The bovine pericardial xenograft. I: Effect of fixation in aldehydes without constraint on the tensile viscoelastic properties of bovine pericardium. *J Biomed Mater Res* 1989;23:457–475.

25. Ramshaw AM, Casagranda F, White JF, Edwards GA, Hunt JA, Williams DF. Effects of mesh modification on the structure of a mandrel-grown biosynthetic vascular prosthesis. *J Biomed Mater Res* 1999;47:309–315.

26. Wakitani S, Saito T, Caplan AI. Myogenic cells derived from rat bone marrow mesenchymal stem cells exposed to 5-aza-cytidine. *Muscle Nerve* 1995;18:1417–1426.

27. Tomita S, Li RK, Weisel RD, Mickle DAG, Kim EJ, Sakai T, et al. Autologous transplantation of bone marrow cells improves damaged heart function. *Circulation* 1999;100(Suppl):II247–II256.

28. Krupnick AS, Kreisel D, Engels FH, Szeto WY, Plappert T, Popma SH, *et al.* A novel small animal model of left ventricular tissue engineering. *J Heart Lung Transplant* 2002;21:233–243.

29. Ozawa T, Mickle DAG, Weisel RD, Koyama N, Ozawa S, Li RK. Optimal biomaterial for creation of autologous cardiac grafts. *Circulation* 2002;106(Suppl I):I176–I182.

30. Chiang HK, Chu CW, Chen GY, Kuo CD. A new 3-D display method for 12-lead ECG. *IEEE Trans Biomed Eng* 2001;48:1195–1202.

31. Miyagawa S, Sawa Y, Taketani S, Kawaguchi N, Nakamura T, Matsuura N, *et al.* Myocardial regeneration therapy for heart failure hepatocyte growth factor enhances the effect of cellular cardiomyoplasty. *Circulation* 2002;105:2556–2561.

32. Prophet EB, Mills B, Arrington JB, Sobin LH. *Laboratory Methods in Histotechnology*. American Registry of Pathology, Washington, DC, USA, 1994, pp. 136–137.

33. Allaire E, Guettier C, Bruneval P, Plissonnier D, Michel JB. Cell-free arterial grafts: morphologic characteristics of aortic isografts, allografts, and xenografts in rats. *J Vasc Surg* 1994;19:446–456.

34. Hsieh PCH, Davis ME, Gannon J, MacGillivray C, Lee RT. Controlled delivery of PDGF-BB for myocardial protection using injectable self-assembling peptide nanofibers. *J Clin Invest* 2006;116:237–248.

35. Bader A, Schiling T, Teebken OE. Tissue engineering of heart valves-human endothelial cell seeding of detergent acellularized porcine valves. *Eur J Cardiothorac Surg* 1998;14:279–284.

36. Courtman DW, Errett BF, Wilson GJ. The role of crosslinking in modification of the immune response elicited against xenogenic vascular acellular matrices. *J Biomed Mater Res* 2001;55:576–586.

37. Li RK, Jia ZQ, Weisel RD. Cardiomyocyte transplantation improves heart function. *Ann Thorac Surg* 1996;62:654–660.

38. Shimizu T, Yamato M, Kikuchi A, Okano T. Cell sheet engineering for myocardial tissue reconstruction. *Biomaterials* 2003;24:2309–2316.

39. Matsubayashi K, Fedak PWM, Mickle DAG, Richard D, Weisel RD, Ozawa T, *et al.* Improved left ventricular aneurysm repair with bioengineered vascular smooth muscle grafts. *Circulation* 2003;108(Suppl II):II219–II225.

40. Izutani H, Gundry SR, Vricella LA, Xu H, Bailey LL. Right ventricular outflow tract reconstruction using a GoreTex membrane monocusp valve in infant animals. *ASAIO J* 2000;46:553–555.

41. Uemura H, Yagihara T, Kawahira Y, Yoshikawa Y. Kitamura S. Total cavopulmonary connection in children with body weight less than 10 kg. *Eur J Cardiothorac Surg* 2000;17:543–549.

42. Macchi E, Cavalieri M, Stilli D, Musso E, Baruffi S, Olivetti G, *et al.* High-density epicardial mapping during current injection and ventricular activation in rat hearts. *Am J Physiol* 1998;275:H1886-H1897.

43. Whitaker D, Papadimitriou JM, Walters M. The mesothelium: its fibrinolytic properties. *J Pathol* 1982;136:291–299.

44. Elmadbouh I, Chen Y, Louedec L, Silberman S, Pouzet B, Meilhac O, *et al.* Mesothelial cells transplantation in the infarct scar induces neovascularization and improves heart function. *Cardiovasc Res* 2005;68:307–317.

45. Scott SM, Barth MG, Gaddy LR, Ahl ET Jr. The role of circulating cells in the healing of vascular prostheses. *J Vasc Surg* 1994;19:585–593.

46. Malone JM, Brendel K, Duhamel RC, Reinert RL. Detergent-extracted small-diameter vascular prostheses. *J Vasc Surg* 1984;1:181–191.

47. Graham LM, Vinter DW, Ford JW, Kahn RH, Burkel WE, Stanley JC. Endothelial cell seeding of prosthetic vascular grafts. Early experimental studies with cultured autologous canine endothelium. *Arch Surg* 1980;115:929–933.

48. Sue E, Matsuno H, Ishisaki A, Kitajima Y, Kozawa O. Lack of plasminogen activator inhibitor-1 enhances the preventive effect of DX-9065a, a selective factor Xa inhibitor, on venous thrombus and acute pulmonary embolism in mice. *Pathophysiol Haemost Thromb* 2003;33:206–213.

49. Shinoka T, Ma PX, Shum-Tim D, Breuer CK, Cusick RA, Zund G. Tissue-engineered heart valves: autologous valve leaflet replacement study in a lamb model. *Circulation* 1996;94(Suppl 9):II164–II168.
50. Wang JS, Shum-Tim D, Galipeau J. Marrow stromal cells for cellular cardiomyoplasty: feasibility and potential clinical advantages. *J Thorac Cardiovasc Surg* 2000;120:999–1005.
51. Davani S, Marandin A, Mersin N, Royer B, Kantelip B, Hervé P, *et al.* Mesenchymal progenitor cells differentiate into an endothelial phenotype, enhance vascular density, and improve heart function in a rat cellular cardiomyoplasty model. *Circulation* 2003;108(Suppl II):II253–II258.
52. Pitterger MF, Martin BJ. Mesenchymal stem cells and their potential as cardiac therapeutics. *Circ Res* 2004;95:9–20.

13 Stem Cells and Heart Valves

Danielle Gottlieb, Fraser W. H. Sutherland
and John E. Mayer, Jr.

Introduction

An estimated 275,000 patients worldwide undergo heart valve replacement each year, and the rate of heart valve replacement has been increasing at a compound annual growth of 2% over the past 20 years.[1,2] However, existing approaches to heart valve replacement in adults require improvement, both because of the morbidity associated with lifelong anti-coagulation for mechanical valves, and the limited durability of existing bioprostheses.[3,4]

Approximately 0.8% of live births are affected by a cardiac anomaly, and due to absence or hypoplasia of the native main pulmonary artery (MPA) and pulmonary valve (PV), many of those requiring surgical correction need reconstruction of the right ventricular outflow tract (RVOT). In children, for whom mechanical and bio-prosthetic heart valves are largely unsuitable, the five-year function of cadaveric homografts has been reported to be as low as 17%.[5]

Despite a continuing reduction in the morbidity and mortality associated with congenital heart surgery, long-term outcomes for such patients are still heavily influenced by the morbidities associated with RVOT reconstruction.[6] As a result, there is an increasing patient population with ongoing valve-related morbidity requiring late reoperative valve replacement surgery. While perioperative heart valve replacement mortality has improved dramatically over decades of experience, reoperative heart valve replacement carries higher perioperative morbidity and mortality.[7-9] In children, multiple open heart operations have implications for neurodevelopmental outcomes and quality of life.[10]

Due to the personal, social, and economic costs for patients and payors associated with repeat valve replacement, tissue engineering has been adopted by many research groups worldwide toward the goal of developing a biocompatible replacement heart valve. The success of tissue engineering for heart valve replacement could transform operative palliation of valve disease to curative treatment.

Heart valve tissue engineering holds innate challenges, as the complex structure–function relationships of heart valve tissue remain incompletely understood. In normal heart valves, valve cells and the extracellular matrix (ECM) they produce serve to function from intrauterine life until the end of life, with 40 million yearly cycles of leaflet opening and closing.[11] Therefore, tissue engineered heart valves have the unique design objective of immediate and perfect function at the time of implantation, followed by subsequent *in vivo* cellular and extracellular maturation in order to maintain perfect function through the time of somatic growth into adulthood.

Though tissue engineering of heart valves represents a new paradigm for heart valve replacement, the market for heart valves has spurred decades of research and development with only moderate long-term success. Due to an ongoing medical need, the heart valve replacement device industry is thriving, with worldwide sales which totaled greater than US$1 billion in 2005.[12] With decades of engineering and clinical experience in valve design and implantation, a review of modes of failure in historic and currently implanted heart valves provides necessary background toward the design of a tissue-engineered valve replacement.

Modes of Failure in Existing Heart Valves

Mechanical valves

The first two clinical implantations of ball valves were performed in 1960 by Dwight Harken in Boston, MA, USA. Early models of the ball-and-cage design underwent a series of design iterations to the current design, which is still in limited use today. One of the strengths of this design is the four-decade follow-up associated with the valve. In 1989, four patients were alive with their original valves, more than 40 years after implantation. Nevertheless, usage has fallen due to newer designs of both mechanical and bioprosthetic valves.

Subsequent ball-and-cage valves including the Starr–Edwards valve and the Smeloff–Cutter valve underwent multiple iterations. The dominant mode of failure was ball variance, or fragmentation of the ball over time. Bleeding, thromboembolism, persistent left ventricular dysfunction, left ventricular free wall rupture, and aortic outlet obstruction were also seen.

The ball-and-cage design was followed by caged disc valves, cageless tilting disc valves, and bileaflet mechanical valves. In caged disc valves, the ball was replaced by a disc, though the caged retaining mechanism was unaltered. Though caged disc valves provided a lower profile, the presence of a disc in the flow path was more obstructive to flow than the spherical poppet of the ball-and-cage valves. Disc occluders also showed wear, and disc escape was found to occur. In all caged designs, the primary design flaw was the cloth covering. Blood flow through the valve caused red blood cell collisions against the cloth, resulting in hemolysis.[13] Red blood cell damage was found to be closely related to flow characteristics in caged disc valves.[14] Cageless tilting disc valves were the next strategy; disc wear and detachment of the occluder were again found, as was variable degrees of occluder movement at the hinge. As a result, authors recommended elective replacement for patients surviving with these valves.[15] A late version of a cageless tilting disc valve was the Bjork–Shiley spherical disc prosthesis, in which the disc was manufactured from pyrolytic carbon. In this design, occluder wear was minimal in recovered valves. However, the largest sized valves with a convex–concave disc failed due to strut fracture resulting in disc embolization. Strut fractures occurred as a result of the closing dynamics of the valve.[16] The last generation of such valves, the Bjork–Shiley monostrut, overcame the welded area of stress which had led to strut fracture.

Bileaflet mechanical valves, including the St. Jude valve, are currently used in clinical practice. Structural valve failure has been rare, though leaflet fracture has been seen.[17] Once current designs were produced, durability has been maintained for mechanical valves, and the principal shortcoming of mechanical prostheses has become the need for anti-coagulation, with its attendant risks of bleeding and thromboembolism.[3] Hypothetical advantages of tissue engineered valves over the currently available mechanical prostheses for adult and pediatric application include the lack of requirement for anti-coagulation and the possibility of growth. However, further investigation of the opening and closing characteristics of native valves is required, and tissue engineered valve designs developed which have comparable mechanical opening and closing characteristics to mechanical valves. In children, somatic growth creates an additional limitation: as growth occurs, a valve of fixed orifice area becomes flow limiting over time.

Tissue valves

The great advantages of tissue valves are their relative freedom from thromboembolic complications and the absence of hemolysis. Variation in the design occurs in the choice of substitute tissue, its method of preservation, and the mode of structural support of the tissue. Early iterations of tissue valves included fascia lata valves, whose failures were attributed to contraction of fibrous tissue, a high incidence of endocarditis, and functional deterioration of the valve at five years.[18]

Three types of bioprosthetic valves are currently available: porcine xenograft valves, bovine pericardial valves, and allograft or homograft valves. Porcine valves are assembled by explanting a pig aortic root *en bloc*, fixing in phosphate-buffered glutaraldehyde, then mounting onto a cloth-covered flexible stent attached to a sewing ring. Bovine pericardial valves are constructed from up to three pieces of glutaraldehyde-treated pericardium, affixed to a supporting stent and sewing ring. For both porcine and bovine valves, there is a variation in the materials used for stent and sewing ring, depending on the valve manufacturer. In addition to fixing the tissue, low-concentration glutaraldehyde serves to reduce the antigenicity of xenogenic tissues. Valves are treated with additional agents, such as sodium dodecyl sulfate, to minimize calcification.

Homograft valves are obtained from cadaveric tissue donors, stored by cryopreservation as entire aortic or pulmonary roots. Primary modes of failure for tissue valves are structural deterioration and calcification. For a 65-year-old man, life expectancy after pericardial valve and porcine valve in one meta-analysis was 10.8 and 10.9 years, the median time to structural valve deterioration was 20.1 and 22.2 years, and lifetime risk of reoperation due to structural valve failure was

18.3% and 14%, respectively. The authors concluded that no appreciable differences are discernible in the long-term outcome of adults receiving porcine or pericardial valves.[19]

Though homograft valves offer superior hemodynamics over artificial valves, long-term degeneration of structural components severely limits the durability of these valves. Additionally, homografts perform poorly in the pulmonary position in children,[20] though they represent the best current technology for this application. It is in this context, then, that surgeons seek alternative technologies to provide the heart valve replacements of the future. The tissue engineering approach requires a change in emphasis from improvement of existing technologies to a redesign of the mode of production of replacement heart valves. As in xenograft and homograft valves, the design principles of longevity, freedom from calcification and tissue degeneration will be required for clinical use of a tissue engineered heart valve. In addition to its basis in engineering design principles, tissue engineering is grounded in principles from heart valve developmental biology as a theoretical underpinning.

Embryologic Development of Native Semilunar Valves

By day 15 of human embryo development, specification of myocardial and endothelial cardiac cells occurs. At day 21, fusion at the midline and formation of a linear heart tube occurs.[21] These events are followed by rightward looping of the heart, resulting in orientation of the cardiac chambers into their final positions, and opposition of two specialized segments of cardiac jelly (extracellular matrix). In these swellings of cardiac jelly, called the endocardial cushions, a subset of endothelial cells delaminate, invade the ECM, and undergo endothelial-to-mesenchymal transformation (EMT). After migration into the deeper regions of the endocardial cushion, these cells proliferate and complete their EMT, becoming valve interstitial cells.[22] The endothelium fills the gaps made by migration of endothelial cells. Signaling pathways driving these cellular events are complex and overlapping; those elucidated to date and thought to govern this process include: VEGF-R, NFATc1, and wnt/β-catenin, BMP/TGF-β, erbB and NF1.[22] The sequence of developmental and morphogenetic events in valvulogenesis outlined here provides a potential sequence through which a tissue engineered heart valve might mature; therefore, an understanding of these events and the resultant valve structure and function relationships are paramount to developing a tissue engineered heart valve replacement.

Structure and Function of the Native Semilunar Heart Valves

Mature semilunar valves show a trilaminar histologic structure, with variation in the cell and ECM composition between layers.[23] The ventricularis, on the ventricular surface, is rich in elastin; the centrally located spongiosa is primarily composed of proteoglycans and their derivative glycosaminoglycans; the outflow surface fibrosa is comprised of densely packed collagens. Two cell types predominate in the native semilunar valve: the valvular interstitial (VIC) and endothelial cell (VEC) pheno-types, which in combination, are thought to be critical for production and repair of ECM, and consequently, for maintenance of valve function.[24]

Aikawa *et al.* described the histology of human pulmonary valve cell and ECM composition throughout life. A consistent process of valve cell and ECM maturation over time was found. At 14 weeks gestation, fetal valves were found to contain high VIC density and to be highly proliferative, with structural homogeneity, coalescing into a trilaminar structure between 20 and 36 weeks gestational age.[24] Early fetal valves were highest in proteolytic enzymes, including SMemb, a non-muscle myosin expressed in activated mesenchymal cells; MMP-1 and MMP-3, matrix metallopro-teinases responsible for degradation of ECM proteins. These tissue turnover enzymes are associated with an activated myofibroblast-like phenotype, and in fetal valves, these activated VICs were distributed homogenously throughout the valve cusps.

Likely due to changes in right versus left heart pressures at birth, the shift from fetal to neonatal circulation was associated with a change in phenotype from acti-vated to quiescent interstitial cells in pulmonary, but not in aortic, valves. Cellularity was seen to decrease through fetal life, and this decrease in VIC cellularity continued through childhood into adulthood. Similarly, VECs were noted to have an activated phenotype in fetal, but not adult, life. Further, the disorganized, thin collagen fibers were found to align, thicken and mature from fetal life to adulthood.

Once thought to be passive structures, increasing evidence shows cardiac valves to be dynamic organs. Mechanistically, the finding that cellular and extracellular components mature is thought to reflect the dramatic changes in flow and biome-chanical loading conditions from fetal to post-natal life.[25,26] It is thought that during the cardiac cycle, forces acting on the valve at the organ level (flow, flexure, tension) are translated to the tissue level (collagen crimping, uncrimping, and orientation) and finally to the cell level, inducing phenotypic change and cell signaling events.[27]

At the cell level, investigators have shown cell–matrix interactions to be recip-rocal and additionally determined by the biochemical microenvironment experi-enced in regions of valve tissue.[28] Several molecules involved in cell signaling between VIC and other cells have been identified, including N-cadherin, connexin-26 and -43, and desmoglein, which are involved in communication from VICs to other valve cell types, including VECs.[28] The cell–cell and cell–matrix interactions are incompletely defined in native valve leaflets, and though not yet well elucidated, complex signaling events are thought to play an important role in matrix deposition, remodeling and in long-term native valve function.

The goal of heart valve tissue engineering, then, is to understand the cells, extracellular matrix, biomechanical loading conditions and biochemical signaling involved in native valve structure and function, and to adopt an approach to engineering tissues which contains the necessary components for this complex system.

Approaches to a Tissue Engineered Heart Valve

Approaches to tissue engineered heart valves (TEHV) can be divided into two main categories, based on the type of scaffold employed: bioresorbable scaffolds (non-woven felt, knit, gel-based and combinations) and decellularized tissue-based scaffolds. The bioresorbable scaffold approach begins with seeding cells and culturing under appropriate *in vitro* biomechanical conditions (static flow, pulsatile flow, cyclic strain) and nutrient medium. As the bioresorbable scaffold degrades, new ECM is synthesized by the seeded cells. Porosity is an important characteristic of bioresorbable scaffolds as it provides a permeable framework for cell migration, nutrient supply, and waste removal. The scaffold degradation process is accompanied by a reciprocal increase in ECM during *in vitro* culture. When a sufficiently stable tissue is formed, *in vivo* implantation follows, placing a newly synthesized tissue, devoid of foreign elements, in the required site.

Proponents of the decellularized tissue method base their approach on the findings that human aortic valve homografts are implanted without tissue-type matching, become acellular after several months *in vivo*, yet retain their mechanical properties for decades. As the exact features endowing the aortic valve with such durability are unknown, it is hypothesized that these features are retained after *ex vivo* decellularization. Scaffolds are either seeded with cells prior to implantation, or implanted without seeded cells, with the expectation that the scaffolds will be populated *in vivo* by appropriate circulating cell populations. Without seeding of cells prior to implantation, however, decellularized matrices are insufficiently endothelialized by circulating cells to resist surface thrombus formation *in vivo*.[27]

Review of literature encompassing the cumulative experience of tissue engineering of heart valves is complex, as the type of scaffold, three-dimensional scaffold structure, cell source and type(s), extent and method of *in vitro* pre-conditioning, and duration of *in vivo* implantation have all varied. A complete review of past investigations is beyond the scope of this chapter; however, a summary of the fundamental elements of synthetic, biomaterial-based TEHV will be considered in detail in the following sections.

Differentiated and Stem Cells for Tissue Engineered Heart Valves

The ideal cell type in TEHV would closely resemble both VIC and VEC functions. The goal, then, is to identify cell types that would make ECM, reduce thrombogenicity, lack antigenicity, and respond to biochemical and mechanical signals in a similar fashion to the native tissue, while maintaining the capability for self-renewal. For pediatric valve replacement applications, cell self-renewal or ongoing *in vivo* repopulation of the tissue with appropriate circulating cells would hypothetically lead to tissue growth.

The first TEHV experiments were performed with differentiated cells from artery or vein, including vascular smooth muscle cells, fibroblasts, and endothelial cells from ovine femoral or carotid arteries.[29] As sacrifice of a limb- or brain-perfusing artery is not clinically feasible, subsequent investigators isolated and seeded smooth muscle cells from human saphenous vein and demonstrated stable production of tissue.[30]

The parallel development of stem cell biology has influenced cell choices in tissue engineering, as the capacity for self-renewal and high-volume cell expansion are highly desirable characteristics which have proven difficult in terminally differentiated cells. Mesenchymal stem cells, also called multipotent mesenchymal stromal cells (MSCs), can be isolated from bone marrow, adipose tissue, synovium, umbilical cord blood and amniotic fluid, and display multipotency, with the ability to differentiate along adipogenic, chondrogenic and osteogenic lineages.[31–35] Cells are isolated by density centrifugation and selected by their propensity to adhere strongly to tissue culture plastic. MSCs form colonies in culture and can be verified by colony-forming (CFU-F) assays.[36] Unlike hematopoetic progenitors, the hierarchy of differentiation remains poorly understood, however, after approximately 20 doublings, cells begin to lose multipotency.[35] Mechanisms governing colony formation, multipotency and self-renewal are active areas of investigation and are currently unknown in detail. To date, MSCs can be identified only by these functional characteristics, accompanied by the absence of endothelial markers such as CD31 or von Willebrand factor forming Weibel–Palade bodies, and the presence of smooth muscle cell cytoskeletal markers such as alpha-smooth muscle actin and calponin[31] (Fig. 1). Due to the lack of specificity of this phenotype, cells documented in the literature as MSCs are likely to represent heterogeneous populations of bone marrow and fetal tissue-derived cells. Further, MSCs are likely to be heterogeneous within cultures, as macrophages, lymphocytes, endothelial cells and smooth muscle cells, also adhere to plastic, likely rendering early preparations impure.[36] Over multiple passages, though surface markers are retained, cultures remain morphologically

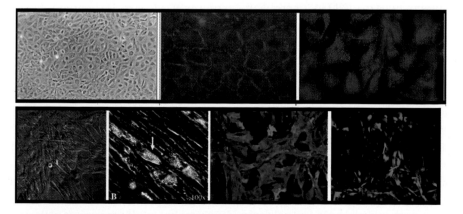

Fig. 1. EPC and MSC phenotypic characterization. **Top panels (L to R)**: cobblestone morphology of endothelial progenitor cells; CD 31+ immunostaining, vWF immunostaining demonstrating Weibel–Palade bodies. **Bottom panels (L to R)**: spindle morphology of mesenchymal progenitor cells; lipid vacuoles demonstrating adipogenesis; immunostaining for αSMA; immunostaining for calponin.

heterogeneous, containing cells ranging from spindle-shaped to polygonal to cuboidal.[36]

Endothelial progenitor cells (EPCs) are found circulating in the blood and represent the precursors to endothelial cells. Endothelial cells are a critical component in the efficient function of cardiovascular tissues, and are central in heart valve tissue development as described earlier. A recent report describes embryologic development in the absence of survivin, an inhibitor of apoptosis and regulator of mitosis in endothelial cells. Mouse embryos lacking the survivin gene experience lethal embryonic defects in cardiogenesis, angiogenesis, and neural tube closure.[37] In addition to its role in development, mature endothelium is involved in the physiologic events of angiogenesis, inflammation, and thrombosis. Circulating EPCs have been identified as cells involved in tissue vascularization of non-cardiovascular tissues, postnatal organ regeneration, and tissue neoangiogenesis.[38] Due to their ability to self-renew, and because their collection does not require sacrifice of a blood vessel, circulating EPCs have received substantial attention in tissue engineering and for their potential applications in myocardial repair after infarction. EPCs have also been used as a single cell type in biopolymer scaffold-based heart valve tissue engineering, in conjunction with MSCs in replacement heart valves, as a mechanism to decrease the thrombogenic potential of decellularized scaffolds in heart valve tissue engineering, and in vascular tissue engineering to serve as a non-thrombogenic surface on dacron grafts and PTFE.[39−44]

Endothelial progenitor cells have been isolated from the mononuclear layer of peripheral blood and umbilical cord blood on gelatin- or fibronectin-coated tissue culture plates.[45] EPCs have also been identified in bone marrow, fetal liver, and skeletal muscle.[38] After isolation, cells can be grown as clonal populations by a limiting dilution technique, or expansion from potentially more heterogeneous colony-based populations. EPCs are identified by their characteristic cobblestone morphology, immunophenotype of CD31 expression on cell–cell membrane borders, CD34 expression, expression of von Willebrand factor in Weibel–Palade bodies, absence of mesenchymal markers, and functional assays, such as their uptake of acetylated low-density lipoprotein and expression of E-selectin in response to lipopolysaccharide[46] (Fig. 1). Additionally, the kinase insert domain receptor (KDR), vascular endothelial growth factor-2 (VEGF-2), and the angiopoietin receptor Tie-2 are expressed by endothelial progenitor cells and can be detected by Western blotting.[47] While these markers establish the endothelial nature of cells, they are also expressed on other endothelial cell types, while the progenitor cell marker CD133 is specific for progenitor cells. In combination, these markers confirm the EPC phenotype.[38]

The finding that human pulmonary valve endothelial progenitor clones treated with TGFβ2 undergo endothelial-to-mesenchymal transformation, expressing alpha-smooth muscle actin and calponin, suggests that adult progenitor cells exhibit plasticity.[48] However, after undergoing EMT, the differentiation potential of these cells was limited, as they did not undergo adipogenesis, osteogenesis, or chondrogenesis after exhibiting a mesenchymal phenotype.

Though incompletely understood, the apparent plasticity of adult progenitor stem cells is a promising characteristic for tissue engineering. All tissues throughout life appear to have a small population of intrinsic repair cells; this finding has led investigators to hypothesize the role for adult progenitor stem cells in organ repair, and has led to their application in tissue engineering.[38,49] Though adult progenitor stem cells may find their appropriate microenvironment *in vivo*, the *ex vivo* isolation and culture of these cells is potentially more complicated. Because non-clonal stem cell populations are heterogeneous and the required signals are incompletely understood, investigators may be expanding different undefined subpopulations, potentially accounting for the variability in results related to cell plasticity and engineered tissue formation found in the tissue engineering literature. Detailed investigations of the signals involved in regulation of phenotype are underway; such data may allow more reproducible results and finely tuned control over steps in engineered tissue development.

Scaffolds: Polymers

Bioresorbable polymers are used as a temporary scaffold for seeding of cells during an *in vitro*, pre-implantation culture period. Ideally, during this period of time, while seeded cells form ECM, scaffolds mimic the complex and anisotropic properties of heart valve tissue. The first generation of scaffolds used in heart valve tissue engineering were highly porous, non-woven felt produced from fibers of polyglycolic acid (PGA).[29] PGA and related polymers continue to be the most widely used for multiple tissue engineering applications.[50–53] The advantage of PGA and related aliphatic polyesters, including poly-l-lactic acid (PLLA) and copolymers, is their safety, biocompatibility, lack of toxicity, and commercial availability. PGA has been used as the commercially available Dexon suture material since the 1970s.[54] Additionally, material properties of these non-woven felts are well established, with reproducible hydrolytic degradation times; PGA alone degrades in two to four weeks, while the majority of fibers of the more hydrophobic PLLA degrade within four to six weeks. These scaffolds lose strength prior to losing mass, which challenges tissue engineers to seed sufficient numbers of matrix-producing cells so as to replace the scaffold strength as it is lost. Several other similar scaffold materials have been utilized with moderate success toward a tissue engineered heart valve: PGA dip-coated with poly-4-hydroxybutyrate (P4HB), polyhydroxyalkanoate; PGA and PLLA have been compounded with a collagen microsponge and used as patches in the canine pulmonary artery.[55–57] Due to the stiffness of non-woven bioresorbable polymer-based scaffolds, and to concerns about the acidity of degradation products, some groups have used hydrogel scaffolds.

Hydrogel-based scaffolds, including collagen, alginate, agarose, gelatin, fibrin, chitosan, polyethylene glycol, hyaluronic acid, and dehydrated sheets of extracellular matrix have formed the basis of many experiments in tissue engineering.[58–65] When gels become solid, cells are trapped in the gel, providing a homogeneous distribution of cells embedded in a temporary matrix. A bileaflet heart valve was produced in the Tranquillo laboratory using collagen-based scaffolds seeded with dermal fibroblasts.[58] Recent advances in drug delivery technologies and microfluidics have resulted in further design of scaffold characteristics, including orchestration of scaffold polymerization with changes in temperature, pH, or photopolymerization; engineering of nano- and micro-scale cell environments in order to direct cell distribution throughout a scaffold, and microencapsulation of growth factors and adhesion peptides on scaffolds for improved cell attachment and proliferation.[66–70] Finally, cells and scaffold have been incorporated and directed in nanoscale fabrication techniques such as electrospinning.[71,72]

The complex anisotropy and three-dimensional structure of heart valves has been a challenge to reproduce in a tissue engineered device. By imaging normal heart valve anatomy derived from computed tomography images, Sodian employed stereolithography to print a three-dimensional model for use as a mold for the thermoplastic polymer P4HB. The mold was used for curing the polymer into three-dimensional valve anatomy.[73] Tranquillo's bileaflet valve was created by solidification of a hydrogel in a bileaflet valve mold.[58]

As improvements in imaging technology yield higher spatial and temporal resolution, anatomic definition of thin, moving structures in the heart such as valve leaflets will be more feasible. With defined anatomic dimensions and an understanding of outflow tract and leaflet motion, tissue engineers will have clear targets for three-dimensional fabrication of heart valves (Fig. 2).

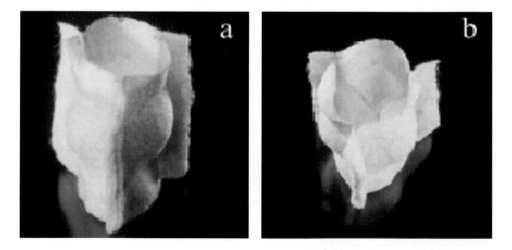

Fig. 2.   Three-dimensional configuration of non-woven scaffold for heart valve tissue engineering.

Tissue Formation: Biochemical and Biophysical Pre-Conditioning

Identification of anatomic targets and engineering appropriate scaffold material properties are individual steps in engineering a heart valve replacement tissue. Another step involves induction of a phenotypically similar cell type to native tissue. However, simply identifying an appropriate cell type is necessary but not sufficient to engineer replacement tissues. Proper cellular orientation and three-dimensional microstructure are also required for tissue function; tissues demonstrate organization of cells and matrix across multiple levels of scale. In this regard, engineered tissues require coordination.

Stem cells hold promise for responding to local cues resulting in assembly of complex tissues. In one study, human embryonic stem (ES) cells injected into the vitreous of newborn mice invaded all retinal layers of the vitreous, and differentiated into appropriate neural cell types in each layer.[74] Hypothetically, in an engineered biochemical and mechanical environment similar to the fetal milieu, progenitor cells will transduce biochemical and biophysical signals to form similarly anisotropic, strong, and flexible tissues to those formed during development.

During heart valve development, the signals involved in cell proliferation, migration, and apoptosis occur through both biochemical and mechanical cues. *In vitro*, mature endothelial cell gene regulation is altered in response to shear stress.[75] And with constant concentrations of soluble growth factors, endothelial and other cells can be switched between growth, differentiation and apoptosis programs by varying the extent to which the cell is spread or stretched.[76] As cells are embedded in and coupled to their extracellular matrix, ECM is a vehicle through which signals must pass. Biochemical signals can be sequestered by ECM, and as cells bind to surrounding ECM via integrin receptors; the binding itself can induce phenotypic changes. Additionally, as in the retinal example, cellular location within the ECM can determine cell phenotype and behavior. Integrin receptors are again involved in mechanotransduction of signals through ECM to alter cellular cytoskeleton and cytoplasm.[76,77] As progenitor cells are capable of differentiating into multiple cell types, their differentiation could potentially be controlled by biophysical signals. When human MSCs are plated onto large tissue islands that promote cell spreading, they efficiently differentiate into bone cells; when plated onto small islands, the same cells in the same culture medium differentiate into adipocytes.[78]

Biophysical signaling is therefore likely to be important in tissue engineering at the cell level, at the tissue organization level, and at the tissue engineered organ level. Since heart valve tissue engineering requires implanted valves to function perfectly at the time of implantation, leaflet motion and coaptation with an adequate but non-regurgitant valve orifice must be engineered prior to implantation. The concept of *in vitro* pre-conditioning for *in vivo* implantation arises from this organ-level requirement. With cell and tissue-level data suggesting that mechanical forces stimulate tissue formation *in vitro*, sterile vessels termed "bioreactors" which apply mechanical force while providing circulating tissue culture medium to developing tissues, have been utilized toward the goal of increasing tissue formation while creating a moving, three-dimensional valve structure.

The optimal pressure for medium perfusion and the ideal pre-conditioning forces required for optimal heart valve tissue formation remain unknown. For engineering a valved conduit, it is likely that pre-conditioning strategies will differ for the leaflet and arterial wall conduit portions of the tissue engineered graft. Several strategies have been investigated.

The first bioreactor created for heart valve tissue engineering occurred in the Mayer laboratory. A pulse duplicator system was created from plexiglass, with pulsatile flow propelled by a respirator (Fig. 3). Pulsatile flow capability ranged from 50 ml/min to 2 l/min; pressures generated ranged from 10 to 240 mm Hg.[79] This system was used for fabrication of valves from PGA dip-coated in polyhydroxy-4-butyrate. Myofibroblasts were isolated from ovine carotid arteries and seeded onto the scaffold for four days in static culture, then seeded with endothelial cells and subjected to gradually increasing flow and pressure conditions (125–750 ml/min at 30–50 mm Hg). After 14 days of pulsatile flow, the resultant tissue was superior to that grown in static culture: DNA and collagen content were substantially greater, tissue was denser and more organized.[55] In this system, heart valve scaffolds composed of PGA dip-coated in P4HB were placed in a novel diastolic duplicator bioreactor, representing only the diastolic phase of the cardiac cycle when leaflets are subjected to dynamic strain. Scaffolds were seeded with vascular smooth muscle cells and underwent dynamic pre-conditioning with transvalvar pressures gradually increasing from 0 to 80 mm Hg over four weeks, and were compared to a group of scaffolds undergoing static culture and a group with pre-strained leaflets constrained by a stent.[80] Both the pre-strain and dynamic conditioning groups resulted in more homogeneous and cellular tissue formation with higher collagen content. Strain at breaking increased with longer culture time. After two weeks, the mechanical behavior of the tissue was linear, reflecting scaffold properties, however after three to four weeks of culture time, the stress–strain profile was non-linear and suggestive of a tissue contribution. A similar pulse duplicator system was later created for vascular conduit scaffold seeding followed by pulsatile flow, with controllable

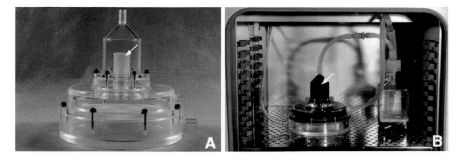

Fig. 3.   Pulse duplicator system (bioreactor) for heart valve tissue engineering. **(A)** Pulse duplicator with mounted heart valve scaffold (arrow). **(B)** Pulse duplicator in standard tissue culture incubator.[55]

culture medium flow rates of 100 ml/min to 3 l/min, pressures ranging from 3 to 150 mm Hg, and shear stresses between 1.12 and 32.45 dyn/cm^2.[81] Small diameter blood vessels have been cultured using smooth muscle and endothelial cells, under conditions of pulsatile radial stress in bioreactors with continuous medium flow and compared to vessels formed without pulsatile pre-conditioning. Groups with and without culture medium supplemented with ascorbic acid, copper sulfate, proline, alanine and glycine were compared. All vessels, regardless of biophysical force applied, ruptured at less than 300 mm Hg without culture medium supplementation. With medium supplementation, pulsed vessels ruptured at >570 mm Hg, suggesting a synergistic effect of biophysical and biochemical signals.[82] Finally, leaflet-sized scaffolds of PGA/PLLA seeded with bone marrow-derived MSCs were exposed to laminar flow, cyclic flexure, or flexure and flow. The combination of cyclic flexure and laminar flow yielded the greatest tissue formation after three weeks of culture, displaying 75% more collagen, suggesting that when combined with tissue culture medium flow, a cyclic flexure pre-conditioning regimen, like pulsatile flow, provides stimulus for increased tissue formation.[83]

In Vivo Experiments

Heart valve tissue engineering has been conceptually established through proof-of-concept experiments in animals. The first *in vivo* tissue engineered heart valve experiment was implantation of a single leaflet into the pulmonary circulation.[29] Cells were isolated from ovine arteries and sorted into endothelial and fibroblast populations. Cells were labeled and seeded onto PGA scaffold and implanted into animals. Labeled cells were visualized in explanted specimens after six hours and after one, six, seven, nine, or 11 weeks. Cells produced extracellular matrix, and leaflets persisted in the circulation at all time points.

A second set of experiments involved implantation of a pulmonary valve composed of autologous ovine endothelial cells and myofibroblasts on PGA coated with P4HB.[55] After 14 days of exposure to gradually increasing flow and pressure conditions in a pulse duplicator, valves were implanted into sheep for one day, four, six, eight, 16, and 20 weeks. Valves demonstrated acceptable hemodynamics and had no evidence of thrombosis, stenosis or aneurysm formation up to 20 weeks. Central valvar regurgitation was noted after 16 weeks. Tissue analysis showed a layered structure with central glycosaminoglycans, collagen on the outflow surface and elastin on the inflow surface. Elastin was detectable in the leaflets by six weeks *in vivo*. Over the 20-week study period, scaffold degradation was coincident with decreased stiffness of the leaflets. Histological, biochemical, and biomechanical parameters were similar to those of native pulmonary artery and leaflet tissue.

Using MSCs on a PGA/PLLA scaffold, investigators fabricated and implanted appropriately-sized heart valves in the pulmonary position of juvenile sheep[84] (Fig. 4). Valves were evaluated by *in vivo* echocardiography; explanted tissues were

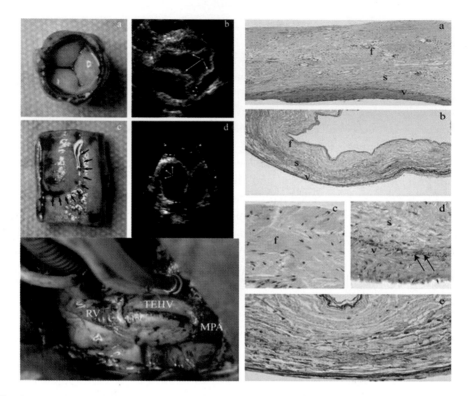

Fig. 4. Tissue engineered heart valve. **Top left panels**: tissue engineered heart valve prior to implantation, and echocardiographic images after implantation. **Bottom left panels**: intraoperative photo of implanted tissue engineered heart valve. **Right panels**: Movat pentchrome stains of low (×100 a, b) and high (×400 c, d, e) magnification sections of tissue engineered heart valve leaflets (**a**) in comparison with normal ovine pulmonary valve leaflets (**b**) and demonstrating rudimentary zona fibrosa, f; zona spongiosa, s; and zona ventrciularis, v in tissue engineered heart valve leaflets (**c, d**) in comparison with normal ovine pulmonary valve leaflet (**e**). Arrows point to a layer of elastin observed on the ventricular side of the tissue engineered heart valve leaflet (**d**).

analyzed by histology and immunostaining. At the time of implantation, echocardiograms demonstrated a maximum instantaneous gradient of 17.2 ± 1.33 mm Hg, with estimates of regurgitation ranging from trivial to mild. Four animals survived the immediate post-operative period. After four months *in vivo*, there were no statistically significant differences in maximum instantaneous gradient, mean gradient or effective orifice area. Histologic analysis at eight post-operative months showed explanted tissue to have a layered organization, with elastin fibers identified on the inflow surface and collagen dominating the outflow surface. Glycosaminoglycans were distributed throughout the remainder of the valve. At the time of implantation, cells uniformly expressed the mesenchymal cell marker α-SMA, but at the time of explantation, these cells were confined to the subendothelial layer, while surface cells expressed von Willebrand factor, suggesting *in vivo* endothelialization of the graft. This study demonstrated the feasibility of implantation of a pulmonary valve using stem cells.

The major hypothetical advantage of TEHV for congenital cardiac surgery is the hypothetical growth that heart valves would undergo during a child's period of maximum somatic growth. Though implanting a tissue engineered pulmonary valve has been possible, questions remain about the *in vivo* growth of TEHV.

Fourteen lambs underwent implantation of tissue engineered pulmonary conduits constructed from committed vascular cells (endothelial and myofibroblast) seeded sequentially onto a PGA scaffold dip-coated in P4HB.[85] Animals were followed for ≤100 weeks, during which time, their body weight doubled. Echocardiography and computed tomography (CT) were performed at defined intervals, and histologic analysis of explanted grafts was performed. Twelve animals were analyzed and sacrificed at post-operative week 20, 50, 80, and 100. At 20 weeks post-implantation, grafts were found to increase in length by 23%, followed by an additional increase in mean length by 18% during the subsequent 80 weeks. By angiography, grafts were found to increase in circumference by 30% between 20 and 80 weeks. No stenosis or aneurysm formation was identified. These results were promising, suggesting that the hypothetical growth potential of tissue engineered cardiovascular structures may be demonstrated *in vivo*.

To date, it is unclear whether the post-operative histologic change observed represents a true increase in stem cell number corresponding with an increase in graft size, or whether changes can be attributed only to influx of inflammatory cells. Additionally, for growth to be demonstrated, increases in size must be accompanied by maintenance of tissue integrity over time. New imaging technologies such as PET scanning, MRI and CT will allow labeling of stem cells for visualization post-implantation accompanied by detailed anatomic monitoring. In combination, imaging and molecular biology techniques for cell labeling will allow further elucidation of the remodeling process.

Heart Valve Tissue Engineering: Present and Future

After a decade of effort toward a tissue engineered heart valve, a clinical need for a more durable valve replacement still exists. Modes of failure of currently available heart valve replacements are well described and provide insight into strengths and weaknesses of previous designs. Though the appropriate cell source and scaffold material for heart valve tissue engineering have been topics of substantial effort, the optimal cell type, scaffold material and *in vitro* culture conditions for a tissue engineered heart valve still remain unknown. Stem cells and adult progenitor cells offer promise toward this end, as they possess the capacity for self-renewal and multipotential differentiation, and therefore, the potential for growth. Evolving imaging technologies will allow a more complete understanding of post-implantation remodeling and the potential for growth. Though many details are yet to be elucidated, bioreactor pre-conditioning has the potential to reduce *in vitro* culture times, thereby facilitating clinical application.

Complex biological events, such as heart valve development, occur through both biochemical and biophysical cues occurring in parallel at the cellular, tissue and organ levels. A detailed understanding of the sequence of necessary developmental events may allow engineering of appropriate microenvironments and three-dimensional structures leading to the goal: a clinically translatable heart valve replacement with the capacity to grow and remain durable for a lifetime of cardiac cycles.

References

1. Schoen F. Pathology of heart valve substitution with mechanical and tissue prostheses. In: Silver M, Gotlieb A, Schoen F (eds.) *Cardiovascular Pathology.* WB Saunders, New York, 2001, pp. 629–677.
2. Society of Cardiac Surgeons of Great Britain and Ireland. United Kingdom Cardiac Surgical Register 1999–2000; 2000.
3. Dalrymple-Hay M, Pearce R, Dawkins S, Alexiou C, Haw M, Livesey S. Mid-term results with 1,503 CarboMedics mechanical valve implants. *J Heart Valve Dis* 2000;9(3):389–395.
4. Schoen F, Levy R. Tissue heart valves: current challenges and future research perspectives. *J Biomed Mater Res* 1999;47(4):439–465.
5. Perron J, Moran AM, Gauvreau K, del Nido PJ, Mayer JE, Jonas RA. Valved homograft conduit repair of the right heart in early infancy. *Ann Thorac Surg* 1999;68(2):542–548.
6. Mayer JE. Uses of homograft conduits for right ventricle to pulmonary artery connections in the neonatal period. *Semin Thorac Cardiovasc Surg* 1995;7(3):130–132.
7. McGrath L, Fernandez J, Laub G, Anderson W, Bailey B, Chen C. Perioperative events in patients with failed mechanical and bioprosthetic valves. *Ann Thorac Surg* 1995;60:S475–S478.
8. Bioglioli P, DiMatteo S, Parolari A, Antona C, Arena V, Sala A. Reoperative cardiac valve surgery: a multivariable analysis of risk factors. *Cardiovasc Surg* 1994;2:216–222.
9. Kojori F, Chen R, Calderone C, Merklinger S, Azakie A, Williams W, *et al.* Outcomes of mitral valve replacement in children: a competing-risks analysis. *J Thorac Cardiovasc Surg* 2004;128(5):703–709.
10. Robbins R, Jr. FB, Malm J. Cardiac valve replacement in children: a twenty-year series. *Ann Thorac Surg* 1988;45(1):56–61.
11. Schoen F, Edwards W. Valvular heart disease: general principles and stenosis. In: Silver M, Gotlieb A, Schoen F (eds.) *Cardiovascular Pathology*, 3rd ed. Churchill Livingstone, New York, 2001, pp. 402–405.
12. Vesely I. Heart valve tissue engineering. *Circ Resh* 2005;97:743–755.
13. Roberts W, Fishbein M, Golden A. Cardiac pathology after valve replacement by disc prosthesis. A study of 61 necropsy patients. *Am J Cardiol* 1975;35(5):740–760.
14. Sallam I, Shaw A, Bain W. Experimental evaluation of mechanical hemolysis with Starr–Edwards, Kay-Shiley and Bjork–Shiley valves. *Scand J Thorac Cardiovasc Surg* 1976;10(2):117–122.
15. Roe B, Fishman N, Hutchinson J, Goodenough S. Occluder disruption of the Wada–Cutter valve prosthesis. *Ann Thorac Surg* 1975;20(3):256–264.
16. Schreck S, Inderbitzen R, Chin H, Wieting D, Smilor M, Breznock E. Dynamics of Bjork–Shiley convexo-concave mitral valves in sheep. *J Heart Valve Dis* 1995;4(Suppl 1):S21–S24.
17. Emery R, Krogh C, Arom K, Emery A, Benyo-Albrecht K, Joyce L. The St. Jude Medical cardiac valve prosthesis: a 25-year experience with single valve replacement. *Ann Thorac Surg* 2005;79(3):776–782.

18. Olsen E, Janabi NA, Salamao C, Ross D. Fascia lata valves: a clinicopathological study. *Thorax* 1975;30(5):528–534.

19. Puvimanasinghe J, Takkenberg J, Eijkemans M, van Herwerden L, Jamieson W, Grunkemeier G. Comparison of Carpentier–Edwards pericardial and supraannular bioprostheses in aortic valve replacement. *Eur J Cardiothorac Surg* 2006;29(3):374–379.

20. Karamlou T, Jang K, Williams W, Caldarone C, Arsdell GV, Coles J, *et al.* Outcomes and associated risk factors for aortic valve replacement in 160 children: a competing-risks analysis. *Circulation* 2005;112(22):3462–3469.

21. Moore K, Persaud T. *The Cardiovascular System. The Developing Human*, 6th ed. WB Saunders, Philadelphia, 1998, pp. 563.

22. Armstrong E, Bischoff J. Heart valve development: endothelial cell signaling and differentiation. *Circ Res* 2004;95(5):459–470.

23. Junqueira L, Carneiro J, Kelley R. *The Circulatory System: Basic Histology*, 8th ed. Appleton & Lange, 1995, p. 488.

24. Aikawa E, Whittaker P, Farber M, Mendelson K, Padera R, Aikawa M, *et al.* Human semilunar cardiac valve remodeling by activated cells from fetus to adult: implications for postnatal adaptation, pathology, and tissue engineering. *Circulation* 2006;113:1344–1352.

25. Butcher J, Tressel S, Johnson T, Turner D, Sorescu G, Jo H, *et al.* Transcriptional profiles of valvular and vascular endothelial cells reveal phenotypic differences: influence of shear stress. *Arterioscler Thromb Vasc Biol* 2006;26(1):69–77.

26. Ku C, Johnson P, Batten P, Sarathchandra P, Chambers R, Taylor P, *et al.* Collagen synthesis by mesenchymal stem cells and aortic valve interstitial cells in response to mechanical stretch. *Cardiovasc Res* 2006;71(3):548–556.

27. Mendelson K, Schoen F. Heart valve tissue engineering: concepts, approaches, progress and challenges. *Ann Biomed Eng* 2006;34(12):1799–1819.

28. Latif N, Sarathchandra P, Taylor P, Antoniw J, Brand N, Yacoub M. Characterization of molecules mediating cell–cell communication in human cardiac valve interstitial cells. *Cell Biochem Biophys* 2006;45(3):255–264.

29. Shinoka T, Breuer C, Tanel R, Zund G, Miura T, Ma P, *et al.* Tissue engineering heart valves: valve leaflet replacement study in a lamb model. *Ann Thorac Surg* 1995;60(Suppl 6):S513–S516.

30. Schnell A, Hoerstrup S, Zund G, Kolb S, Sodian R, Visjager J, *et al.* Optimal cell source for cardiovascular tissue engineering: venous vs. aortic human myofibroblasts. *Thorac Cardiovasc Surg* 2001;49(4):221–225.

31. Pittenger M, Mackay A, Beck S, Jaiswal R, Douglas R, Mosca J, *et al.* Multilineage potential of adult human mesenchymal stem cells. *Science* 1999;284(5411):143–147.

32. Breymann C, Schmidt D, Hoerstrup S. Umbilical cord cells as a source of cardiovascular tissue engineering. *Stem Cell Rev* 2006;2(2):87–92.

33. Marion N, Mao J. Mesenchymal stem cells and tissue engineering. *Methods Enzymol* 2006;420:339–361.

34. Baksh D, Song L, Tuan R. Adult mesenchymal stem cells: characterization, differentiation, and application in cell and gene therapy. *J Cell Mol Med* 2004;8(3):301–316.

35. Muraglia A, Cancedda R, Quarto R. Clonal mesenchymal progenitors from human bone marrow differentiate *in vitro* according to a hierarchical model. *J Cell Sci* 2000;113:1161–1166.

36. Javazon E, Beggs K, Flake A. Mesenchymal stem cells: paradoxes of passaging. *Exp Hematol* 2004;32:414–425.

37. Zwerts F, Lupu F, Vriese AD, Pollefeyt S, Moons L, Altura R, *et al.* Lack of endothelial cell survivin causes embryonic defects in angiogenesis, cardiogenesis, and neural tube closure. *Blood* 2007;109(11):4742–4752.

38. Yu Y, Flint A, Mulliken J, Wu J, Bischoff J. Endothelial progenitor cells in infantile hemangioma. *Blood* 2004;103(4):1373–1375.

39. Wilhelm C, Bal L, Smirnov P, Galy-Fauroux I, Clement O, Gazeau F, *et al.* Magnetic control of vascular network formation with magnetically labeled endothelial progenitor cells. *Biomaterials* 2007.

40. Thebaud N, Pierron D, Bareille R, Visage CL, Letourneur D, Bordenave L. Human endothelial progenitor cell attachment to polysaccharide-based hydrogels: a pre-requisite for vascular tissue engineering. *J Mater Sci Mater Med* 2007;18(2):339–345.

41. Melero-Martin J, Khan Z, Picard A, Wu X, Paruchuri S, Bischoff J. *In vivo* vasculogenic potential of human blood-derived endothelial progenitor cells. *Blood* 2007;109(11):4761–4768.

42. Cebotari S, Lichtenberg A, Tudorache I, Hilfiker A, Mertsching H, Leyh R, *et al.* Clinical application of tissue engineered human heart valves using autologous progenitor cells. *Circulation* 2006;114(Suppl I):I132–I137.

43. Sreerekha P, Krishnan L. Cultivation of endothelial progenitor cells on fibrin matrix and layering on dacron/polytetrafluoroethylene vascular grafts. *Artif Organs* 2006;30(4):242–249.

44. Losordo D, Schatz R, White C, Udelson J, Veereshwarayya V, Durgin M, *et al.* Intramyocardial transplantation of autologous CD34+ stem cells for intractable angina: a phase I/IIa double-blind, randomized controlled trial. *Circulation* 2007;115(25):3165–3172.

45. Schmidt D, Asmis L, Odermatt B, Kelm J, Breymann C, Gossi M, *et al.* Engineered living blood vessels: functional endothelia generated from human umbilical cord-derived progenitors. *Ann Thorac Surg* 2006;82(4):1465–1471.

46. Urbich C, Dimmeler S. Endothelial progenitor cells: characterization and role in vascular biology. *Circ Res* 2004;95(4):343–353.

47. Kaushal S, Amiel G, Guleserian K, Shapira O, Perry T, Sutherland F, *et al.* Functional small-diameter neovessels created using endothelial progenitor cells expanded *ex vivo*. *Nat Med* 2001;7(9):1035–1040.

48. Paruchuri S, Yang J, Aikawa E, Melero-Martin J, Khan Z, Loukogeorgakis S, *et al.* Human pulmonary valve progenitor cells exhibit endothelial/mesenchymal plasticity in response to vascular endothelial growth factor-A and transforming growth factor-beta2. *Circ Res* 2006;99(8):861–869.

49. Visconti R, Ebihara Y, LaRue A, Fleming P, McQuinn T, Masuya M, *et al.* An *in vivo* analysis of hematopoietic stem cell potential: hematopoietic origin of cardiac valve interstitial cells. *Circ Res* 2006;98(5):690–696.

50. Bailey M, Wang L, Bode C, Mitchell K, Detamore M. A comparison of human umbilical cord matrix stem cells and temporomandibular joint condylar chondrocytes for tissue engineering temporomandibular joint condylar cartilage. *Tissue Eng* 2007;13(8):2003–2010.

51. Rohman G, Pettit J, Isaure F, Cameron N, Southgate J. Influence of the physical properties of two-dimensional polyester substrates on the growth of normal human urothelial and urinary smooth muscle cells *in vitro*. *Biomaterials* 2007;28(14):2264–2274.

52. Roh J, Brennan M, Lopez-Soler R, Fong P, Goyal A, Dardik A, *et al.* Construction of an autologous tissue-engineered venous conduit from bone marrow-derived vascular cells: optimization of cell harvest and seeding techniques. *J Pediatr Surg* 2007;42(1):198–202.

53. Dahl S, Rhim C, Song Y, Niklason L. Mechanical properties and compositions of tissue engineered and native arteries. *Ann Biomed Eng* 2007;35(3):348–355.

54. Gao J, Niklason L, Langer R. Surface hydrolysis of poly(glycolic acid) meshes increases the seeding density of vascular smooth muscle cells. *J Biomed Mater Res* 1998;42:417–424.

55. Hoerstrup S, Sodian R, Daebritz S, Wang J, Bacha E, Martin D, *et al.* Functional living trileaflet heart valves grown *in vitro*. *Circulation* 2000;102(9 Suppl 3):III44–III49.

56. Sodian R, Hoerstrup S, Sperling J, Daebritz S, Martin D, Moran A, *et al.* Early *in vivo* experience with tissue-engineered trileaflet heart valves. *Circulation* 2000;102(19 Suppl 3):III22–III29.

57. Iwai S, Sawa Y, Ichikawa H, Taketani S, Uchimura E, Chen G, *et al.* Biodegradable polymer with collagen microsponge serves as a new bioengineered cardiovascular prosthesis. *J Thorac Cardiovasc Surg* 2004;128(3):472–479.

58. Neidert M, Tranquillo R. Tissue-engineered valves with commissural alignment. *Tissue Eng* 2006;12(4):891–903.
59. Hannouche D, Terai H, Fuchs J, Terada S, Zand S, Nasseri B, *et al.* Engineering of implantable cartilaginous structures from bone marrow-derived mesenchymal stem cells. *Tissue Eng* 2007;13(1):87–89.
60. Roman J, Cabanas M, Pena J, Doadrio J, Vallet-Regi M. An optimized beta-tricalcium phosphate and agarose scaffold fabrication technique. *J Biomed Mater Res* 2007;84(1):99–107.
61. Haraguchi T, Okada K, Tabata Y, Maniwa Y, Hayashi Y, Okita Y. Controlled release of basic fibroblast growth factor from gelatin hydrogel sheet improves structural and physiological properties of vein graft in rat. *Arterioscler Thromb Vasc Biol* 2007;27(3):548–444.
62. Mol A, van Lieshout M, Dam-de Veen CG, Neuenschwander S, Hoerstrup S, Baaijens F, Bouten CV. Fibrin as a cell carrier in cardiovascular tissue engineering applications. *Biomaterials* 2005;26(16):3113–3121.
63. Wang W. A novel hydrogel crosslinked hyaluronan with glycol chitosan. *J Mater Sci Mater Med* 2006;17(12):1259–1265.
64. Ferreira L, Gerecht S, Fuller J, Shieh H, Vunjak-Novakovic G, Langer R. Bioactive hydrogel scaffolds for controllable vascular differentiation of human embryonic stem cells. *Biomaterials* 2007;28(17):2706–2717.
65. Mironov V, Kasyanov V, Shu XZ, Eisenberg C, Eisenberg L, Gonda S, *et al.* Fabrication of tubular tissue constructs by centrifugal casting of cells suspended in an *in situ* crosslinkable hyaluronan-gelatin hydrogel. *Biomaterials* 2005;26(36):7628–7635.
66. Park H, Cannizzaro C, Vunjak-Novakovic G, Langer R, Vacanti C, Farokhzad O. Nanofabrication and microfabrication of functional materials for tissue engineering. *Tissue Eng* 2007;13:1867–1877.
67. Burdick J, Khademhosseini A, Langer R. Fabrication of gradient hydrogels using a microfluidics/photopolymerization process. *Langmuir* 2004;20(13):5153–5156.
68. Silva E, Mooney D. Spatiotemporal control of vascular endothelial growth factor delivery from injectable hydrogels enhances angiogenesis. *J Thromb Haemost* 2007;5(3):590–598.
69. Matsumoto T, Mooney D. Cell instructive polymers. *Adv Biochem Eng Biotechnol* 2006;102:113–137.
70. Bacakova L, Filova E, Kubies D, Machova L, Proks V, Malinova V, *et al.* Adhesion and growth of vascular smooth muscle cells in cultures on bioactive RGD peptide-carrying polylactides. *J Mater Sci Mater Med* 2007;18(7):1317–1323.
71. Lee S, Yoo J, Lim G, Atala A, Stitzel J. *In vitro* evaluation of electrospun nanofiber scaffolds for vascular graft application. *J Biomed Mater Res A* 2007;83(4):999–1008.
72. Goldberg M, Langer R, Jia X. Nanostructured materials for applications in drug delivery and tissue engineering. *J Biomater Sci Polym Ed* 2007;18(3):241–268.
73. Sodian R, Fu P, Lueders C, Szymanski D, Fritsche C, Gutberlet M, *et al.* Tissue engineering of vascular conduits: fabrication of custom-made scaffolds using rapid prototyping techniques. *Thorac Cardiovasc Surg* 2005;53(3):144–149.
74. Lamba D, Karl M, Ware C, Reh T. Efficient generation of retinal progenitor cells from human embryonic stem cells. *Proc Natl Acad Sci USA* 2006;103(34):12769–12774.
75. Resnick N, Jr. MG. Hemodynamic forces are complex regulators of endothelial gene expression. *FASEB J* 1995;9(10):874–882.
76. Ingber D. Mechanical control of tissue morphogenesis during embryological development. *Int J Dev Biol* 2006;50:255–266.
77. Stegemann J, Hong H, Nerem R. Mechanical, biochemical, and extracellular matrix effects on vascular smooth muscle cell phenotype. *J Appl Physiol* 2005;98(6):2321–2327.
78. Ingber D, Levin M. What lies at the interface of regenerative medicine and developmental biology? *Development* 2007;134(14):2541–2547.

79. Hoerstrup S, Sodian R, Sperling J, Vacanti J, Jr. JM. New pulsatile bioreactor for *in vitro* formation of tissue engineered heart valves. *Tissue Eng* 2000;6(1):75–79.

80. Mol A, Driessen N, Rutten M, Hoerstrup S, Bouten C, Baaijens F. Tissue engineering of human heart valve leaflets: a novel bioreactor for a strain-based conditioning approach. *Ann Biomed Eng* 2005;33(12):1778–1788.

81. Sodian R, Lemke T, Fritsche C, Hoerstrup S, Fu P, Potapov E, *et al.* Tissue-engineering bioreactors: a new combined cell–seeding and perfusion system for vascular tissue engineering. *Tissue Eng* 2002;8(5):863–870.

82. Niklason L, Gao J, Abbott W, Hirschi K, Houser S, Marini R, *et al.* Functional arteries grown *in vitro*. *Science* 1999;284(5413):489–493.

83. Engelmayr G, Sales V, Mayer JE, Sacks M. Cyclic flexure and laminar flow synergistically accelerate mesenchymal stem cell-mediated engineered tissue formation: implications for engineered heart valve tissues. *Biomaterials* 2006;27(36):6083–6095.

84. Sutherland F, Perry T, Yu Y, Sherwood M, Rabkin E, Masuda Y, *et al.* From stem cells to viable autologous semilunar heart valve. *Circulation* 2005;111(21):2783–2791.

85. Hoerstrup S, Cumming Mrcs I, Lachat M, Schoen F, Jenni R, Leschka S, *et al.* Functional growth in tissue-engineered living, vascular grafts: follow-up at 100 weeks in a large animal model. *Circulation* 2006;114(Suppl 1):I159–I166.

14 Potential Applications of Combined Stem Cell and Gene Therapy for the Treatment of Infarcted Heart

Husnain Khawaja Haider and Muhammad Ashraf

Introduction

Congestive heart failure remains the leading cause of morbidity and death. It is characterized by extensive cardiomyocytes apoptosis and fibrous replacement of myocardial tissue after infarction episode. The changes result in increased wall stress in the viable myocardium thus initiating a progression of cellular and physiological alterations which end up in left ventricular remodeling. The conventional surgical and pharmacological interventions for management of the disease process are insufficient and warrant alternative strategies to reduce the incidence of congestive heart failure. Heart transplantation is considered as the gold standard treatment modality the success of which is limited by lack of appropriate number of donors. Stem cell transplantation and therapeutic gene delivery have emerged as alternative potential treatment methods. Although the existence of resident cardiac stem cells (CSCs) in myocardial niches has been reported,[1,2] and a subpopulation of cardiomyocytes has been identified which may re-enter into cell cycle after myocardial infarction,[3] their capacity to compensate for massive cardiomyocyte loss and amelioration of the vicious effects of ventricular remodeling are limited. Nevertheless, their presence has brought a paradigm shift in the long-standing dogma about the heart being terminally differentiated organ which is unable to repair itself in the event of injury.

In heart cell therapy, the donor cells are engrafted in the infarcted heart to compensate for the functioning cardiomyocytes loss, induce angiogenesis, and stimulate matrix remodeling to preserve left ventricular contractile function.[4-6] Despite the initial promise of stem cell therapy for myocardial repair, the approach still faces formidable challenges ranging from choice of cells to methodological defects.[7] Beside other factors, prognostic outcome in terms of left ventricular function has shown a direct proportional relation with the total number of engrafted donor cells and their post-transplantation survival which is influenced by multiple factors in the ischemic heart.

In order to enhance the therapeutic efficacy of cellular cardiomyoplasty, multiple remedial measures have been adopted. A common perception is that pre-treatment of the host myocardium with angiogenic growth factor protein or gene overexpression prior to cell transplantation makes it more favorable for the engrafted cells to proliferate and integrate with the host cells.[8–10] Similarly, pharmacological (i.e. growth factors, cytokines) or physical (i.e. heat shock or ischemic pre-conditioning) treatment of cells prior to transplantation makes them resistant to tissue ischemia which may otherwise adversely affect cell survival and lower the consequent functional benefits of cellular cardiomyoplasty.[11,12] Alternatively, genetic modification of the cells represents a promising new avenue of investigation which is performed to achieve sustained expression of cytoprotective or angiogeneic recombinant proteins. The uniqueness of the combined cell and gene therapy approach is attributed to the fact that it facilitates an inexpensive, safe, rapid, and site specific transgene delivery *in vivo* which may in turn, intricately influence the effectiveness of cellular cardiomyoplasty. However, many problems and unresolved issues may need to be addressed in future research before an appropriate clinical application of the cell-based gene delivery is appreciated. This chapter will give an overview of the different types of cells and vector systems which have been used for *ex vivo* gene delivery to the infarcted heart, and will discuss both the limitations and innovations of the combined cell and gene therapy approach.

Stem Cells for Cellular Cardiomyoplasty

The use of stem cells for the treatment of heart muscle diseases remains the ultimate aspiration in cardiovascular therapeutics. The transplanted stem cells differentiate to adopt cardiac phenotype and if possible, restore regional blood flow in the ischemic myocardium by participation in angiogenesis. In order to achieve these goals, cells from diverse sources and with varying regenerative ability have been used (Table 1). Earlier results clearly favored the use of *embryonic, fetal and adult cardiomyocytes* for transplantation.[13–15] Based on their capability to engraft easily without fusion and electrically couple with the host myocardium, cardiomyocytes provide

Table 1. Cells used for heart cell therapy.

Cell type	Angiogenesis	Myogenesis	References
Bone marrow stem cells	+	+	4, 68, 75
Cardiac stem cells	+	+	2, 23, 24
Embryonic stem cells	+	+	20, 22, 79
Endothelial cells and EPC	+	−	27, 100
Cardiomyocytes	−	+	13–15
Fibroblasts	+	−	71, 72
Smooth muscle cells	−	+	101, 102
Skeletal myoblast	+	+	5, 26, 66
Stable cell lines	−	+	103, 104

an optimal cell type for heart cell therapy. However, difficulty in availability and the ethical issues involved, their use has remained restricted to the studies in experimental animal models. In order to overcome these issues, research is underway to derive cardiomyocytes from embryonic stem cells (ESCs).[16,17]

ESCs derived from the inner mass of blastocysts are pluripotent. They are capable of differentiation into various cell lineages.[18] Their cardiogenic differentiation is manifested by the expression of cardiac specific proteins and transcription factors. They beat spontaneously and rhythmically *in vitro* cell culture conditions.[19] Furthermore, the ESC-derived cardiomyocytes show action potential which is characteristically similar to atrial and ventricular cardiomyocytes.[20] More importantly, they can be propagated for multiple doublings *in vitro* without changing their phenotype.[21] These attributes advocate that the ESC-derived cardiomyocytes establish a renewable source of donor cardiomyocytes suitable for heart cell therapy. There is overwhelming evidence about their histological and electrophysiological integration with the host cardiomyocytes post-transplantation.[17] Despite these encouraging attributes, their purification from the undifferentiated ESCs in cell culture remains a challenge and is currently being performed by genetic markers.[22] The teratogenic potential of the ESC-derived cardiomyocytes is another concern in their clinical application. This may be overcome by ensuring that the entire population of hESCs is either differentiated or fully committed to well defined cardiac lineage prior to transplantation.

Resident CSCs have recently been identified and characterized in animal and human hearts.[23] Human CSCs express the stem cell antigens c-kit, MDR1 and Sca-1, either alone or in various combinations. The expression of a distinct stem cell epitope has a functional significance with respect to their growth, self-renewal, and differentiation potential. As CSCs are inherently programmed to create heart muscle and coronary vessels, their stimulation or transplantation for active participation in the cardiac repair process may rescue the failing heart.[24] Importantly, their use is without ethical issues and procedural complications. Moreover, CSC should be more effective in making new myocardium than stem/progenitor cells from other organs including the bone marrow.[25]

Skeletal muscle-derived myoblasts (SkMs) and *bone marrow-derived stem cells (BMSCs)* are the only cells which have been used in the clinical studies after extensive *in vitro* as well as *in vivo* characterization in experimental animal models.[26,27] SkMs are inherently myogenic and get mobilized in response to muscle injury. Likewise, there is compelling evidence that BMSCs get mobilized in response to tissue ischemia and show tropism for the ischemic heart.[28] Notwithstanding their important characteristics (Table 2), BMSCs are still under intense scrutiny for their potential to repair injured myocardium due to their plastic nature.[29] Whereas the major shortcoming of SkMs is their inability to electrically couple with the host cardiomyocytes,[30] the cardiomyogenic potential of BMSCs remains unsure.[31−33] Furthermore, the underlying mechanism of their cardiac function improvement is contentious. Various cell re-programming measures have been adopted to overcome

Table 2.   Characteristics of SkMs and BMSCs.

Characteristics	SkMs	BMSCs
Safety and feasibility for engraftment	+	+
Easy availability from autologous source	+	+
No ethical issues	+	+
More resistant to ischemia	+	−
Form functioning muscle	+	+
Adoption of cardiac phenotype	−	+
Develop gap junctions	−	+
Promote angiogenesis via paracrine effects	+	+

these deficiencies, one of these being genetic modulation of the cells prior to engraftment.[34–36]

Gene Therapy for Myocardial Repair

Gene therapy in the broader perspective is intended as a corrective measure which involves transgene/s delivery into the cell to supplement or replace a malfunctioning or attenuated gene resulting from a mutation or setting in of a disease process. The expression product of the transgene is intended to restore the normal function of the cells and helps to overcome the adverse effects resulting from the abnormal endogenous gene expression.[37,38] In other cases, transgene delivery is meant for silencing the endogenous gene involved in the disease process (i.e. inhibiting cell proliferation in the arterial wall).[39] Cytostatic and cytotoxic strategies have been employed to prevent cells from entering into the undesired mitotic cycle and to destroy the dividing cells.[40,41] More recently, gene therapy approach is being employed for controlled gene expression in stem cells to influence their fate.[42] One of the most commonly intended therapeutic effect from transgene delivery to the heart is the overexpression of pro-angiogenic growth factors to achieve angiogenesis which alleviates myocardial ischemia via enhanced regional blood flow.[43] Likewise, delivery of the transgenes encoding for survival factors achieved cardiac protection and anti-apoptotic effects.[44,45] Moreover, gene therapy targeting the underlying biological processes affecting the injured cardiomyocytes, including modification of cellular contractile signaling by the normal function of the β-adrenergic receptor cascade.[38,46] and regulatory pathways such as intracellular calcium signaling[37,47] present promising new objectives for biologic therapy of heart disease.

The success of an experimental or clinical gene therapy procedure is critically dependent on the efficiency of gene transfer and expression. By and large, gene delivery to the heart is achieved by using various gene delivery vectors to compensate for the poor penetrability of the naked DNA plasmid across the cell membrane which translates into a low level of transgene expression (Table 3). The pattern of transgene

Table 3. Vector systems used for transgene delivery into cells.

Naked plasmids

Physical vectors
    (a) Microinjection   (b) Particle bombardment "gene gun"
    (c) Electroporation   (d) Sonoporation
    (e) Laser irradiation

Non-viral vectors
    (a) Calcium phosphate   (b) DEAE dextran
    (c) Cationic polyplexes
         — Polylysin vectors
         — Polycation
    (d) Cationic lipoplexes
         — Liposomal
         — Non-liposomal
    (e) Cationic bioplexes

Viral vectors
    (a) Retrovirus   (b) Adenovirus
    (c) Adeno-associated virus   (d) Lentivirus
    (e) Other virus

Table 4. Factors effecting gene transfer into cells.

Type of the target cell or tissue
Type of the delivery vector
Presence and absence of serum proteins
Temperature and medium of reaction
Size and residual charge of the vector-transgene complex

expression varies with the choice of delivery vectors and its *in vivo* pharmaco-kinetics is influenced by various physical, chemical, and biological factors (Table 4). The currently available gene transfer vectors are less than ideal in their character-istics for clinical applications. Hence, there is a need to design a vector which may overcome these limitations to accomplish optimal gene transfer with minimum undesired effects (Table 5). Of the commonly used vectors, replication deficient viral vectors based on RNA and DNA viruses have been extensively used for thera-peutic gene delivery.[48] Particular viral gene delivery systems have been derived from retrovirus, adenovirus, adeno-associated virus, poxvirus and herpes virus. Despite their distinctive advantages, the risks associated with their use (Table 6) have forced the researchers to design alternative gene delivery strategies including non-viral vectors. Regardless of their low transfection efficiency and short term expression, the use of non-viral gene delivery system is safe and flexible in terms of the plasmid DNA size which can be delivered.[49] Some of the frequently used non-viral vectors for gene delivery to the heart include lipopolyplexes, liposomes, and polaxamine

Table 5.   Advantages and limitations of viral and non-viral vectors.

	Advantages	Limitations
Viral vectors	Higher transfection efficiency	Restricted size of gene delivery Immune response Lack of target specificity Problems in large scale production Safety concerns Probability of long term integration with the host genome Potentially tumorigenic
Non-viral vectors	Ease of manipulation and low cost Large scale production Flexible in gene size insertion Lack of safety concerns Non-immunogenic	Low transfection efficiency Short term expression

Table 6.   Desirable characteristics of an ideal vector.

Non-toxic and non-immunogenic
High efficiency of transgene delivery
Should not interfere with self-renewal and differentiation potential of stem cells
Regulatable expression of transgene
Ease of delivery and site-specific targeting
Reliable and optimal duration of gene expression
Free from replication competent infectious viral particles
High capacity gene insertion size
Stable, cost effective, and ease of production on large scale

nanospheres.[50–52] We have successfully designed a polyethyleneimine (PEI)-based nanoparticle system to deliver human VEGF$_{165}$ into human SkMs.[53] PEI has cationic nature and has strong DNA compaction capacity, effective DNA protection and with an intrinsic endosomolytic activity which contribute to superior gene transfection efficiency.[54] We achieved more than 11% transfection efficiency of SkMs with the peak level expression (25 ng/ml) on days 4–6 after transfection. At four weeks after transplantation into an experimental rat heart model of permanent coronary artery ligation, blood vessel density, and regional blood flow (ml/min/g) in the left ventricle were significantly higher in the transfected cell transplanted group of animals as compared with the controls. Our results clearly signify the safety and effectiveness of the nanoparticle-based non-viral vectors for gene delivery to the heart.

Combining Cellular Cardiomyoplasty with Gene Delivery

Until recently, cell therapy and gene therapy were considered as "rival approaches." The recent progress in both these potential therapeutic strategies has encouraged the researchers for their combined application. The cell-based therapeutic gene delivery is a safer alternative to direct injection of the viral vectors, use of which is not without biological consequence. Cell-mediated gene transfer is useful for sustained local protein delivery and provides a more efficient option as compared with the direct plasmid injection or non-viral vector delivery approach. The results of our recent study vividly support the feasibility and effectiveness of combining Ang-1 delivery with SkMs transplantation for myocardial angiogenesis and improvement of left ventricular contractile function.[55]

Cell-Based Gene Delivery to Alleviate Cellular Apoptosis

The effectiveness of stem cell therapy is flawed by the poor survival of the donor cells after engraftment. In the absence of a delineated molecular mechanism, apoptosis is considered to play a major role in the massive donor cell death which ensues during the acute phase after heart cell therapy.[56] This emphasizes the need for strategies to limit the factors such as mechanical stress, bouts of ischemia and ischemia-reperfusion, and more importantly the host inflammatory response mediators and pro-apoptotic factors in the ischemic heart to improve donor cell survival. Upregulated expression of heat shock proteins stabilizes various aspects of cell metabolism and function and can be cytoprotective.[57] Combining of cell transplantation with angiogenic growth factor gene delivery can improve the cell graft survival in the heart. A multimodal approach of bone marrow cell-based delivery of insulin-like growth factor (IGF) and VEGF transgenes resulted in reduced apoptosis of the transplanted cells.[58] Genetic modification of donor cells for overexpression of the cell survival signaling mediators significantly prevent their apoptosis after transplantation. Akt and its downstream mediator Bcl.xl have a central role in cell survival signaling,[59] their gene transfer exerted a powerful cardio-protection after transient ischemia, inhibited cardiomyocyte death and improved function of the surviving cardiomyocytes.[44,60] Transplantation of mesenchymal stem cells (MSCs) expressing Akt in a rat heart model restored four-fold greater myocardial volumes as compared with non-transduced MSCs.[45] These results were duplicated by Lim *et al.* in a porcine heart model of myocardial infarction.[60] Similar effects were achieved by adenoviral transduction of human Bcl2 transgene into cardiomyoblasts which attenuated their death during early phase after transplantation into ischemic rat hearts.[61] We have recently reported that co-overexpression of Ang-1 and Akt in MSCs (Fig. 1) produced powerful cytoprotection against oxygen and glucose deprivation *in vitro* and improved their survival *in vivo* after transplantation.[62] These cells also showed differentiation into myocytes and promoted neovascularization

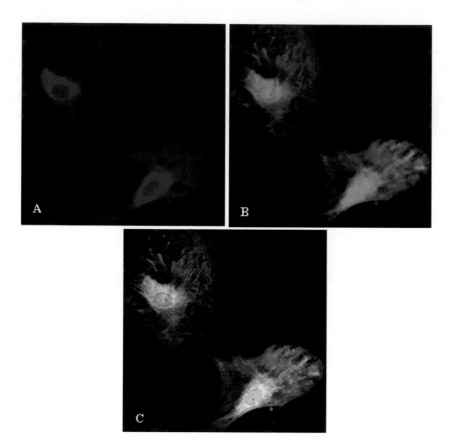

Fig. 1. Bone marrow-derived mesenchymal stem cells co-overexpressing Ang-1 (red fluorescence; **A**) and Akt (green fluorescence; **B**). The cells were transduced with adenoviral vectors encoding for Ang-1 and Akt. (**C**) represents merged image of (A) and (B).

(Fig. 2). In a recently concluded study, we have exploited pharmacological pre-conditioning approach to trigger cell survival signaling in SkMs.[63] We hypothesized that the powerful cytoprotective nature of pre-conditioning using pre-conditioning mimetics would enhance their resistance to ischemia-induced apoptosis. Real-time PCR for *sry*-gene four days after transplantation showed a significant 1.93-fold higher survival of SkMs in preconditioned SkMs transplanted animals as compared with non-preconditioned SkMs transplanted animals. Currently, we are assessing the versatility of this approach using BMSCs and decipher the underlying mechanism of cytoprotection.

A distinctive approach of cell manipulation has been adopted by Tara *et al.* to prevent their apoptotic death under nutrient and oxidant stress.[64,65] A super anti-apoptotic protein FNK was developed and used for transduction of the cells. The novel protein FNK was previously constructed from an anti-apoptotic factor Bcl-xl by site-directed mutagenesis and was transduced into the cells using protein transfer domain (PTD) of HIV-Tat protein. They have explicitly demonstrated better

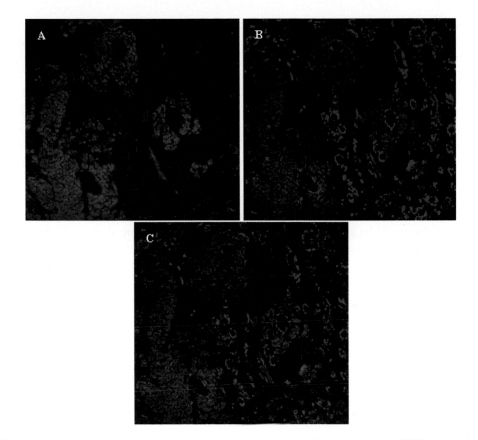

Fig. 2.   Fluorescent immunostaining of the rat heart tissue for the expression of vWillbrand Factor VIII (red fluorescence; **A**) at four weeks after mesenchymal stem cell transplantation co-overexpressing Ang-1 and Akt. The tissue architecture was revealed by immunostaining for actinin (blue fluorescence; **B**). Extensive angiogenesis was observed in the peri-infarct region (**C**; merged image of A and B).

survival of bone marrow mononuclear cells (BM-MNCs) treated with PTD-FNK under oxidant stress and serum starvation conditions *in vitro* and depicted higher engraftment efficiency in a rat hind-limb ischemia model.

Cell-Based Gene Delivery for Myogenesis and Electromechanical Integration

One of the expected consequences of cell-based therapeutic gene delivery to the heart is that the donor cells undergo myogenic differentiation. SkMs in this regard provide an automatic choice and has been extensively studied for *ex vivo* gene delivery in the experimental animal models.[55,66] Nevertheless, a major discredit of SkMs is their failure to integrate with host cardiomyocytes which hampers their

synchronous beating with the host myocardium. In order to overcome this deficiency, SkMs have been transduced to overexpress gap junction proteins, especially
connexin-43.[67] On the other hand, BMSCs provide a preferred alternative with the
ability to adopt cardiomyogenic phenotype. Their use for heart cell therapy has
accomplished encouraging results in terms of their myogenic differentiation and
integration with the host tissue.[45,62,68] A more elegant approach is to re-program
these cells by genetic modulation for directed differentiation to adopt cardiac phenotype. Myocardin is transcriptional regulator of cardiac and smooth muscle development which transactivates both smooth muscle and heart muscle specific genes.
In a recent study, primary human MSCs and fibroblasts were transduced with human
adenovirus vectors encoding for human myocardin gene.[69] Although the upregulated
expression of myocardin alone strongly induced both smooth and cardiac muscle
genes, there might be a need for additional factors to fully commit the cells to either
cardiac or smooth muscle fates. An identical approach has previously been adopted
for fibroblasts which are non-myogenic cells and play a key role in growth factor
secretion, matrix deposition and matrix degradation, and therefore are important in
many pathologic processes. Gene transfer of muscle-specific MyoD family of transcription factors into fibroblasts induced their myogenic differentiation.[70] A recent
study has shown that MyoD transduced cardiac fibroblasts underwent myogenic differentiation post-engraftment in an infarcted heart.[71] These cells expressed myosin
heavy chain (fast skeletal isoform) in the rat heart, with very few of them staining
positively for randomly distributed connexin-43. However, forced expression of
MyoD did not change their electrically inert nature and prevented their excitability
and electrical coupling with the host tissue. In order to overcome this deficiency,
fibroblasts were transduced with third-generation lentivirus vectors for simultaneous
overexpression of MyoD and connexin-43.[72] The genetically modified fibroblasts
may provide an alternative source for heart cell therapy in future.

Cell-Based Gene Delivery for Angiogenesis

Angiogenesis involves sprouting and growth of capillaries from the pre-existing
blood vessels and forms an essential component of various physiological as well as
pathological conditions that occur during embryonic development and throughout
postnatal life. The cascade of events during angiogenesis maintains a delicate
balance between pro- and anti-angiogenic factors. In the patients suffering from
coronary artery disease, angiogenesis is induced as a natural response to regional
myocardial ischemia. Indeed, copious amount of angiogenic growth factors are produced in the ischemic heart but the expression level slowly tapers off over a short
period of time. Therefore, the angiogenic response in the ischemic heart is invariably
insufficient to compensate for the deteriorating blood supply. These findings clearly
suggest that there may be a natural feedback mechanism which maintains a delicate balance between the pro- and anti-angiogenic factors. Impaired angiogenic

response has also been observed in certain physiological as well as pathological conditions like aging, diabetes, and hypercholestremia. In such cases, supplementation of proper angiogenic factor is beneficial and may potentially augment the natural angiogenic cascade leading to the symptomatic relief.

Stem cell-derived neovascularization

One of the suggested mechanisms of functional improvement after heart cell therapy is that the donor cells help the healing process through angiogcnesis by adoption of endothelial phenotype.[73] They also release paracrine factors which promote donor cell survival and induce neovascularization.[74,75] Different cells vary in their angiogenic potential.[6,76–79] For example, peripheral blood mononuclear cells (PBMNCs) and BMMNCs show similar levels of VEGF and basic fibroblast growth factor, whereas interleukin-1β and tumor necrosis factor-α were higher in PBMNCs. The results from human clinical trials have shown significantly improved regional perfusion of left ventricle due to the paracrine factors released from stem cell after transplantation.[80,81] Despite these encouraging results, there are issues in relation to cellular angiogenesis which need to be addressed. The choice and the source of donor cells are particularly important factors, especially in the old patients. The pathophysiological conditions and the age-related impairment of angiogcnesis are associated with altered expression of VEGF, VEGF receptors (Flt-1 and Flk-1) as well as angiopoietin-1 (Ang-1) and angiopoietin-2 (Ang-2).[82] Therefore, a more efficient approach may be to combine stem cell transplantation with gene therapy.

Angiogenesis Using Genetically Engineered Cells

SkMs-mediated angiogenic gene delivery

Primary SkMs were the first to be used as a platform for the transgene delivery encoding for angiogenic growth factors in the ischemic heart. As early as 1995, Koh *et al.* genetically modified SkMs for the delivery of recombinant transforming growth factor-β (rTGF-β) to the heart.[83] C2C12 SkMs were transfected with a construct consisting of an inducible metallothionein promoter fused to a modified TGF-β cDNA. After intra-cardiac delivery into syngenic C3Heb/FeJ hosts, the genetically modified cells produced viable grafts as long as three months after implantation. Regions of apparent neovascularization were observed in the bordering areas of the cell graft overexpressing TGF-β. Although the study failed to provide any evidence of improved heart function, the results clearly depicted SkMs as the potential transgene carriers for recombinant molecules to the heart. In a later study, SkMs were transfected with human VEGF$_{165}$ gene using liposome complexed with hemagglutinatinin virus of Japan.[84] At a transduction efficiency of >95%, 25-fold increase in VEGF expression was observed in the culture medium of VEGF-transfected SkMs as compared with the control non-transfected SkMs. Injection of these cells into a rat model of acute myocardial infarction gave increased myocardial VEGF level and a resultant increase in angiogenesis. Infarct size was significantly reduced and both systolic and diastolic functions were best preserved after VEGF transduced SkMs transplantation as compared with the control groups. Askari *et al.* reported that *ex vivo* delivery of VEGF was more efficient and gave improved cardiac function as compared with the non-transfected cells.[85] Most of these studies however used SkMs derived from animals. Unlike these studies, our research group has studied the potential of human SkMs to deliver VEGF$_{165}$ into porcine infarcted heart.[66] The cells were transduced with replication deficient adenoviral vector carrying VEGF$_{165}$ gene. Enzyme-linked immunosorbent assay showed that the level of secreted VEGF$_{165}$ in the culture medium of VEGF-transfected SkMs was significantly higher as compared with the null-adenovirus transfected SkMs. The transfected cells actively secreted biologically active VEGF$_{165}$ for 18 days with a peak level expression (37 ng/ml) on day 7. Transplantation of these genetically modified cells increased the myocardial vascular density as well as improved pig heart function. Our results showed the cardioprotective effects of VEGF-induced vasodilatation in the early phase after transplantation, followed by angiogenesis which salvaged the host myocardium with functional benefits from the donor cell-derived neomyogenesis.[53] Bearing in mind the complexity of angiogenic cascade, it is obvious that a single growth factor (i.e. VEGF) will be less efficient to complete the biological process of angiogenesis alone. Therefore, the multiple growth factor delivery approach may be more pertinent in creating a sustainable angiogenic effect with clinically relevant outcome. The challenge is to find complementary partner for VEGF in angiogenesis. The angiopoietin family of growth factors and their receptors are critical regulators of vascular structure maturity[86] and have an indispensable role in coronary artery development.[87] Hence, we designed a novel bicistronic vector encoding for VEGF and Ang-1.[88] Due to their unique complementary properties, we observed that the combination not only enhanced angiogenesis but also gave healthy and mature vascular network formation in the ischemic heart.

Cardiomyocyte-mediated gene delivery

Some of the untransfected donor cell types including cardiomyocytes induce a limited degree of angiogenesis after transplantation. Gene modification of cardiomyocytes to overexpress angiogenic growth factors is expected to enhance their angiogenic potential post-transplantation. However, adult cardiomyocytes are quite resistant to gene transfer. Both viral as well as non-viral gene delivery strategies have been developed and optimized for genetic modification of cardiomyocytes.[89] Transfection efficiency to the level of 25%–30% has been achieved in cardiomyocytes with lipid vector-VEGF$_{165}$ plasmid DNA complex. Intracellular VEGF expression was 6.1-fold greater in transfected cells as compared with the untransfected cells and the concentrations of VEGF secreted into the supernatant was 3.8-fold greater in transfected cells.[90] In another study, transduction of cardiomyocytes with the IGF carrying adenoviral vector resulted in a dose- and time-dependent increase in IGF-I protein expression which significantly abrogated cell apoptosis induced by ischemia-reoxygenation, ceramide, or heat shock.[91] These cardiomyocytes were later used in a Matrigel plug model of *in vivo* angiogenesis in nude mice. They provided a sustained upregulated expression of IGF-1 which produced enhanced angiogenesis and prevented the host cardiomyocytes apoptosis. Similar observations were reported with VEGF transfected heart cells and SkMs after transplantation in the ischemic heart in comparison with their non-transfected counterparts.[92] The transfected cells showed significantly improved survival and also enhanced angiogenesis at the site of the cells graft.

Bone marrow cell-based gene delivery

As discussed earlier, bone marrow-derived cells have distinct advantages of easy availability, high *in vitro* proliferative and self-renewal capability and multilineage differentiation potential. Moreover, they can also secrete a broad spectrum of angiogenic cytokines. Therefore, it may be reasonable to expect that bone marrow cells genetically modified for augmented expression of angiogenic growth factors may be beneficial for the treatment of ischemic heart.[93,94] In a series of experiments, Yau *et al.* sought to augment the angiogenic and myogenic effects of bone marrow cell transplantation by simultaneous overexpression of VEGF and IGF-I, which also induced myocytes hypertrophy and inhibited apoptosis.[58] They also observed greater restoration of left ventricular function due to synergistic interaction between overly expressed transgenes. We hypothesized that the consequence of intramyocardial delivery of genetically modified bone marrow cells overexpressing growth factors and cytokines might be amplified by increasing the supply of circulating progenitor cells via pharmacological mobilization of bone marrow cells.[95] We observed that granulocyte colony stimulating factor induced mobilization of stem cells from their bone marrow niches and increased their number in the peripheral circulation. In order to provide homing signal for the blood borne stem cells into the infarcted heart, we intramyocardially transplanted MSCs overexpressing VEGF. Our results showed significantly higher angiogenic response and improvement in cardiac function in the experimental animal group as compared with the control group.

Future Directions

Both cell and gene therapy approaches have entered into clinical phase with their respective therapeutic advantages. The strategy of combining gene delivery with cell transplantation using genetically engineered cells is advantageous in terms of localized and regulatable delivery of the gene of interest with significantly reduced incidence of untoward effects. However, different aspects of cell-based gene delivery need to be optimized and studied in-depth for optimum benefits. The choice of the cell type is domineering for *ex vivo* delivery of the therapeutic genes to the heart. Besides reaping the beneficial effects of *ex vivo* gene delivery such as sustained and localized overexpression of the transgene product, the choice of the cell type will be influenced by the desired outcome of the cellular cardiomyoplasty procedure. For example, if improvement of the systolic function is the ultimate goal, bone marrow cells should better be used as the carriers of the transgene as they can transdifferentiate to adopt cardiac phenotype. However, if improvement of the diastolic function is the prime consideration, SkMs may be the best choice to serve as the carriers of the transgene as their myogenic differentiation would exert scaffolding effect and prevent left ventricular dilatation. ESC-derived cardiomyocytes may provide an effective alternative as they will better integrate with the host myocardium as compared with the other cells, thus alleviating the problem of arrhythmia.

Likewise, further investigative work needs to be carried out on the basic issues such as the optimal number of cells, the route of administration and time of injection after infarction episode. The purification of donor cells for each patient and their propagation *in vitro* to get the required number is time consuming and expensive. Off-the-shelf availability of stem cells for genetic modification will aid in overcoming the logistic concerns as well as make the whole procedure cost effective and more efficient. The source of the donor cells (autologous versus allogenic or xenogenic) will be of significance in this regard. Indeed, transduction of autologous cells can also modify their immunogenicity thus prompting concerns about their acceptance by the host immune system. Cell encapsulation technology may provide an answer to this problem. Encapsulation of the genetically modified cells within a microporous polymer membrane will isolate the transplanted tissue from the host and allow the eventual retrieval of engineered cells.[96] Immunoisolation of the donor cells in this manner improves their protection from cell contact-mediated rejection by the host's immune system.

Transfer of therapeutic gene to stem cells may be inefficient due to cell quiescence or because of the insufficient viral receptor density. A number of investigators have attempted to alter viral vectors to allow their entrance into target cells via more specific or more abundant receptors. A safer option may be to implicate nanotechnology for gene transfection into stem cells.[53,97] This will ensure more successful gene transfer option into the cells without compromising their immune status. Another key consideration in cellular gene modification, especially in the case of stem cells, is that transgene transfer should be without interference with their differentiation potential.[98] The strategy of peptide-based gene transfer into stem cells provides another efficient alternative to the viral vectors.[99] We, therefore, contemplate that the use of gene therapy, cell transplantation, and hematogenous cytokine administration will benefit increasing number of patients with ischemic heart disease. However, long term studies monitoring side effects to establish safety and effectiveness are required.

References

1. Urbanek K, Cesselli D, Rota M, Nascimbene A, De Angelis A, Hosoda T, Bearzi C, Boni A, Bolli R, Kajstura J, Anversa P, Leri A. Stem cell niches in the adult mouse heart. *Proc Natl Acad Sci USA* 2006;103:9226–9231.
2. Leri A, Kajstura J, Anversa P. Cardiac stem cells and mechanisms of myocardial regeneration. *Physiol Rev* 2005;85:1373–1416.
3. Anversa P, Kajstura J, Leri A, Bolli R. Life and death of cardiac stem cells: a paradigm shift in cardiac biology. *Circulation* 2006;113:1451–1463.
4. Orlic D, Kajstura J, Chimenti S, Jakoniuk I, Anderson SM, Li B, Pickel J, McKay R, Nadal-Ginard B, Bodine DM, Leri A, Anversa P. Bone marrow cells regenerate infarcted myocardium. *Nature* 2001;410:701–705.
5. Taylor DA, Atkins BZ, Hungspreugs P, Jones TR, Reedy MC, Hutcheson KA, Glower DD, Kraus WE. Regenerating functional myocardium: improved performance after skeletal myoblast transplantation. *Nat Med* 1998;4:929–933.
6. Jackson KA, Majka SM, Wang H, Pocius J, Hartley CJ, Majesky MW, Entman ML, Michael LH, Hirschi KK, Goodell MA. Regeneration of ischemic cardiac muscle and vascular endothelium by adult stem cells. *J Clin Invest* 2001;107:1395–1402.
7. Anversa P, Leri A, Rota M, Bearzi C, Urbanek K, Kajstura J, Bolli R. Concise review: stem cells, myocardial regeneration and methodological artifacts. *Stem Cells* 2007;25:589–601.
8. Retuerto MA, Schalch P, Patejunas G, Carbray J, Liu N, Esser K, Crystal RG, Rosengart TK. Angiogenic pretreatment improves the efficacy of cellular cardiomyoplasty performed with fetal cardiomyocyte implantation. *J Thorac Cardiovasc Surg* 2004;127:1041–1051.
9. Bartunek J, Croissant JD, Wijns W, Gofflot S, de Lavareille A, Vanderheyden M, Kaluzhny Y, Mazouz N, Wlllemsen P, Penicka M, Mathieu M, Homsy C, De Bruyne B, McEntee K, Lee IW, Heyndrickx GR. Pretreatment of adult bone marrow mesenchymal stem cells with cardiomyogenic growth factors and repair of the chronically infarcted myocardium. *Am J Physiol Heart Circ Physiol* 2007;292:H1095–H1104.
10. Shintani S, Kusano K, Ii M, Iwakura A, Heyd L, Curry C, Wecker A, Gavin M, Ma H, Kearney M, Silver M, Thorne T, Murohara T, Losordo DW. Synergistic effect of combined intramyocardial CD34+ cells and VEGF2 gene therapy after MI. *Nat Clin Pract Cardiovasc Med* 2006;3 (Suppl 1):S123–S128.
11. Sakakibara Y, Nishimura K, Tambara K, Yamamoto M, Lu F, Tabata Y, Komeda M. Prevascularization with gelatin microspheres containing basic fibroblast growth factor enhances the benefits of cardiomyocyte transplantation. *J Thorac Cardiovasc Surg* 2002;124:50–56.
12. Muller-Ehmsen J, Whittaker P, Kloner RA, Dow JS, Sakoda T, Long TI, Laird PW, Kedes L. Survival and development of neonatal rat cardiomyocytes transplanted into adult myocardium. *J Mol Cell Cardiol* 2002;34:107–116.
13. Reinecke H, Zhang M, Bartosek T, Murry CE. Survival, integration, and differentiation of cardiomyocyte grafts: a study in normal and injured rat hearts. *Circulation* 1999;100:193–202.
14. Li RK, Weisel RD, Mickle DA, Jia ZQ, Kim EJ, Sakai T, Tomita S, Schwartz L, Iwanochko M, Husain M, Cusimano RJ, Burns RJ, Yau TM. Autologous porcine heart cell transplantation improved heart function after a myocardial infarction. *J Thorac Cardiovasc Surg* 2000;119: 62–68.
15. Roell W, Lu ZJ, Bloch W, Siedner S, Tiemann K, Xia Y, Stoecker E, Fleischmann M, Bohlen H, Stehle R, Kolossov E, Brem G, Addicks K, Pfitzer G, Welz A, Hescheler J, Fleischmann BK. Cellular cardiomyoplasty improves survival after myocardial injury. *Circulation* 2002;105: 2435–2441.
16. Wei H, Juhasz O, Li J, Tarasova YS, Boheler KR. Embryonic stem cells and cardiomyocyte differentiation: phenotypic and molecular analyses. *J Cell Mol Med* 2005;9:804–817.

17. Johkura K, Cui L, Suzuki A, Teng R, Kamiyoshi A, Okamura S, Kubota S, Zhao X, Asanuma K, Okouchi Y, Ogiwara N, Tagawa Y, Sasaki K. Survival and function of mouse embryonic stem cell-derived cardiomyocytes in ectopic transplants. *Cardiovasc Res* 2003;58:435–443.
18. Srivastava D, Ivey KN. Potential of stem-cell-based therapies for heart disease. *Nature* 2006;441:1097–1099.
19. Xu C, Police S, Rao N, Carpenter MK. Characterization and enrichment of cardiomyocytes derived from human embryonic stem cells. *Circ Res* 2002;91:501–508.
20. Kehat I, Kenyagin-Karsenti D, Snir M, Segev H, Amit M, Gepstein A, Livne E, Binah O, Itskovitz-Eldor J, Gepstein L. Human embryonic stem cells can differentiate into myocytes with structural and functional properties of cardiomyocytes. *J Clin Invest* 2001;108:407–414.
21. Amit M, Carpenter MK, Inokuma MS, Chiu CP, Harris CP, Waknitz MA, Itskovitz-Eldor J, Thomson JA. Clonally derived human embryonic stem cell lines maintain pluripotency and proliferative potential for prolonged periods of culture. *Dev Biol* 2000;227:271–278.
22. Kolossov E, Bostani T, Roell W, Breitbach M, Pillekamp F, Nygren JM, Sasse P, Rubenchik O, Fries JW, Wenzel D, Geisen C, Xia Y, Lu Z, Duan Y, Kettenhofen R, Jovinge S, Bloch W, Bohlen H, Welz A, Hescheler J, Jacobsen SE, Fleischmann BK. Engraftment of engineered ES cell-derived cardiomyocytes but not BM cells restores contractile function to the infarcted myocardium. *J Exp Med* 2006;203:2315–2327.
23. Beltrami AP, Barlucchi L, Torella D, Baker M, Limana F, Chimenti S, Kasahara H, Rota M, Musso E, Urbanek K, Leri A, Kajstura J, Nadal-Ginard B, Anversa P. Adult cardiac stem cells are multipotent and support myocardial regeneration. *Cell* 2003;114:763–776.
24. Dawn B, Stein AB, Urbanek K, Rota M, Whang B, Rastaldo R, Torella D, Tang XL, Rezazadeh A, Kajstura J, Leri A, Hunt G, Varma J, Prabhu SD, Anversa P, Bolli R. Cardiac stem cells delivered intravascularly traverse the vessel barrier, regenerate infarcted myocardium, and improve cardiac function. *Proc Natl Acad Sci USA* 2005;102:3766–3771.
25. Linke A, Muller P, Nurzynska D, Casarsa C, Torella D, Nascimbene A, Castaldo C, Cascapera S, Bohm M, Quaini F, Urbanek K, Leri A, Hintze TH, Kajstura J, Anversa P. Stem cells in the dog heart are self-renewing, clonogenic, and multipotent and regenerate infarcted myocardium, improving cardiac function. *Proc Natl Acad Sci USA* 2005;102:8966–8971.
26. Dib N, Michler RE, Pagani FD, Wright S, Kereiakes DJ, Lengerich R, Binkley P, Buchele D, Anand I, Swingen C, Di Carli MF, Thomas JD, Jaber WA, Opie SR, Campbell A, McCarthy P, Yeager M, Dilsizian V, Griffith BP, Korn R, Kreuger SK, Ghazoul M, MacLellan WR, Fonarow G, Eisen HJ, Dinsmore J, Diethrich E. Safety and feasibility of autologous myoblast transplantation in patients with ischemic cardiomyopathy: four-year follow-up. *Circulation* 2005;112:1748–1755.
27. Bartunek J, Vanderheyden M, Vandekerckhove B, Mansour S, De Bruyne B, De Bondt P, Van Haute I, Lootens N, Heyndrickx G, Wijns W. Intracoronary injection of CD133-positive enriched bone marrow progenitor cells promotes cardiac recovery after recent myocardial infarction: feasibility and safety. *Circulation* 2005;112:I178–I183.
28. Wang Y, Haider H, Ahmad N, Zhang D, Ashraf M. Evidence for ischemia induced host-derived bone marrow cell mobilization into cardiac allografts. *J Mol Cell Cardiol* 2006;41:478–487.
29. Goodell MA, Jackson KA, Majka SM, Mi T, Wang H, Pocius J, Hartley CJ, Majesky MW, Entman ML, Michael LH, Hirschi KK. Stem cell plasticity in muscle and bone marrow. *Ann N Y Acad Sci* 2001;938:208–220.
30. Reinecke H, MacDonald GH, Hauschka SD, Murry CE. Electromechanical coupling between skeletal and cardiac muscle. Implications for infarct repair. *J Cell Biol* 2000;149:731–740.
31. Balsam LB, Wagers AJ, Christensen JL, Kofidis T, Weissman IL, Robbins RC. Haematopoietic stem cells adopt mature haematopoietic fates in ischaemic myocardium. *Nature* 2004;428:668–673.

32. Murry CE, Soonpaa MH, Reinecke H, Nakajima H, Nakajima HO, Rubart M, Pasumarthi KB, Virag JI, Bartelmez SH, Poppa V, Bradford G, Dowell JD, Williams DA, Field LJ. Haematopoietic stem cells do not transdifferentiate into cardiac myocytes in myocardial infarcts. *Nature* 2004;428:664–668.

33. Nygren JM, Jovinge S, Breitbach M, Sawen P, Roll W, Hescheler J, Taneera J, Fleischmann BK, Jacobsen SE. Bone marrow-derived hematopoietic cells generate cardiomyocytes at a low frequency through cell fusion, but not transdifferentiation. *Nat Med* 2004;10:494–501.

34. Abraham MR, Henrikson CA, Tung L, Chang MG, Aon M, Xue T, Li RA, O'Rourke B, Marbán E. Antiarrhythmic engineering of skeletal myoblasts for cardiac transplantation. *Circ Res* 2005;97:159–167.

35. Stagg MA, Coppen SR, Suzuki K, Varela-Carver A, Lee J, Brand NJ, Fukushima S, Yacoub MH, Terracciano CM. Evaluation of frequency, type, and function of gap junctions between skeletal myoblasts overexpressing connexin-43 and cardiomyocytes: relevance to cell transplantation. *FASEB J* 2006;20:744–746.

36. Li TS, Hayashi M, Ito H, Furutani A, Murata T, Matsuzaki M, Hamano K. Regeneration of infarcted myocardium by intramyocardial implantation of *ex vivo* transforming growth factor-beta-preprogrammed bone marrow stem cells. *Circulation* 2005;111:2438–2445.

37. Logeart D, Vinet L, Ragot T, Heimburger M, Louedec L, Michel JB, Escoubet B, Mercadier JJ. Percutaneous intracoronary delivery of SERCA gene increases myocardial function: a tissue Doppler imaging echocardiographic study. *Am J Physiol Heart Circ Physiol* 2006;291:H1773–H1779.

38. del Monte F, Harding SE, Dec GW, Gwathmey JK, Hajjar RJ. Targeting phospholamban by gene transfer in human heart failure. *Circulation* 2002;105:904–907.

39. Sugano M, Tsuchida K, Hata T, Makino N. RNA interference targeting SHP-1 attenuates myocardial infarction in rats. *FASEB J* 2005;19:2054–2056.

40. Harrell RL, Rajanayagam S, Doanes AM, Guzman RJ, Hirschowitz EA, Crystal RG, Epstein SE, Finkel T. Inhibition of vascular smooth muscle cell proliferation and neointimal accumulation by adenovirus-mediated gene transfer of cytosine deaminase. *Circulation* 1997;96:621–627.

41. Chang MW, Ohno T, Gordon D, Lu MM, Nabel GJ, Nabel EG, Leiden JM. Adenovirus-mediated transfer of the herpes simplex virus thymidine kinase gene inhibits vascular smooth muscle cell proliferation and neointima formation following balloon angioplasty of the rat carotid artery. *Mol Med* 1995;1:172–181.

42. Hough SR, Clements I, Welch PJ, Wiederholt KA. Differentiation of mouse embryonic stem cells after RNA interference-mediated silencing of OCT4 and Nanog. *Stem Cells* 2006;24:1467–1475.

43. Yla-Herttuala S, Martin JF. Cardiovascular gene therapy. *Lancet* 2000;355:213–222.

44. Huang J, Ito Y, Morikawa M, Uchida H, Kobune M, Sasaki K, Abe T, Hamada H. Bcl-xL gene transfer protects the heart against ischemia/reperfusion injury. *Biochem Biophys Res Commun* 2003;311:64–70.

45. Mangi AA, Noiseux N, Kong D, He H, Rezvani M, Ingwall JS, Dzau VJ. Mesenchymal stem cells modified with Akt prevent remodeling and restore performance of infarcted hearts. *Nat Med* 2003;9:1195–1201.

46. Tevaearai HT, Walton GB, Keys JR, Koch WJ, Eckhart AD. Acute ischemic cardiac dysfunction is attenuated via gene transfer of a peptide inhibitor of the beta-adrenergic receptor kinase (betaARK1). *J Gene Med* 2005;7:1172–1177.

47. Ennis IL, Li RA, Murphy AM, Marban E, Nuss HB. Dual gene therapy with SERCA1 and Kir2.1 abbreviates excitation without suppressing contractility. *J Clin Invest* 2002;109:393–400.

48. Young LS, Searle PF, Onion D, Mautner V. Viral gene therapy strategies: from basic science to clinical application. *J Pathol* 2006;208:299–318.

49. Müller OJ, Katus HA, Bekeredjian R. Targeting the heart with gene therapy-optimized gene delivery methods. *Cardiovasc Res* 2007;73:453–462.
50. Bull DA, Bailey SH, Rentz JJ, Zebrack JS, Lee M, Litwin SE, Kim SW. Effect of Terplex/VEGF-165 gene therapy on left ventricular function and structure following myocardial infarction. VEGF gene therapy for myocardial infarction. *J Control Release* 2003;93:175–181.
51. Miyagawa S, Sawa Y, Taketani S, Kawaguchi N, Nakamura T, Matsuura N, Matsuda H. Myocardial regeneration therapy for heart failure: hepatocyte growth factor enhances the effect of cellular cardiomyoplasty. *Circulation* 2002;105:2556–2561.
52. Fukuyama N, Tanaka E, Tabata Y, Fujikura H, Hagihara M, Sakamoto H, Ando K, Nakazawa H, Mori H. Intravenous injection of phagocytes transfected *ex vivo* with FGF4 DNA/biodegradable gelatin complex promotes angiogenesis in a rat myocardial ischemia/reperfusion injury model. *Basic Res Cardiol* 2007;102:209–216.
53. Lei Y, Haider HKh, Shujia J, Tan R, Toh WC, Law PK, Tan WB, Su LP, Zhang W, Ge R, Sim EKW. Transplantation of nanoparticle based skeletal myoblasts overexpressing VEGF165 for cardiac repair. *Circulation* 2006;114:II389.
54. Lungwitz U, Breunig M, Blunk T, Gopferich A. Polyethylenimine-based non-viral gene delivery systems. *Eur J Pharm Biopharm* 2005;60:247–266.
55. Ye L, Haider HK, Jiang S, Tan RS, Toh WC, Ge R, Sim EK. Angiopoietin-1 for myocardial angiogenesis: a comparison between delivery strategies. *Eur J Heart Fail* 2007;9:458–465.
56. Suzuki K, Murtuza B, Beauchamp JR, Smolenski RT, Varela-Carver A, Fukushima S, Coppen SR, Partridge TA, Yacoub MH. Dynamics and mediators of acute graft attrition after myoblast transplantation to the heart. *FASEB J* 2004;18:1153–1155.
57. Kim HP, Morse D, Choi AM. Heat-shock proteins: new keys to the development of cytoprotective therapies. *Expert Opin Ther Targets* 2006;10:759–769.
58. Yau TM, Kim C, Li G, Zhang Y, Weisel RD, Li RK. Maximizing ventricular function with multimodal cell-based gene therapy. *Circulation* 2005;112:I123–I128.
59. Walsh K. Akt signaling and growth of the heart. *Circulation* 2006;113:2032–2034.
60. Lim SY, Kim YS, Ahn Y, Jeong MH, Hong MH, Joo SY, Nam KI, Cho JG, Kang PM, Park JC. The effects of mesenchymal stem cells transduced with Akt in a porcine myocardial infarction model. *Cardiovasc Res* 2006;70:530–542.
61. Kutschka I, Kofidis T, Chen IY, von Degenfeld G, Zwierzchoniewska M, Hoyt G, Arai T, Lebl DR, Hendry SL, Sheikh AY, Cooke DT, Connolly A, Blau HM, Gambhir SS, Robbins RC. Adenoviral human BCL-2 transgene expression attenuates early donor cell death after cardiomyoblast transplantation into ischemic rat hearts. *Circulation* 2006;114:I174–I180.
62. Jiang S, Haider H, Idris NM, Salim A, Ashraf M. Supportive interaction between cell survival signaling and angiocompetent factors enhances donor cell survival and promotes angiomyogenesis for cardiac repair. *Circ Res* 2006;99:776–784.
63. Idris NM, Haider H, Jiang S, Ashraf M. Pharmacologically preconditioned skeletal myoblasts are resistant to oxidative stress and promote angiomyogenesis via release of paracrine factors in the infarcted heart. *Circ Res* 2007;100:545–555.
64. Tara S, Miyamoto M, Asoh S, Ishii N, Yasutake M, Takagi G, Takano T, Ohta S. Transduction of the anti-apoptotic PTD-FNK protein improves the efficiency of transplantation of bone marrow mononuclear cells. *J Mol Cell Cardiol* 2007;42:489–497.
65. Haider H, Ashraf M. Developing "Super cells" by modification with super anti-apoptotic factor. *J Mol Cell Cardiol* 2007;42:478–480.
66. Haider H, Ye L, Jiang S, Ge R, Law PK, Chua T, Wong P, Sim EK. Angiomyogenesis for cardiac repair using human myoblasts as carriers of human vascular endothelial growth factor. *J Mol Med* 2004;82:539–549.

67. Suzuki K, Brand NJ, Allen S, Khan MA, Farrell AO, Murtuza B, Oakley RE, Yacoub MH. Overexpression of connexin 43 in skeletal myoblasts: relevance to cell transplantation to the heart. *J Thorac Cardiovasc Surg* 2001;122:759–766.

68. Tomita S, Li RK, Weisel RD, Mickle DA, Kim EJ, Sakai T, Jia ZQ. Autologous transplantation of bone marrow cells improves damaged heart function. *Circulation* 1999;100:II247–II256.

69. van Tuyn J, Knaan-Shanzer S, van de Watering MJ, de Graaf M, van der Laarse A, Schalij MJ, van der Wall EE, de Vries AA, Atsma DE. Activation of cardiac and smooth muscle-specific genes in primary human cells after forced expression of human myocardin. *Cardiovasc Res* 2005;67:245–255.

70. Lattanzi L, Salvatori G, Coletta M, Sonnino C, Cusella De Angelis MG, Gioglio L, Murry CE, Kelly R, Ferrari G, Molinaro M, Crescenzi M, Mavilio F, Cossu G. High efficiency myogenic conversion of human fibroblasts by adenoviral vector-mediated MyoD gene transfer. An alternative strategy for *ex vivo* gene therapy of primary myopathies. *J Clin Invest* 1998;101:2119–2128.

71. Etzion S, Barbash IM, Feinberg MS, Zarin P, Miller L, Guetta E, Holbova R, Kloner RA, Kedes LH, Leor J. Cellular cardiomyoplasty of cardiac fibroblasts by adenoviral delivery of MyoD *ex vivo*: an unlimited source of cells for myocardial repair. *Circulation* 2002;106:I125–I130.

72. Kizana E, Ginn SL, Allen DG, Ross DL, Alexander IE. Fibroblasts can be genetically modified to produce excitable cells capable of electrical coupling. *Circulation* 2005;111:394–398.

73. Tomita S, Mickle DA, Weisel RD, Jia ZQ, Tumiati LC, Allidina Y, Liu P, Li RK. Improved heart function with myogenesis and angiogenesis after autologous porcine bone marrow stromal cell transplantation. *J Thorac Cardiovasc Surg* 2002;123:1132–1140.

74. Uemura R, Xu M, Ahmad N, Ashraf M. Bone marrow stem cells prevent left ventricular remodeling of ischemic heart through paracrine signaling. *Circ Res* 2006;98:1414–1421.

75. Tang J, Xie Q, Pan G, Wang J, Wang M. Mesenchymal stem cells participate in angiogenesis and improve heart function in rat model of myocardial ischemia with reperfusion. *Eur J Cardiothorac Surg* 2006;30:353–361.

76. Kim EJ, Li RK, Weisel RD, Mickle DA, Jia ZQ, Tomita S, Sakai T, Yau TM. Angiogenesis by endothelial cell transplantation. *J Thorac Cardiovasc Surg* 2001;122:963–971.

77. Fuchs S, Baffour R, Zhou YF, Shou M, Pierre A, Tio FO, Weissman NJ, Leon MB, Epstein SE, Kornowski R. Transendocardial delivery of autologous bone marrow enhances collateral perfusion and regional function in pigs with chronic experimental myocardial ischemia. *J Am Coll Cardiol* 2001;37:1726–1732.

78. Kamihata H, Matsubara H, Nishiue T, Fujiyama S, Amano K, Iba O, Imada T, Iwasaka T. Improvement of collateral perfusion and regional function by implantation of peripheral blood mononuclear cells into ischemic hibernating myocardium. *Arterioscler Thromb Vasc Biol* 2002;22:1804–1810.

79. Levenberg S, Golub JS, Amit M, Itskovitz-Eldor J, Langer R. Endothelial cells derived from human embryonic stem cells. *Proc Natl Acad Sci USA* 2002;99:4391–4396.

80. Dohmann HF, Perin EC, Takiya CM, Silva GV, Silva SA, Sousa AL, Mesquita CT, Rossi MI, Pascarelli BM, Assis IM, Dutra HS, Assad JA, Castello-Branco RV, Drummond C, Dohmann HJ, Willerson JT, Borojevic R. Transendocardial autologous bone marrow mononuclear cell injection in ischemic heart failure: postmortem anatomicopathologic and immunohistochemical findings. *Circulation* 2005;112:521–526.

81. Tse HF, Thambar S, Kwong YL, Rowlings P, Bellamy G, McCrohon J, Bastian B, Chan JK, Lo G, Ho CL, Lau CP. Safety of catheter-based intramyocardial autologous bone marrow cells implantation for therapeutic angiogenesis. *Am J Cardiol* 2006;98:60–62.

82. Qian HS, de Resende MM, Beausejour C, Huw LY, Liu P, Rubanyi GM, Kauser K. Age-dependent acceleration of ischemic injury in endothelial nitric oxide synthase-deficient

mice: potential role of impaired VEGF receptor 2 expression. *J Cardiovasc Pharmacol* 2006;47: 587–593.

83. Koh GY, Kim SJ, Klug MG, Park K, Soonpaa MH, Field LJ. Targeted expression of transforming growth factor-beta 1 in intracardiac grafts promotes vascular endothelial cell DNA synthesis. *J Clin Invest* 1995;95:114–121.

84. Suzuki K, Murtuza B, Smolenski RT, Sammut IA, Suzuki N, Kaneda Y, Yacoub MH. Cell transplantation for the treatment of acute myocardial infarction using vascular endothelial growth factor-expressing skeletal myoblasts. *Circulation* 2001;104:I207–I212.

85. Askari A, Unzek S, Goldman CK, Ellis SG, Thomas JD, DiCorleto PE, Topol EJ, Penn MS. Cellular, but not direct, adenoviral delivery of vascular endothelial growth factor results in improved left ventricular function and neovascularization in dilated ischemic cardiomyopathy. *J Am Coll Cardiol* 2004;43:1908–1914.

86. Markkanen JE, Rissanen TT, Kivela A, Yla-Herttuala S. Growth factor-induced therapeutic angiogenesis and arteriogenesis in the heart–gene therapy. *Cardiovasc Res* 2005;65:656–664.

87. Ward NL, Van Slyke P, Sturk C, Cruz M, Dumont DJ. Angiopoietin 1 expression levels in the myocardium direct coronary vessel development. *Dev Dyn* 2004;229:500–509.

88. Niagara MI, Haider H, Ye L, Koh VS, Lim YT, Poh KK, Ge R, Sim EK. Autologous skeletal myoblasts transduced with a new adenoviral bicistronic vector for treatment of hind limb ischemia. *J Vasc Surg* 2004;40:774–785.

89. Xu P, Li SY, Li Q, Ren J, Van Kirk EA, Murdoch WJ, Radosz M, Shen Y. Biodegradable cationic polyester as an efficient carrier for gene delivery to neonatal cardiomyocytes. *Biotechnol Bioeng* 2006;95:893–903.

90. Yau TM, Fung K, Weisel RD, Fujii T, Mickle DA, Li RK. Enhanced myocardial angiogenesis by gene transfer with transplanted cells. *Circulation* 2001;104:I218–I222.

91. Su EJ, Cioffi CL, Stefansson S, Mittereder N, Garay M, Hreniuk D, Liau G. Gene therapy vector-mediated expression of insulin-like growth factors protects cardiomyocytes from apoptosis and enhances neovascularization. *Am J Physiol Heart Circ Physiol* 2003;284:H1429–H1440.

92. Yau TM, Kim C, Ng D, Li G, Zhang Y, Weisel RD, Li RK. Increasing transplanted cell survival with cell-based angiogenic gene therapy. *Ann Thorac Surg* 2005;80:1779–1786.

93. Matsumoto R, Omura T, Yoshiyama M, Hayashi T, Inamoto S, Koh KR, Ohta K, Izumi Y, Nakamura Y, Akioka K, Kitaura Y, Takeuchi K, Yoshikawa J. Vascular endothelial growth factor-expressing mesenchymal stem cell transplantation for the treatment of acute myocardial infarction. *Arterioscler Thromb Vasc Biol* 2005;25:1168–1173.

94. Yang J, Zhou W, Zheng W, Ma Y, Lin L, Tang T, Liu J, Yu J, Zhou X, Hu J. Effects of myocardial transplantation of marrow mesenchymal stem cells transfected with vascular endothelial growth factor for the improvement of heart function and angiogenesis after myocardial infarction. *Cardiology* 2007;107:17–29.

95. Wang Y, Haider HK, Ahmad N, Xu M, Ge R, Ashraf M. Combining pharmacological mobilization with intramyocardial delivery of bone marrow cells over-expressing VEGF is more effective for cardiac repair. *J Mol Cell Cardiol* 2006;40:736–745.

96. Rinsch C, Quinodoz P, Pittet B, Alizadeh N, Baetens D, Montandon D, Aebischer P, Pepper MS. Delivery of FGF-2 but not VEGF by encapsulated genetically engineered myoblasts improves survival and vascularization in a model of acute skin flap ischemia. *Gene Ther* 2001;8:523–533.

97. Jeong JH, Kim SH, Kim SW, Park TG. Intracellular delivery of poly(ethylene glycol) conjugated antisense oligonucleotide using cationic lipids by formation of self-assembled polyelectrolyte complex micelles. *J Nanosci Nanotechnol* 2006;6:2790–2795.

98. Liu B, Daviau J, Nichols CN, Strayer DS. *In vivo* gene transfer into rat bone marrow progenitor cells using rSV40 viral vectors. *Blood* 2005;106:2655–2662.

99. Haitao P, Qixin Z, Xiaodong G. A novel synthetic peptide vector system for optimal gene delivery to bone marrow stromal cells. *J Pept Sci* 2007;13:154–163.

100. Kawamoto A, Gwon HC, Iwaguro H, Yamaguchi JI, Uchida S, Masuda H, Silver M, Ma H, Kearney M, Isner JM, Asahara T. Therapeutic potential of *ex vivo* expanded endothelial progenitor cells for myocardial ischemia. *Circulation* 2001;103:634–637.
101. Kim C, Li RK, Li G, Zhang Y, Weisel RD, Yau TM. Effects of cell-based angiogenic gene therapy at 6 months: persistent angiogenesis and absence of oncogenicity. *Ann Thorac Surg* 2007;83:640–646.
102. Li RK, Jia ZQ, Weisel RD, Merante F, Mickle DA. Smooth muscle cell transplantation into myocardial scar tissue improves heart function. *J Mol Cell Cardiol* 1999;31:513–522.
103. Koh GY, Klug MG, Soonpaa MH, Field LJ. Differentiation and long-term survival of C2C12 myoblast grafts in heart. *J Clin Invest* 1993;92:1548–1554.
104. Watanabe Y, Sawaishi Y, Tada H, Sato E, Suzuki T, Takada G. Immunoliposome-mediated gene transfer into cultured myotubes. *Tohoku J Exp Med* 2000;192:173–180.

15 Optimal Expansion and Differentiation of Cord Blood Stem Cells Using Design of Experiments

Mayasari Lim, Hua Ye, Nicki Panoskaltsis
and Athanasios Mantalaris

Introduction

The demand for well-defined cellular products that are of clinical grade has increased significantly due to the growth in the use of these cells for the treatment of various medical conditions. Therefore, the *ex vivo* expansion and directed differentiation of HSCs has recently been the main focus of research. However, due to the complexity of HSC cultures, process characterization cannot be easily achieved. Specifically, the challenges faced in process characterization can be attributed to several factors, such as the heterogeneity in cell populations in terms of cell types and maturational stage, the transient kinetics, the high sensitivity and specificity to extrinsic factors, and the interaction between these factors all of which affect cellular behavior. Therefore, the lack of detailed understanding of these process characteristics impedes good process control and results in compromising quality, yield and purity of the final product. The utilization of DOE to identify the critical factors and their interactions, which will result in quantitative process characterization, is therefore essential for optimal bioprocess control. In addition, real-time process monitoring will also become increasingly important in providing information about the on-going process. Consequently, disturbances from the optimal dynamic equilibrium of the system can be quickly detected and appropriate adjustments can be made so that the system will recover back into its optimal operating conditions. The combination of these bioprocessing tools will, therefore, enable the maintenance of an optimal dynamic equilibrium that is robust in terms of achieving high-product quality, high yield, and high reproducibility of stem cell cultures. Such tight control of the final product quality is essential in order to meet the high standards required for the clinical application of the cellular products generated by stem cell cultures.

Hematopoiesis

Hematopoiesis is the process of blood cell formation; it takes place in the bone marrow in adults where all blood lineages are produced prior to entering circulation. Blood cells, in general, have a short lifespan varying from several days to a few months. Specifically, white blood cells have a lifespan of 13–21 days while red blood cells (RBCs) usually last up to 120 days. As blood cells die and are removed from circulation, blood is replenished as required by the body via hematopoiesis. Hematopoiesis, which is a highly complex and highly regulated process, includes the self-renewal, commitment, proliferation, differentiation, and programmed cell death (apoptosis) of the HSCs and their progeny. HSCs reside in the stem cell niches of the bone marrow and give rise to all blood cells. They are distinguished by their unique properties of self-renewal, which enables them to maintain the source of mature blood cells for the lifetime of the host, and multipotency, which allows them to differentiate into different types of blood cells, or lineages; each blood cell lineage having a distinct and specific function (Fig. 1). Red blood cells (erythrocytes), for example, carry oxygen from the lungs to the tissues and transports carbon dioxide from the tissues back to the lungs. White blood cells, which consist largely of granulocytes, monocytes, and lymphocytes, are cells of the immune system. These cells function to protect the host by digesting foreign bodies and micro-organisms, and by producing antibodies and other proteins such as cytokines to neutralize bacteria and viruses and to mobilize the immune system. Platelets (thrombocytes) are the smallest elements of the blood that are derived from megakaryocytic fragments and are mainly responsible for initiating the blood clotting process to prevent blood loss from injured blood vessels.

The site of hematopoiesis, which is comprised of hematopoietic cells, stromal cells, growth factors, and extracellular matrix (ECM) is called the hematopoietic inductive micro-environment (HIM). The bone marrow stromal cells, namely fibroblasts, macrophages, endothelial cells, and adipocytes provide the three-dimensional growth environment and support structure for the hematopoietic cells, and secrete growth factors that control many of the hematopoietic cell functions, such as proliferation and differentiation. Growth factors (cytokines) are glycoproteins that recognize and interact with specific cell-surface receptors to regulate the various hematopoietic processes. Some of the growth factors, such as erythropoietin (EPO), are lineage-specific whilst others, such as stem cell factor (SCF), interleukin-3 (IL-3), and Flt-3 ligand (FL), work across lineage boundaries. Non-lineage-specific growth factors often do not work independently but rather tend to work in conjunction with other growth factors. ECM macromolecules which are secreted by stromal cells include collagen and adhesion proteins, such as fibronectin and laminin, and help to localize both growth factors and hematopoietic cells in their specific three-dimensional micro-environmental niches. With all of these elements intact, regulated cellular processes in the HIM can therefore occur through a series of cell-to-cell and cell-to-matrix interactions in an extremely well-balanced physical, chemical, and cellular micro-environment.

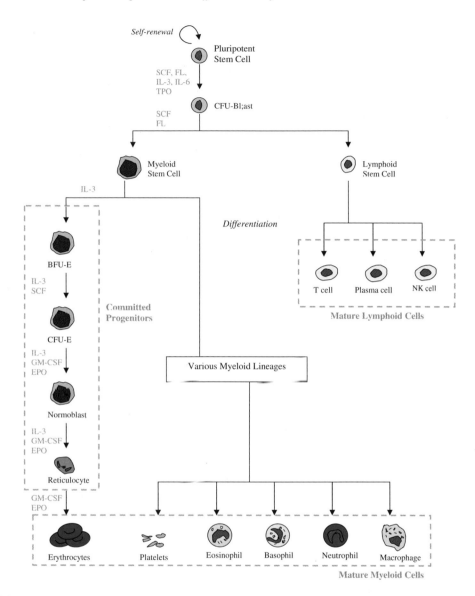

Fig. 1. Schematic of the hematopoietic tree depicting the most important growth factors involved in erythropoiesis.

Ex vivo expansion and differentiation of hematopoietic cells

In order to successfully expand and differentiate hematopoietic cells *ex vivo*, knowledge and control of the various process parameters is essential. These parameters include factors which constitute the physicochemical environment, nutrients and metabolites, and growth factors/cytokines. Each of these factors affects subsets of cells and cell populations differently, depending on the concentration gradient of the factor, metabolic activity of the cell, and abundance of the substrate.

Physicochemical culture parameters such as pH, oxygen and carbon dioxide tension, and temperature affect the cell culture environment and therefore, affect cell behavior and cellular activities. The proliferation and differentiation of different blood lineages operate at unique physicochemical optima. Despite this differential requirement, most current cell culture conditions are maintained at pH 7.40, 20% oxygen tension, 5% carbon dioxide tension and 37°C, regardless of the cultivated cells being studied. As an example, the processes of megakaryopoiesis (differentiation and maturation of megakaryocytic progenitors) and erythropoiesis (differentiation and maturation of erythroid progenitors) are optimal at pH 7.60.[1,2] In contrast, granulopoiesis requires a lower pH and the process is optimized at pH 7.21. The lower pH conditions enhance granulocyte colony stimulating factor receptor (G-CSFR) expression and, therefore, granulocyte proliferation and differentiation.[3] Oxygen demands in hematopoietic cell cultures also vary according to the cell lineage and maturational stage. Low-oxygen tension (5%) is better for progenitor cell expansion and granulocyte differentiation, whereas high-oxygen tension (20%) is most conducive to the growth of mature megakaryocytes and erythrocytes.[4] Oxygen tension modulates cytokine production, surface marker expression and transcription factor production, and may therefore be responsible for the differences found in the expansion and differentiation of cell subtypes.[5] With regards to temperature requirements, most cell cultures appear to grow very effectively at 37°C, although a higher temperature (39°C) enhances megakaryopoiesis in CD34-enriched cord blood cultures.[6] Since these findings indicate unique differences in the physicochemical requirements of different types of bone-marrow derived cells during differentiation *ex vivo*, the data also suggest the existence of physicochemical gradients *in vivo* that are partially responsible for regulating hematopoiesis in the bone marrow micro-environment.

Nutrients, such as glucose and glutamine, are the main energy sources required for cell growth and metabolism. Serum, which is frequently used in cell cultures, provides a wealth of such resources to the cells but is a poorly defined supplement. Therefore, it may contain proteins that could be harmful to a human recipient, and is not ideal for clinical applications. Instead, a defined, serum-free media is required for clinical applications. Waste metabolites, such as lactate and ammonia, are constantly being produced by cells as they metabolize nutrients. The accumulation of these metabolites can become harmful to the cells when in excess, and may eventually cause cells to die. Periodic removal of waste metabolites from the cell culture is therefore necessary to prevent cell death. Similarly, a lack of glucose (nutrient) levels in the cell culture can also result in reduced cell viability. A balance of nutrient and metabolite concentrations is therefore essential for maintaining healthy cell growth and cellular activities. Process monitoring and bioprocess control of key nutrient and metabolite levels is therefore critical in maintaining a good bioprocess operation.

Growth factors and cytokines in hematopoiesis determine the survival, proliferation, and differentiation of cells by providing the necessary signals, both positive and negative, that influence cellular behavior. Growth factors can be lineage-specific, acting only on a specific lineage of interest, or non-lineage specific, acting on multiple lineages to perform various functions. Lineage-specific growth factors and cytokines that have been identified include EPO which control the process of erythropoiesis,[7,8] GM-CSF which activates differentiation of the myeloid pro-genitors and maturation of granulocytes, and IL-5 which differentiates cells into eosinophils.[9] Non-lineage-specific growth factors include early acting cytokines, such as SCF and FL, and mid-acting growth factors, such as thrombopoietin (TPO), IL-3, IL-6 and IL-11.[10,11] These cytokines are found to be responsible for the sur-vival and expansion of hematopoietic stem and progenitor cells. When these growth factors or cytokines are used to activate or influence cells *in vitro*, it is essential to determine and optimize the concentrations and combinations of cytokines required for specific types of cell culture, as these conditions differ from those that may occur *in vivo*.

To date, most studies in cell culture have used the dose–response method of optimizing cytokine concentrations, wherein effects from each factor are investi-gated individually. This method can result in misleading observations as it does not take into account the effects and interactions between different growth factors or other factors of interest. As a consequence, dose–response studies tend to give false optima for the process and thus provide results for suboptimal bio-process conditions. Process influences from physicochemical parameters, nutrients, metabolites, and growth factors, and their corresponding interactions with each other for hematopoietic cell cultures have not yet been fully understood or quan-tified. Process operation of most hematopoietic cell cultures are therefore cur-rently estimates based on what is known in the literature from less-than-ideal dose–response methods of optimization. This estimation of the process is one of the main factors that limits bioprocess control. DOE is widely used in process characterization and optimization in many industries, and is particularly useful in the manufacturing and production of food, chemicals, materials, and semicon-ductors. The application of DOE in identifying and quantifying the effects of all process parameters in hematopoietic cell cultures at both an individual and inter-active level is therefore essential if the process of interest is to be understood and optimized.

Design of Experiments

DOE is an effective and efficient methodology for investigating a process system and obtaining the maximum amount of information about a system with minimum effort (i.e. experimentation).[12] DOE can be used in comparison studies, to perform screening experiments, and to conduct response surface or regression modeling. Specifically, one of the distinct advantages of using DOE is the ability of the design to reveal both individualistic and interactive effects for a process using the minimum number of experimental runs. This saves significant time and operation costs, but more importantly, it reveals the "true" picture of the process of interest. Traditional experimentation methods, such as dose–response studies, do not often paint the "true" picture of the process under investigation as they only examine changes due to one factor while other corresponding process factors remain unaltered. This analysis limits the experimental results and can sometimes lead to misleading results as the optima obtained from the dose–response experiments are often not the true optima of the process.[12] Process interactions cannot be revealed from simple dose–response studies; they do not provide a complete descriptive relationship between the process parameters. The advantage of DOE in such studies is the application of response surface modeling, which yields surface plots of two or more process parameters using the minimum number of experiments (Fig. 2). Although surface response plots can technically be generated from dose–response methods, they require a significantly higher number of experiments to be performed than does the DOE method. Furthermore, DOE is able to quickly screen a large number of factors with minimum experimentation and correctly identify the influencing factors as well as avoiding

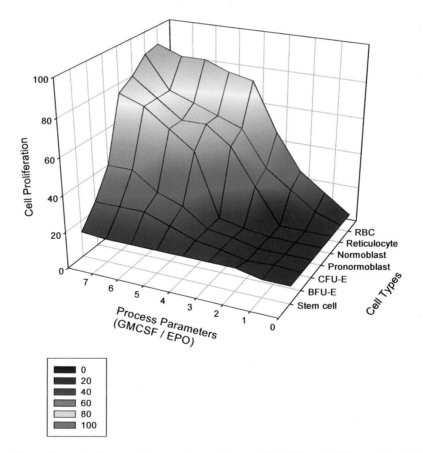

Fig. 2. Three-dimensional map of the combined effects of GM-CSF and EPO on cell proliferation of stem cells, progenitors, precursors and mature cells of the erythroid lineage.

false positives, which is of critical importance in establishing reproducible and controlled bioprocesses. Several types of designs can be employed when performing screening experiments. They include the fractional factorial design and the Plackett–Burman design. A creative and properly designed experiment can provide a wealth of information elucidating the process of interest in a significantly short period of time with a small number of experiments.

To date, the use of DOE in stem cell bioprocessing has been limited; much of HSC research still utilizes traditional dose–response methods. However, a handful of investigators have been quite successful in using DOE for stem cell culture studies. These include the use of fractional factorial designs to perform screening experiments,[13,14] and the use of central-composite or full factorial designs to perform process characterization.[15–17] However, full process characterization, which considers all the process factors including physicochemical parameters, nutrient, metabolite, and growth factor concentrations, has not yet been established. Due to the complexity of their interactions, process investigation of these factors in combination is not straightforward. The successful application of DOE methodologies to resolve the complexity in stem cell bioprocessing will not only yield invaluable information for HSC cultures but also provide a novel approach for other cell culture studies.

In studying a complex system, a systematic approach should be taken to obtain useful information and to determine the optimal operating conditions for the process of interest (Fig. 3). The approach we propose involves screening, characterization, and optimization steps using the appropriate experimental design(s) in each phase.[18] The objective in the first stage, *process screening*, is to identify factors that have a significant influence on the process and correctly discriminate against factors that have little or no influence at all. Hence, in the screening step, the goal of DOE is to efficiently and quickly cover the whole range of the experimental space identifying the critical factors. In the second stage, *process characterization*, the goal is to obtain more detailed and quantitative description about the process by way of

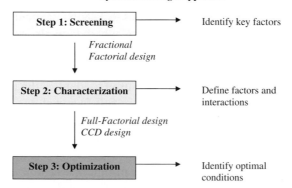

Fig. 3. Systematic design approach for conducting DOE experiments.

three-dimensional surface response plots to describe process characteristics. This is achieved by designing experiments for the parameters identified during the screening stage so that the full interactions of the critical parameters are revealed. In the third stage, *process optimization*, the generation of the three-dimensional plots reveals the optimal operating regimens for the critical process parameters investigated and yields the best conditions for process optimization. The outcome of this systematic framework is the reproducible and statistically valid elucidation of bioprocesses that do not require *a priori* assumptions for the process of interest.

DOE designs

Two-level factorial designs are the basic building blocks for most designs and are therefore the most commonly applied design method. Factorial designs are very effective in studying both main and interactive effects of a process rendering them extremely useful for response surface studies or modeling.[19] A basic two-level factorial design involves all the possible combinations of the high and low values of the various process factors under investigation and yields the total number of experiments required for the design. Specifically, in a two-factor factorial design, the design points relate to the four corners of a square, representing full coverage of the experimental space, and yields four (2^2) experimental runs. Similarly, a three-factor factorial design yields eight (2^3) experimental runs. As a general rule, for any n number of factors in an experimental study, a 2^n number of experimental runs are required. Basic two-level factorial designs are mainly used to study linear relationships between process factors and their responses but are still capable for revealing both main and interactive effects independently. Other variations of the factorial design for the purpose of screening and response surface modeling for second-order models will be discussed in the following sections.

Since stem cell bioprocesses involve a multitude of parameters that need to be investigated, this increase in the number of factors in an experimental study results in a rapid increase in the number of experimental runs required to perform a full factorial design. For instance, in a seven-factor experiment the required number of experimental runs would be 2^7, which equates to 128 experiments. This is a significantly large number of experiments to perform, which is costly and time-consuming, thus making full factorial designs an impractical option. Often in these situations, when a large number of factors (more than five) are involved, screening is performed and two-level fractional factorial designs are used to identify the influencing parameters. Fractional factorial designs derive from full factorial designs, where only a fraction of the full design is used to perform the screening. The assumption made in these designs is that higher-order interactions are often negligible, therefore only the main factors and low-order interactions are estimated in the design. The number of experiments are reduced in this case as factors in the design, both the main and interactive effects, are aliased with each other. The effectiveness of the fractional factorial design depends on the resolution of the design; these include Resolution III,

IV, and V.[12,19] A Resolution III design is one that has no main effects aliased with another main effect but they are aliased with two-factor interactions. Hence, Resolution III designs allow the estimation of the main effects, but they may be biased by two-factor interactions. A Resolution IV design has no main effects aliased with another main effect or a two-factor interaction, but two-factor interactions may be aliased with each other. In practice that means that Resolution IV designs allow the estimation of the main effects unbiased by two-factor interactions, but the two-factor interactions may not be estimable. Finally, a Resolution V design is one that has no main effects or two-factor interactions aliased with each other. Hence, Resolution V designs allow the estimation of main effects and two-factor interactions. Higher-order interactions are usually assumed to be negligible. In conclusion, a design with a higher resolution represents a more thorough design since fewer factors are aliased with each other, but this also means that they require a greater number of experiments. In most screening studies, a Resolution IV design is sufficient to perform accurate screening as the main factors are not aliased with any two-factor interactions, which are often significant in most processes.

A very common and widely applicable design used for characterization studies in order to build second-order response surface models is the central composite design (CCD). CCD derives from two-level factorial designs but also adds center-points and axial points into the basic 2^n design. Center-points are not only used to estimate the curvature in a second-order model, but are also used as replicates for estimating pure experimental error. Axial points are used mainly to estimate the quadratic terms in a second-order model.[19] The position of the axial point depends on α, the axial distance, which varies according to the extent of the region of operation and the process region of interest. The value of α can range from 1 to $\sqrt{n}$, where n is the number of factors in the experiment. The choice of α value affects the design rotatability, which relates to the robustness and stability of the design. A rotatable design is one that provides equal variance in the estimation of predicted values at any two locations in the design space.[19] A CCD with an α value of $\sqrt{n}$ is referred to as a spherical design and is nearly rotatable or considered rotatable. A CCD with an α value of 1 is a cubodial or face-centered design. Though they are not rotatable, face-centered designs are often applied in many experimental studies where design ranges are restricted and are very effective in providing full process characterization for a defined process regime. Other available designs include the Box–Behnken design, other spherical response surface method (RSM) designs, and Equiradial designs.[19] A Box–Behnken design is an independent quadratic design that uses three levels on each factor, and unlike a factorial design, it takes midpoints on the edges of the process space and a center-point to create the design study.

Case study: *in-vitro* erythropoiesis

In order to illustrate the application of the proposed systematic design framework, utilizing DOE methodologies, we investigated the expansion and differentiation of human cord blood stem cells (CD34$^+$) towards the erythroid lineage. As shown in Fig. 1, the process of erythropoiesis, which is transient in nature and involves the expansion and maturation of hematopoietic stem cells, progenitors, precursors, and mature cells, is affected by numerous growth factors that act at particular stages of the erythropoietic process.[20,21] Specifically, the stem cell (CD34$^+$) and early progenitor populations (BFU-E) are affected by early-acting growth factors, such as SCF, whereas late progenitors (CFU-E) and committed precursors (normoblasts) are affected by mid- to late-acting growth factors, such as granulocyte/macrophage-colony stimulating factor (GM-CSF) and EPO, respectively. EPO is a lineage-dominant cytokine, which is specific to red blood cell maturation and is the key initiator for erythrocyte differentiation and maturation. In our case study, we limited our investigation on the effects of different growth factors on the yield of red blood cell formation in terms of total cell number expanded and the percentage of RBCs in the final population as achieved at day 10 of the culture. Specifically, we sought to determine which of the most prominent growth factors observed in the literature to affect erythropoiesis are significant and to establish the optimal concentrations that should be used in culture. The growth factors under investigation included early-acting cytokines, such as SCF, FL, interleukin-6 (IL-6), mid-acting cytokines, such as IL-3 and GM-CSF, and late-acting growth cytokines, such as TPO and EPO.

Cell culture conditions

Cryopreserved human umbilical cord blood units were supplied from the London Cord Blood Bank in accordance with regulatory and ethical policies. Mononuclear cells (MNCs) were isolated from cord blood using Ficoll–Hypaque (GE–Healthcare) density centrifugation following the manufacturer's instructions. MNCs were then incubated with anti-CD34 monoclonal antibodies conjugated to microbeads (Miltenyi Biotec) for 30 minutes at 4°C before being subjected to microbead selection using high-gradient magnetic field and mini-MACS columns (Miltenyi Biotec). Freshly isolated CD34$^+$ cells were then cultivated in Iscove's Modified Dulbecco's Medium (IMDM; Gibco) containing 10% (v/v) fetal bovine serum (FBS; Gibco) plus the growth factor cocktail at the specific concentrations as determined by DOE design run. Cells were cultivated in 24-well plates (Corning) at 5×10^4 cells/ml and were fed every three days starting from day 4 after initial seeding by replacing all of the old culture medium. Culture conditions were maintained at 37°C, 5% carbon dioxide and 20% oxygen in a fully humidified chamber. Recombinant human SCF, IL-3, IL-6, FL, TPO, GM-CSF (all from Biosource), and EPO (R&D Systems) were used in the cell culture DOE investigations.

DOE experimental design

Growth factors that have been identified in the scientific literature as the most important in influencing erythropoiesis were the process factors of interest studied,[8,22] namely SCF, IL3, IL6, FL, TPO, GM-CSF and EPO, which gives a total of seven process factors. For the screening experiment, a Resolution IV fractional factorial design was used, which yielded a total of 16 experiments. The screening experiment, as it will be discussed in the Results section, revealed that only SCF and EPO were significant. For the characterization experiment, a face-centered CCD was used with three center replicates. Hence, the factors studied for the characterization experiment, as determined from the screening experiments, were SCF and EPO. Therefore, for a two-factor CCD design with three center replicates, 11 experimental were required. The DOE design software that was used in generating the experiments and analyzing the data were Design Expert (StatEase) and MODDE7 (Umetrics).

Cell analysis

Cell enumeration was performed during each passage; however, as described above, for the DOE screening and characterization studies, the optimization algorithm was based only on day 10 of the culture (final cell expansion). Briefly, cells were stained with eryothrosin-B stain solution (ATCC) to determine cell numbers and viability, as per the manufacturer's instructions. Surface expression for CD45, CD71 and glycophorin-A (GPA) were analyzed via flow cytometry. Cells were incubated at 4°C for 30 minutes with directly labeled human CD45-FITC, CD71-PE and GPA-PC5 (BD Bioscience). After staining, the cells were washed twice with PBS supplemented with 2% (v/v) FBS and 0.1% (w/v) sodium azide prior to fixation with 1% (w/v) para-formaldehyde. Acquisition took place within one day following fixation. For the DOE analysis, the final day (day 10) expression of GPA on day 10 was used. Data was acquired on an Epics Altra (Beckman Coulter) flow cytometer with 30,000 events being captured for each analysis. Data analysis was performed with WinList software (Verity).

Other cellular analyses (data not shown) included studies in morphology, fluorescence microscopy, and clonogenic assays (colony-forming unit assays; CFUs). Cell morphology was determined using Giemsa–Wright staining for nuclear/cytoplasmic differentiation and Cresyl-blue staining for reticulocyte identification, respectively. Fluorescence microscopy was used to verify cellular surface antigen expression of CD45 and CD71. Clonogenic assays for the erythroid (CFU-E and BFU-E) and macrophage, granulocyte, granulocyte-macrophage, and multilineage (CFU-GM and CFU-GEMM) cells using Methocult H4434 (Stem Cell Technologies) were performed. Briefly, 1×10^3 cells for cord blood CD34$^+$ cells and 500–10,000 cells for cultured cord blood cells at days 4, 7, 10, and 13 were plated in duplicates. Assays were carried out for 14–16 days in a 37°C, 5% carbon dioxide, 20% oxygen fully humidified chamber before being scored. Two independent scores were obtained.

Results and Discussion

The coefficient plot from the screening experiments shows that SCF and EPO have a large effect on both total cell number and the percentage of erythroid cell formation

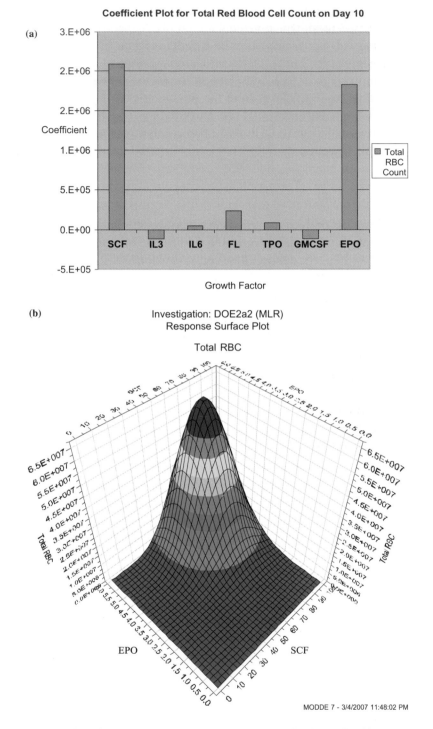

Fig. 4. **(a)** Coefficient plot from the screening DOE experiment for the total red blood cell numbers evaluated at day 10 of culture. **(b)** Response surface plot from the characterization DOE experiment for the total red blood cell numbers evaluated at day 10 of culture.

when compared with that of the other factors (Fig. 4a). From these results, we can conclude that SCF and EPO are the only two growth factors that have a significant influence on the yield of red blood cells formed on the final day (day 10) of cell culture. Having identified the contributing factors, a characterization study was performed using only SCF and EPO as the two main factors of interest. Employing a face-centered CCD design with three center replicates 11 experimental runs were generated to perform this study. From the characterization experiments, a three-dimensional response surface plot was obtained showing the influences of both SCF and EPO on the total number of RBCs on day 10 (Fig. 4b). The area shaded in red in Fig. 4b indicates the region of concentrations of SCF and EPO for optimal expansion of RBCs over a ten-day period. Therefore, the concentrations selected from this optimal process regimen were 75 ng/ml of SCF and 4.5 U/ml of EPO. In addition, in a separate study, cord blood stem cells were cultivated with the optimal concentrations of SCF and EPO for a period of 16 days, and resulted in a significant increase in cell numbers during the initial ten-day period and a significant rapid increase in CD71 and GPA expression. After only four days of culture, almost 90% of the cells already expressed CD71, indicating erythroid precursors. By day 10 of the culture, a 1250-fold total cell expansion along with 96% CD71 and 86% GPA expression (Table 1) was obtained. Morphologic assessment of the cells confirmed the presence of red blood cell maturation, with a small number of enucleated red blood cells being observed on day 10 of the culture (Fig. 5a). Flow cytometry plots showed a strong CD71 expression and weak GPA expression at day 4 of the culture (Fig. 5b). In contrast, at day 10 of the culture, strong GPA expression was observed in the majority of the cultivated cells, substantiating the production and maturation of erythrocytes. Therefore, our DOE study has established that significant improvements in the expansion and maturation of red blood cells from human cord blood stem cells can be generated. In conclusion, utilizing this design approach could enable the process optimization for the expansion and/or differentiation of hematopoietic cells into other cell lineages of interest and the investigation of a higher number of important culture parameters that are usually excluded for current analyses.

Table 1. Results obtained utilizing the optimal DOE design (75 ng/ml SCF and 4.5 U/ml EPO in 10% FBS + IMDM).

Day	Fold expansion	CD45+	CD71+	CD235a+
4	4.6	62.8%	89.8%	42.5%
7	160	24.8%	97.8%	79.2%
10	1257	13.0%	96.3%	86.5%
13	3000	16.0%	89.9%	87.1%
16	6361	26.6%	89.4%	90.8%

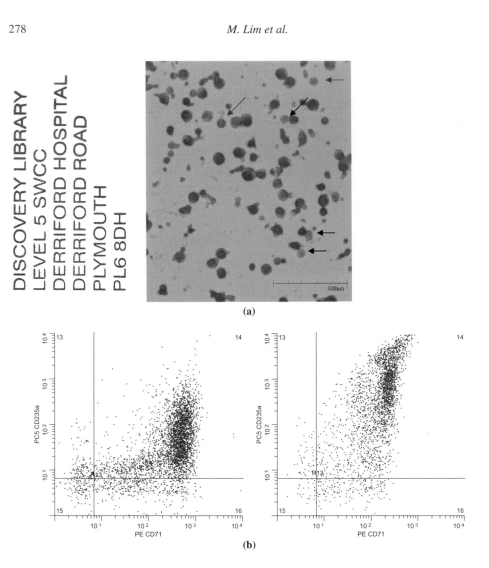

Fig. 5. **(a)** Cell morphology at day 10 of culture (Giemsa–Wright stain; 20× magnification). Signs of red blood cell formation with the nucleus being expelled (black arrows) and enucleated red cells (red arrows) are evident. **(b)** Flow cytometry results for the expression of the erythrocytic surface markers of CD71 and CD235a (GPA) at day 4 (left panel) and at day 10 (right panel).

DOE Application in Stem Cell Bioprocessing

The ultimate goal in utilizing DOE to investigate complex cell culture systems is to characterize the influence of all culture parameters on different cell types and their respective cellular functions. Using a DOE design approach, three-dimensional surface plots could be generated that describe the influence of each culture parameter on the different cell types, such as stem cells, progenitors, and precursors, for various cell functions, such as proliferation, differentiation, and cell death (Fig. 2), which would be represented in the form of x-, y-, and z-axes in a three-dimensional plot.

Culture parameters, as previously discussed, would include physicochemical factors, nutrients and metabolites, and growth factors. Figure 2 includes a non-exhaustive list of defined culture parameters of interest. Hematopoietic cell cultures are transient in nature and are comprised of cell populations that are heterogeneous both in terms of the cell type and in terms of the maturation stage, including immature stem cells, progenitor cells, precursor cells and finally mature cells. Each cell population is influenced by a different and unique set of culture parameters in a potential distinctly different manner. Cellular function, as included in our DOE analysis, consists of proliferation, differentiation and programmed cell death (apoptosis). The cellular behavior of each cell population is significantly influenced by the culture environment. For instance, the proliferation of hematopoietic cells at different maturation stages towards the erythrocytic lineage is influenced in a dissimilar manner by physicochemical parameters, such as pH, and growth factors, such as SCF, EPO, and GM-CSF. Specifically, pH can significantly influence hematopoietic stem cell cultures; a pH around 7.4 is optimal for stem cells whereas a pH around 7.6 is optimal for mature red blood cells.[1] Similarly, mid- to late-acting growth factors, such as GM-CSF and EPO, have a great influence on red blood cell precursors and mature red blood cells, respectively, whereas they have no or little influence on stem/progenitor cells. Such information-rich data can result in the generation of a three-dimensional map of how different process parameters influence the various hematopoietic cell populations. Figure 2 shows the three-dimensional map of the combined effects of the mid-acting GM-CSF and the late acting EPO growth factors. Specifically, it shows that the mid-acting GM-CSF cytokine has significant influence on the proliferation of erythroid progenitors and precursors while it has little or no influence on stem cells or mature red blood cells, in accordance with published studies.[23] In contrast, EPO stimulates erythroid precursors and mature erythrocytes, yet has no influence on stem cells.[8,23,24] Although, the three-dimensional surface plots of these factors are not truly representative of the actual process, since no process characterization on such study has been performed, the use of DOE could result in the generation of a complete and quantitative process representation, as that shown in Fig. 2.

Conclusion

Stem cell bioprocesses, including hematopoietic stem cell cultures, are complex and highly dynamic and influenced by a multitude of physicochemical, cellular, and molecular factors. Due to the multiplicity and complexity of the process factors involved, a complete characterization of these bioprocesses has yet to be established. Consequently, this lack of understanding of the characteristics of the bioprocess results in sub-optimal operation, which compromises the quality and yield of the final product. DOE should become an essential tool in process development and process characterization of stem cell bioprocesses. It enables a thorough understanding of cell culture processes by identifying the true factors that influence the culture and, with the aid of on-line monitoring, could assist in process control. Furthermore, DOE facilitates process characterization by quantifying the parameters, and their interactions, as well as determining optimal operating conditions. This control ensures the production of a high-quality and high-yield product. In addition, tight process control will then result in the high reproducibility of the bioprocess operation. Achieving reproducibility and yield of the desired cellular product is essential in meeting (and exceeding) the high-clinical standards. Therefore, the focus of stem cell bioprocessing, which aims to generate products that are suitable and applicable for therapeutic purposes, should ultimately endeavor to meet this goal.

References

1. McAdams TA, Miller WM, Papoutsakis TE. pH is a potent modulator of erythroid differentiation. *Br J Hematol* 1998;103:317–325.
2. Yang H, Miller WM, Papoutsakis ET. Higher pH promotes megakaryocytic maturation and apoptosis. *Stem Cells* 2002;20:320–328.
3. Hevehan DL, Wenning LA, Miller WM, Papoutsakis TE. Dynamic model of *ex vivo* granulocytic kinetics to examine the effects of oxygen tension, pH and IL-3. *Exp Hematol* 2000;28:1016–1028.
4. Hevehan DL, Papoutsakis TE, Miller WM. Physiologically significant effects of pH and oxygen on granulopoiesis. *Exp Hematol* 2000;28:267–275.
5. Mostafa SS, Papoutsakis TE, Miller WM. Oxygen tension modulates the expression of cytokine receptors, transcription factors, and lineage-specific markers in culture human megakaryocytes. *Exp Hematol* 2001;29:873–883.
6. Proulx C, Dupuis N, St-Amour I, Boyer L, Lemieux R. Increased megakaryopoiesis in cultures in CD34-enriched cord blood cells maintained at 39°C. *Biotechnol Bioeng* 2004;88(6):675–680.
7. Flores-Guzman P, Guiterrez-Rodriguez M, Mayani H. *In vitro* proliferation, expansion and differentiation of a CD34+ cell-enriched hematopoietic cell population from human umbilical cord blood in response to recombinant cytokines. *Arch Med Res* 2002;33:107–114.
8. Sakatoku H, Inoue S. *In vitro* proliferation and differentiation of erythroid progenitors of cord blood. *Stem Cells* 1997;15(4):268–274.
9. Shearer WT, Rosenwasser LJ, Bochner BS, Martinez-Moczygemba M, Huston DP. Biology of common beta receptor-signaling cytokines: IL-3, IL-5 and GM-CSF. *J Allergy Clin Immunol* 2003;112(4):653–665.
10. Alcorn MJ, Holyoake TL. Ex vivo expansion of hematopoietic progenitor cells. *Blood Rev* 1996;10(3):167–176.
11. Banu N, Deng B, Lyman SD, Avraham H. Modulation of hematopoietic progenitor development by Flt3-ligand. *Cytokine* 1998;11(9):679–688.
12. Montgomery DC. *Design and Analysis of Experiments*, 5th ed. John Wiley and Sons, Inc., Arizona, 2001.
13. Yao C-L, Chu I-M, Hsieh T-B, Hwang S-M. A systematic strategy to optimize *ex vivo* expansion medium for human hematopoietic stem cells derived from umbilical cord blood mononuclear cells. *Exp Hematol* 2004;32:720–727.
14. Yao C-L, Liu C-H, Chu I-M, Hsieh T-B, Hwang S-M. Factorial designs combined with the steepest ascent method to optimize serum-free media for *ex vivo* expansion of human hematopoietic stem cells. *Enzyme Microbial Technol* 2003;33(4):343–352.

15. Cortin V, Garnier A, Pineault N, Lemieux R, Boyer L, Proulx C. Efficient *in vitro* megakaryocyte maturation using cytokine cocktails optimized by statistical experimental design. *Exp Hematol* 2005;33:1182–1191.
16. Zandstra PW, Conneally E, Petzer AL, Piret JM, Eaves CJ. Cytokine manipulation of primitive human hematopoietic cell self-renewal. *Proc Natl Acad Sci USA* 1997;94:4698–4703.
17. Zandstra PW, Petzer AL, Eaves CJ, Piret JM. Cellular determinants affecting the rate of cytokine depletion in cultures of human hematopoietic cells. *Biotechnol Bioeng* 1997;54(1):58–66.
18. Lim M, Ye H, Panoskaltsis N, Drakakis EM, Yue X, Cass AEG, *et al.* Intelligent bioprocessing of hematopoietic cell cultures using monitoring and design of experiments. *Biotechnol Adv* 2007;25(4):353–368.
19. Myers RH, Montgomery DC. *Response Surface Methodology: Process and Product Optimization Using Designed Experiments*, 2nd ed. John Wiley and Sons, Inc., 2002.
20. Beutler E. Production and destruction of erythrocytes. In: Beutler E, Lichtman MA, Coller BS, Kipps TJ, Seligsohn U (eds.) *Williams Hematology*. McGraw-Hill, New York, 2001, pp. 355–368.
21. Kufe DW, Pollock RE, Weichselbaum RR, Bast Jr RC, Gansler TS, Holland JF, *et al. Cancer Medicine*, 6th ed. BC Decker Inc., Hamilton, Canada, 2003.
22. Mantalaris A, Keng P, Bourne P, Chang AYC, Wu DJH. Engineering a human bone marrow model: a case study on *ex vivo* erythropoiesis. *Biotechnol Prog* 1998;14:126–133.
23. Battaglia A, Fattorossi A, Pierelli L, Bonanno G, Marone M, Ranelletti FO, *et al.* The fusion protein MEN 11303 (granulocyte-macrophage colony stimulating factor/erythropoietin) acts as a potent inducer of erythropoiesis. *Exp Hematol* 2000;28(5):490–498.
24. Klinken PS. Red blood cells. *Int J Biochem Cell Biol* 2002;34(12):1513–1518.

Index